Lecture Notes in Civil Engineering

Volume 479

Lecture Notes in Civil Engineering (LNCE) publishes the latest developments in Civil Engineering—quickly, informally and in top quality. Though original research reported in proceedings and post-proceedings represents the core of LNCE, edited volumes of exceptionally high quality and interest may also be considered for publication. Volumes published in LNCE embrace all aspects and subfields of, as well as new challenges in, Civil Engineering. Topics in the series include:

- Construction and Structural Mechanics
- Building Materials
- Concrete, Steel and Timber Structures
- Geotechnical Engineering
- Earthquake Engineering
- Coastal Engineering
- Ocean and Offshore Engineering; Ships and Floating Structures
- Hydraulics, Hydrology and Water Resources Engineering
- Environmental Engineering and Sustainability
- Structural Health and Monitoring
- Surveying and Geographical Information Systems
- Indoor Environments
- Transportation and Traffic
- Risk Analysis
- Safety and Security

To submit a proposal or request further information, please contact the appropriate Springer Editor:

- Pierpaolo Riva at pierpaolo.riva@springer.com (Europe and Americas);
- Swati Meherishi at swati.meherishi@springer.com (Asia—except China, Australia, and New Zealand);
- Wayne Hu at wayne.hu@springer.com (China).

All books in the series now indexed by Scopus and EI Compendex database!

Babu T. Jose · Dipak Kumar Sahoo ·
Anand J. Puppala · C. N. V. Satyanarayana Reddy ·
Benny Mathews Abraham · Ravikiran Vaidya
Editors

Proceedings of the Indian Geotechnical Conference 2022 Volume 4

Geotechnics: Learning, Evaluation, Analysis and Practice (GEOLEAP)

Springer

Editors
Babu T. Jose
Cochin University of Science
and Technology
Kochi, Kerala, India

Dipak Kumar Sahoo
Cochin University of Science
and Technology
Kochi, Kerala, India

Anand J. Puppala
Zachry Department of Civil
and Environmental Engineering
Texas A&M University
College Station, TX, USA

C. N. V. Satyanarayana Reddy
Civil Engineering Department
Andhra University College of Engineering
Visakhapatnam, India

Benny Mathews Abraham
Department of Civil Engineering
Albertian Institute of Science
and Technology
Kochi, Kerala, India

Ravikiran Vaidya
Geo Dynamics
Vadodara, Gujarat, India

ISSN 2366-2557 ISSN 2366-2565 (electronic)
Lecture Notes in Civil Engineering
ISBN 978-981-97-1752-1 ISBN 978-981-97-1753-8 (eBook)
https://doi.org/10.1007/978-981-97-1753-8

This Springer imprint is published by the registered company Springer Nature Singapore Pte Ltd.
The registered company address is: 152 Beach Road, #21-01/04 Gateway East, Singapore 189721, Singapore

Preface

The Indian Geotechnical Society, Kochi (IGS-Kochi) Chapter, in association with Cochin University of Science and Technology (CUSAT), India, organized the Indian Geotechnical Conference (IGC-2022) at Kochi during 15–17 December 2022. The main theme of the conference was "**GEOLEAP—GEO**TECHNICS: **L**EARNING, **E**VALUATION, **A**NALYSIS AND **P**RACTICE".

The sub-theme of the conference include:

1. Behaviour of Soils and Soil—Structure Interaction
2. Soil Stabilization, Ground Improvement and Land Reclamation
3. Shallow and Deep Foundations
4. Geotechnical, Geological and Geophysical Investigation
5. Rock Engineering, Tunnelling and Underground Structures
6. Slope Stability, Landslides and Liquefaction
7. Earth Retaining Structures and Deep Excavations
8. Geosynthetics Engineering
9. Geo-Environmental Engineering, Sustainable Geotechnics and Landfill Design
10. Geohydrology, Dam and Embankment Engineering
11. Earthquake Geotechnical Engineering
12. Transportation Geotechnics
13. Forensic Geotechnical Engineering and Retrofitting of Geotechnical Structures
14. Offshore Geotechnics, Marine Geology and Subsea Site Investigation
15. Computational, Analytical and Numerical Modelling
16. Reliability in Geotechnical Engineering.

The proceedings of this conference consist of selected papers presented at the conference, organized into five volumes. The keynote and theme lectures presented during IGC 2022 are being published as a special issue of the *Indian Geotechnical Journal.*

We sincerely thank all the authors who have contributed their papers to the conference proceedings. We also thank all the theme editors and reviewers who have been instrumental in giving their valuable inputs for improving the quality of the final papers. Finally, we thank the Springer team for their support and full cooperation for publishing these five volumes of IGC-2022 proceedings.

Kochi, India

Dr. Babu T. Jose
Chairman IGC-2022

Contents

About the Editors

Dr. Babu T. Jose completed his B.Tech. from NIT Warangal in 1967 with Nehru Memorial Gold Medal, M.Tech. from IIT Madras in 1972 and Ph.D. from Cochin University of Science and Technology in 1990. He was a lecturer in NIT, Calicut, for 13 years (1967–1980). Currently, he is an emeritus professor at Cochin University of Science and Technology since 2004. Dr. Jose has commendable experience in research. He has published more than 200 research papers, and on the consultancy front, he has successfully completed more than 500 projects. On the administrative side, he was a principal at the School of Engineering, Cochin University, during 1980–1995 and a director of two engineering colleges from 1985 to 2004. Dr. Babu T. Jose is the founder chairman of IGS, Kochi Chapter. He was the chairman of the organizing committee of IGC in 2011 and the chief editor of its proceedings.

Dr. Dipak Kumar Sahoo has been a full-time faculty member since 2001, and currently a professor since 2009, in the School of Engineering, Cochin University of Science and Technology, Kochi. Prior to academics, he was working in the Fertilizers and Chemicals Travancore Limited, a Central Public Sector Company, in India for more than ten years as a consulting engineer to many public and private sector companies for setting up their green-field petrochemical complexes in different parts of India. He received his B.Sc. Engineering (Civil) from CET Bhubaneswar in 1987, M.Tech. in Construction Engineering and Management from Cochin University of Science and Technology in 1998, Ph.D. in Structural Engineering from the Indian Institute of Technology Roorkee in 2009 and LL.B. from the Cochin University of Science and Technology in 2015.

Dr. Anand J. Puppala currently serves as A. P. Wiley and Florence Chair of Zachry Civil and Environmental Engineering at Texas A&M University and has been an Associate Director of Center for Infrastructure Renewal (CIR) since 2019. Dr. Puppala was the chair of Soil Mechanics section (AFS00) of the Transportation Research Board (TRB). He also chaired American Society of Civil Engineers (ASCE)'s Geotechnical Institute's (GI) "Engineering Geology and Site Characterization" committee and TRB committee on "Soil and Rock Instrumentation."

Dr. Puppala is the current chair of ISSMGE's Technical Committee 307 on Sustainability in Geotechnical Engineering. He has given several keynotes and invited talks worldwide.

Dr. C. N. V. Satyanarayana Reddy is a professor of Civil Engineering at College of Engineering, Andhra University, Visakhapatnam, and has 29 years of teaching, research, and consultancy experience. His expertise and research interests are in the areas of reinforced soils, ground improvement, landfills, soil retention in excavations, deep foundations, and forensic geotechnical engineering. He obtained B.Tech. (Civil Engineering) degree securing first rank from Nagarjuna University, M.Tech. (Geotechnical Engineering) degree from IIT Madras, M.E. (Structures) from Andhra University, and Ph.D. from NIT Warangal. He has guided 9 Ph.D. scholars and 105 M.Tech. Dissertations so far. He has more than 110 publications in various national and international journals and seminar and conference proceedings. He served as a director of Andhra University Development Centre during the period 2013–2017.

Dr. Benny Mathews Abraham graduated from Regional Engineering College Calicut in 1982. He completed his M.Tech. from IIT Madras in 1984 and Ph.D. from Cochin University in the year 1994. He joined Cochin University as a lecturer in 1985 and became a professor in the year 2001. He was the head of Division of Civil Engineering, CUSAT, for 13 years and was the dean of Faculty of Architecture, CUSAT for three years. He was the chairman of Board of Studies and Academic Council at CUSAT. He retired from Cochin University service in May 2020, and currently, he is the professor and head of the department at Albertian Institute of Science and Technology, Kalamassery. He has more than 80 publications out of which 25 are in peer-reviewed international journals. He won the IGS best research paper awards in 1989, 1991, and 2016.

Mr. Ravikiran Vaidya Principal Engineer of Geodynamics, has been instrumental in popularizing the concept of Deep Foundation Testing in India. He has created a deep foundation testing industry and today Geo Dynamics is India's premier testing company in this field. Jointly with Dr. Jaykumar Shukla, he has also indigenized and successfully demonstrated bi-directional load testing even for large-capacity monopiles which is yet another milestone for the country. Thermal integrity profiling and instrumentation have been his other contributions. He has worked on almost all the prestigious projects in the country and in more than 20 countries worldwide. He has written more than 21 technical papers in conferences and journals. He was the co-editor of the Indian Geotechnical Conference, Vizag, in 2020. He has guided several master's theses, and his work has been used by students in their Ph.D. thesis.

Soil Stabilization, Ground Improvement and Land Reclamation

Natural Rubber Latex for Reducing Soil Brittleness Induced by Post-compaction Moisture Reduction

U. Veena and Naveen James

1 Introduction

Post-compaction drying of soil results in very high strength due to the development of matric suction. At the same time, the reduction in moisture content causes the soil mass to act as a brittle material in which rupture occurs within a small strain without giving necessary warning. Most of the widely adapted chemical stabilisation techniques result in a further increase in the brittleness of soil [1]. In the last few decades, a number of studies have been performed to improve the ductility of soil by means of geosynthetic reinforcement [2–5] and natural or synthetic fibre reinforcement [6–9]. Even though geosynthetic and fibre reinforcement was proved to be effective in reducing soil brittleness, their manufacturing and field deployment are associated with complicated processes and demand skilled labour. The inclusion of rubber tyre chips is a reliable method of increasing ductility [10–13]. Since there is a chance of groundwater pollution due to the heavy metal leachate from tyre chips [14–16], environmental studies are required before implementing the technique in the field.

In this context, the need for an environment-friendly material which can be introduced into soil by a simple procedure made the authors think about the applicability of natural rubber latex (NRL) for inducing ductility in soil. Being a material of biotic origin, NRL can be considered as a sustainable and environment-friendly admix. It was observed that NRL could improve the ductility of freshly compacted soil appreciably [17]. The present paper investigates the efficacy of NRL treatment in maintaining the improved ductility in soil subjected to post-compaction drying. The

U. Veena (✉) · N. James
Department of Civil Engineering, Indian Institute of Technology Ropar, Rupnagar, Punjab 140001, India
e-mail: veenau1010@gmail.com

B. T. Jose et al. (eds.), *Proceedings of the Indian Geotechnical Conference 2022 Volume 4*, Lecture Notes in Civil Engineering 479,
https://doi.org/10.1007/978-981-97-1753-8_1

experimental investigation was conducted by performing unconfined compression tests of untreated and NRL-treated soil samples after various drying durations.

2 Materials and Methodology

2.1 Material Selection and Characterisation

The soil used for the experimental study was collected from a field nearby IIT Ropar. After air drying and crushing the lumps, the soil was sieved through the 2.36 mm sieve to remove foreign objects. The index properties of the processed soil were calculated by conducting laboratory experiments following the corresponding IS codes. The grain size distribution of the soil is shown in Fig. 1a. Figure 1b shows the location of the soil in the plasticity chart [18]. Based on the fine fraction (fraction of soil passing 75 μm sieve) and its position in the plasticity chart, the soil is classified as MI-CI. The index properties of soil are listed in Table 1. In order to understand the compaction characteristics, the light compaction test was carried out as per IS 2720 (Part 7): 1980 [19]. The maximum dry density and optimum moisture content obtained from the light compaction test are also given in Table 1.

The natural rubber latex is a hydrosol of fine rubber particles obtained by tapping Hevea Brasiliensis trees. For industrial use, the NRL is centrifuged and preserved with a small amount of ammonia. The NRL used in the present study was supplied by Ms Asiatic Rubber Industries, Kottayam, Kerala. The basic characteristics of the NRL used were tested as per the relevant IS codes and are given in Table 2.

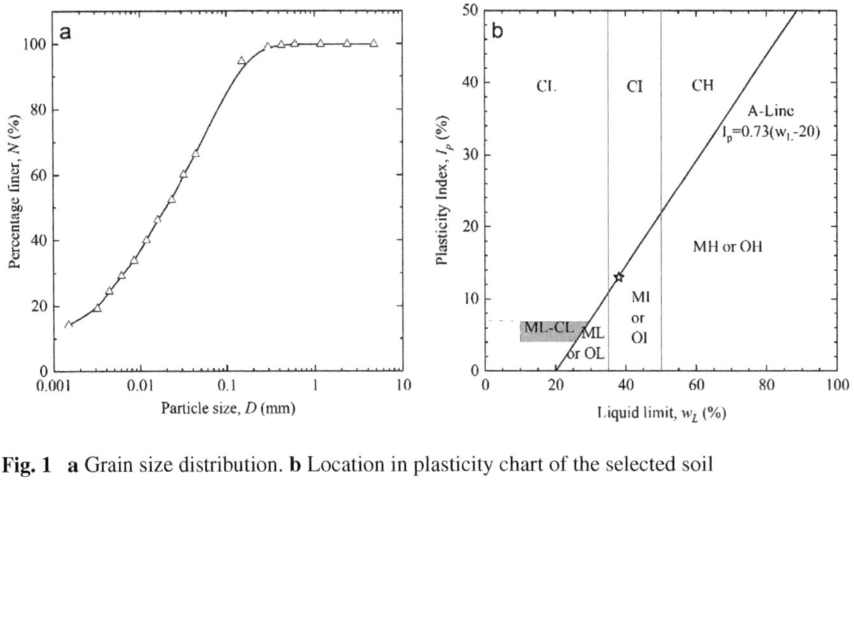

Fig. 1 **a** Grain size distribution. **b** Location in plasticity chart of the selected soil

Table 1 Index properties of selected soil

Property	Value
Specific gravity, G	2.63
Sand size fraction (%)	21
Silt size fraction (%)	63
Clay size fraction (%)	16
Liquid limit, *wL*	38%
Plastic limit, *wP*	25%
Plasticity index, *Ip*	13%
IS classification	CI-MI
γd max (g/cm^3)	1.79
OMC (%)	15

Table 2 Properties of natural rubber latex used for the study

Parameter	Value
Dry rubber content (% by mass)	60.1
Total solids content (% by mass)	61.8
Coagulum content (% by mass)	0.015
Sludge content (% by mass)	BDL*
Alkalinity as ammonia (% by mass)	0.74
Koh number	0.86
Mechanical stability time (s)	726
Volatile fatty acid number	0.05
Copper content (Ppm of total solids)	1.4
Manganese content (Ppm of total solids)	BDL

* Below detection limit

2.2 *Experimental Work*

A series of UCS tests were performed to evaluate strength-deformation characteristics of local soil with and without NRL treatment after various drying periods. The sample preparation method and the details of UCS tests are explained below.

Sample Preparation. All the samples were prepared with a diameter of 38 mm and a height of 76 mm. The target dry density of soil was chosen as 95% of maximum dry density, which is considered as the presumable field density that can be achieved with currently available types of equipment [20]. The moulding water content of the untreated sample was selected as that corresponding to 95% of γd max in the light compaction curve on the dry of optimum. In the case of NRL-treated samples, instead of water, NRL was mixed with dry soil such that the resulting water content was the same as that of untreated samples. The NRL content, which gives the target moulding water content, was determined by calculating the water contents of samples

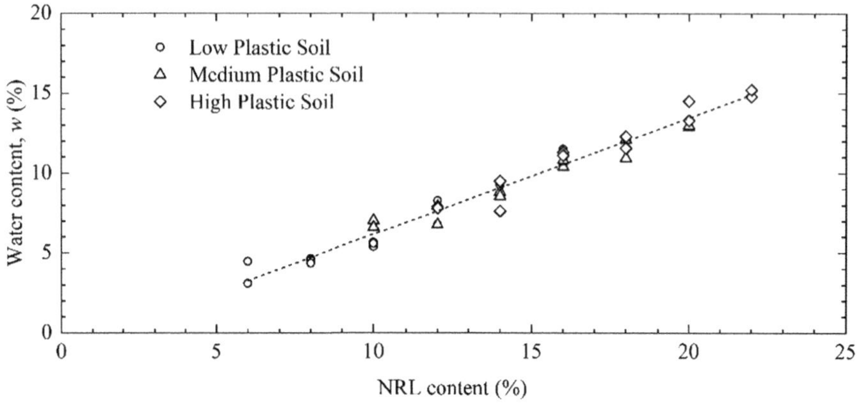

Fig. 2 Variation of moulding water content with NRL content [17]

prepared with various percentages of NRL, as explained by Veena and James [17]. Figure 2 shows the variation of sample moulding water content with NRL content in different types of soils. Here, the NRL content indicates the quantity of NRL as a percentage of soil dry weight. The parameters chosen for sample preparation are given in Table 3.

The untreated and NRL-treated compacted samples were subjected to air drying by keeping them open to atmospheric conditions. UCS tests were carried out after different periods of 0, 1, 2, 4, 7, 10, and 14 days. The daytime room temperature was found to vary between 30 °C and 35 °C during the drying period. A couple of NRL-treated samples were kept for drying for 21 and 28 days to perceive the influence of further drying on their performance. In order to understand the effect of drying temperature, two sets of NRL-treated samples were dried under the controlled temperature of 25 °C and 50 °C for durations varying from 1 to 28 days and tested.

Unconfined Compressive Strength (UCS) Test. For comparative load-deformation studies of cohesive soils, the UCS test is widely accepted as it is quick, cheap, and simple [21–23]. The UCS tests were carried out on all the samples as per IS: 2720 (Part 10) 1991 [24]. The tests were conducted with a strain rate of 1.25 mm/min, and the load applied on the sample was recorded at every 0.5 mm displacement.

Table 3 Sample preparation parameters

Parameter	Untreated sample	NRL-treated sample
Dry density, γd (kg/m^3)	1698	1698
Water content, w (%)	11	–
NRL content (%)	–	18

3 Results and Discussions

3.1 Unconfined Compressive Strength

Figure 3a shows the strength variation of untreated and NRL-treated soil with the drying period of up to 14 days. It is well-known that compacted, unsaturated fine-grained soil strength is primarily controlled by suction [25–27]. Hence, it is expected that the soil specimens will exhibit an increase in strength with drying due to the increase in suction. In untreated soil, it can be seen that after four days, the strength remains almost constant, whereas, in NRL-treated soil, the strength increases till seven days and remains constant after that. It is worth noting that the increase in strength is slow in NRL-treated soil compared with untreated soil. Figure 3b shows the strength development in NRL-treated soil when dried under different conditions. When dried at 25 °C, the slow rate of drying retards the development of suction and, thus, the strength of the soil. After 21 days, these samples also attain the strength of samples dried at room temperature. The samples dried at 50 °C show a different trend of strength development. Due to the fast drying rate, these samples achieved a very high strength within two days but lost the strength after that. This can be attributed to the softening and expansion of rubber content when subjected to high temperature for a long time, causing reduction in the number of interparticle contacts, which is the most critical factor that influences cohesion [28].

Figure 4 depicts the variation of soil strength with the sample's water content during the test. The strength of all the soil samples was found to have an exponential correlation with the water content during the test irrespective of the presence of NRL, except the NRL-treated sample, dried at 50 °C. This result infers that under normal temperature conditions, the influence of NRL on the strength of dry soil is practically

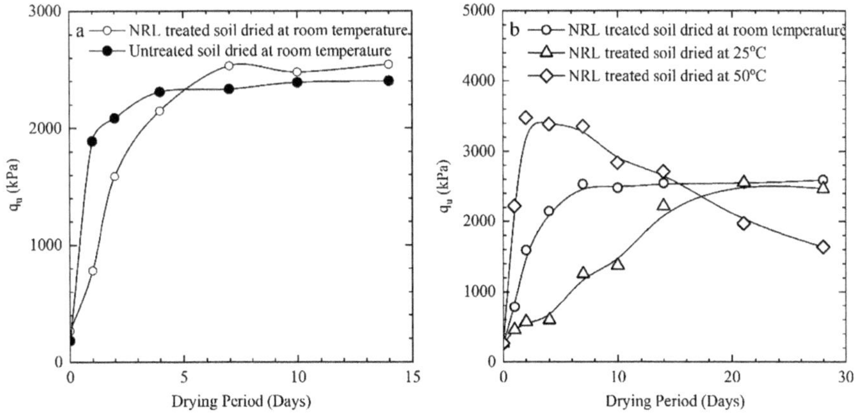

Fig. 3 Variation of strength with drying period. **a** Untreated and NRL-treated soil at room temperature. **b** NRL-treated soil under different conditions

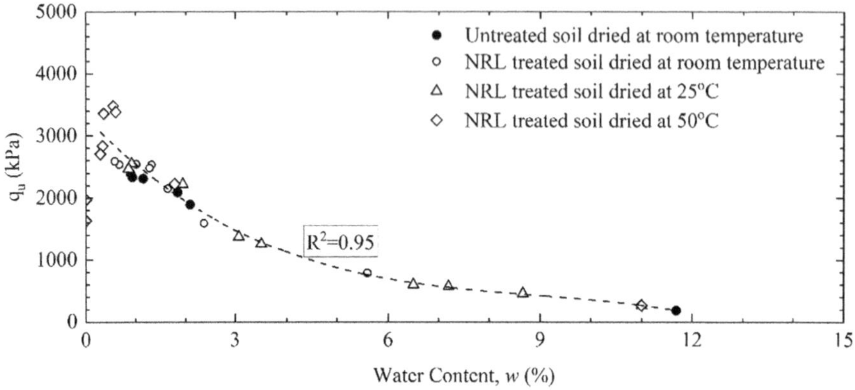

Fig. 4 Variation of UCS values with water content of sample during the test

nil. As mentioned earlier, the variation in behaviour of NRL-treated sample dried at 50 °C can be attributed to the expansion of rubber.

3.2 *Ductility Characteristics*

In the present study, more emphasis is given to the ductility characteristics of the soils. Here, the term ductility indicates the ability of the material to deform without failure. Figure 5a shows the stress–strain curves of untreated and NRL-treated samples dried at room temperature for various time durations. It can be seen that even though there is no considerable improvement in strength, the failure occurs at a higher strain in NRL-treated soil as compared to untreated soil. Figure 5b shows the stress–strain curves of NRL-treated samples subjected to different drying conditions after two days and 28 days. It can be seen that the strain associated with peak stress is comparable in all these samples except the sample dried at 50 °C for 28 days, which shows a reduced strength and failed at a lower strain.

Deformability Index. For quantifying the improvement in ductility due to the NRL treatment, the parameter deformability index [29, 30] is used. The deformability index (*ID*) can be calculated by using Eq. 1.

$$I_D = \frac{\varepsilon_{\text{treated}}}{\varepsilon_{\text{untreated}}} \tag{1}$$

Here, εuntreated and εtreated indicate the strain associated with peak stress in untreated and NRL-treated soil samples. The values of the deformability index calculated for all the NRL-treated samples are plotted against the drying period in Fig. 6a. It should be noted that the ID values of most of the NRL-treated samples are nearly 2. It shows that NRL-treated soil can sustain double the strain than the untreated soil

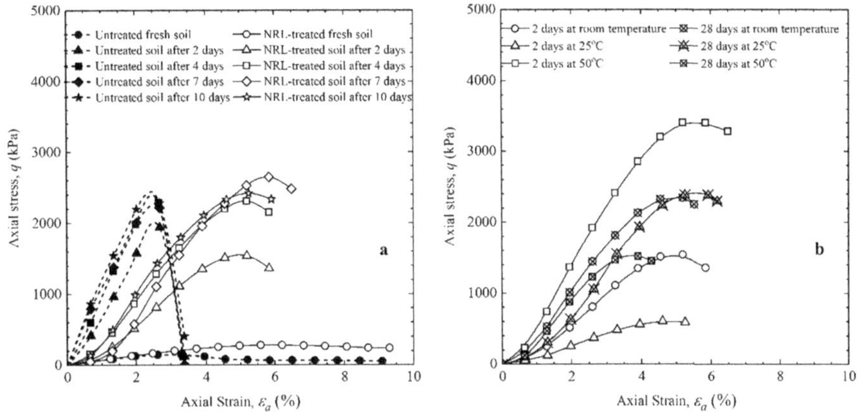

Fig. 5 **a** Stress–strain curves of untreated and NRL-treated soils dried at room temperature. **b** Stress–strain curves of NRL-treated soil dried under different conditions

before failure, even in the dry state. In order to distinguish the influence of NRL in improving the ductility from that of the lubrication effect of sample moisture content, the failure strain was plotted against sample water content during the test in Fig. 6b. Interestingly, it is evident that even at the same water content, NRL-treated samples are able to carry almost two times the strain taken by the untreated sample before failure.

Energy Absorption Capacity. The variation in soil's strength and strain capacity is reflected in the parameter absorbed strain energy as it is calculated as the area under the stress–strain curve [9]. The strain energy absorbed by all the samples before failure is plotted against the drying period in Fig. 7a and against the water

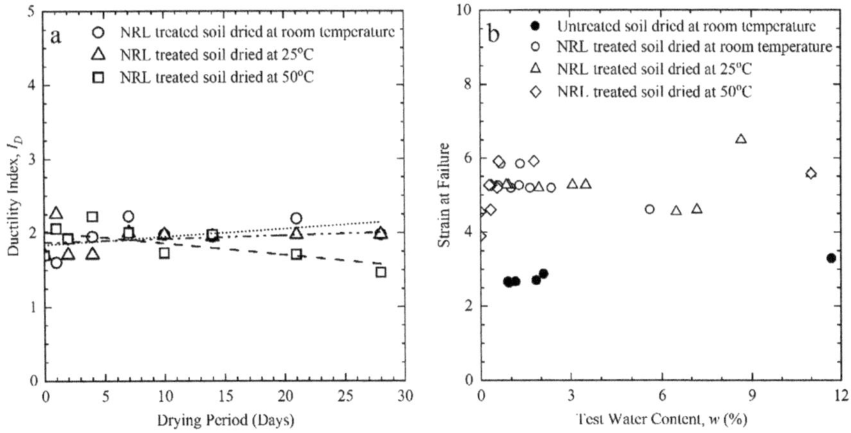

Fig. 6 **a** Deformability index of NRL-treated samples. **b** Strain at failure versus test water content

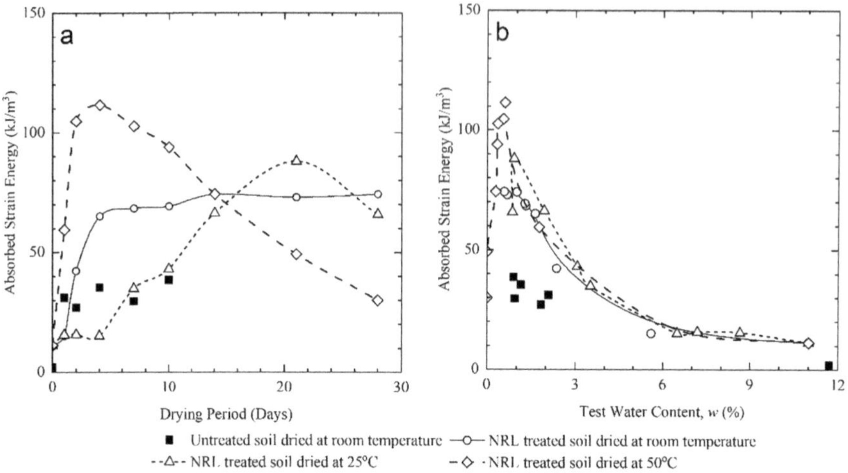

Fig. 7 Strain energy absorbed by the samples before failure plotted against. **a** Drying period and **b** water content during the test

content during the test in Fig. 7b. Irrespective of the comparable strength, the energy absorption capacity of NRL-treated soil samples was found to be higher than that of untreated soil. It was seen in Sect. 3.1 that NRL-treated soil dried at 50°C became weaker than the untreated soil after a period of 14 days. It is noteworthy that even these samples show an energy absorption capacity comparable to that of untreated soil samples. This represents the improved ductility acquired by the soil as a result of NRL treatment.

Post-failure Behaviour. To get an insight into the post-failure behaviour of dry soil, an untreated soil sample and an NRL-treated soil sample were subjected to the UCS test to a higher value of post-failure strain. Samples dried for 14 days were chosen for this test. Figure 8 shows the stress–strain curves obtained from the tests. The stress–strain curve of untreated soil shows a pure brittle behaviour indicated by a substantial post-failure strength loss, whereas in NRL-treated soil, nearly 22% of the peak strength was preserved after the rupture. From the peak strength *qp* and residual strength *qr*, the brittleness index *IB* [31], a parameter to quantify the soil brittleness, can be calculated using Eq. 2.

$$I_B = \frac{q_p - q_r}{q_p} \tag{2}$$

A higher value of *IB* (≈1) indicates brittle failure, and a lower value indicates ductile failure. The brittleness index of dried untreated soil sample calculated from Fig. 8 is 0.99, while that of dried NRL-treated soil sample is only 0.78. This result manifests the potential of NRL treatment in reducing the soil brittleness induced by post-compaction moisture reduction.

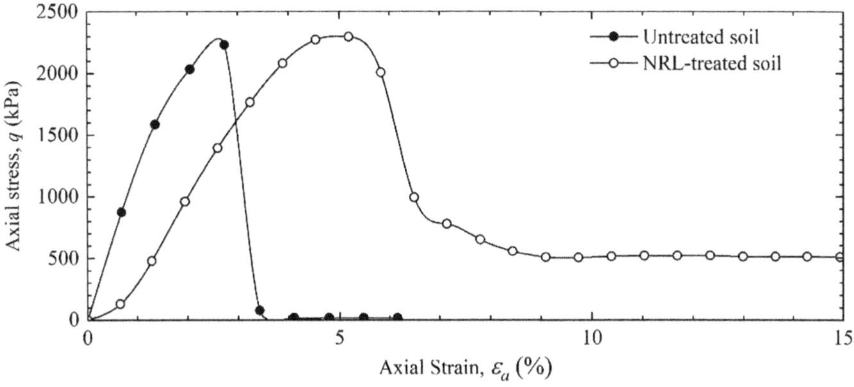

Fig. 8 Stress–strain curves showing post-failure behaviour of soil samples

4 Conclusion

In the present study, the effect of NRL treatment on load-deformation characteristics of soil samples subjected to post-compaction drying was investigated. Untreated and NRL-treated soil samples were prepared and dried for different periods, and UCS tests were carried out. The following conclusions were drawn from the experimental study.

- Development of strength of a compacted sample after drying is mainly controlled by matric suction. The presence of NRL does not cause considerable improvement in strength. When dried at a higher temperature, the presence of rubber resulted in softening of soil.
- In NRL-treated samples, failure occurred at a higher strain level, nearly double that in untreated samples. This indicates that NRL improved the ductility of soil, and hence, NRL treatment can help to avoid brittleness-induced failure of dry soil structures to some extent.
- The energy absorption capacity was found to be much more in NRL-treated soil, indicating improved ductility of dry NRL-treated soil.

References

1. Dhar S, Hussain M (2019) The strength behaviour of lime-stabilised plastic fibre-reinforced clayey soil. Road Mater Pavement Des 20:1757–1778. https://doi.org/10.1080/14680629.2018.1468803
2. Ashmawy AK, Bourdeau PL (1996) Response of a woven and a nonwoven geotextile to monotonic and cyclic simple tension. Geosynth Int 3:493–515. https://doi.org/10.1680/gein.3.0072

3. Azadegan O, Li J, Hadi Jafari S, Ren G (2013) Geogrid reinforced lime cement-treated granular soils. Appl Mech Mater 1090–1094. https://doi.org/10.4028/www.scientific.net/AMM.330.1090
4. Jahandari S, Mojtahedi SF, Zivari F, Jafari M, Mahmoudi MR, Shokrgozar A, Kharazmi S, Vosough Hosseini B, Rezvani S, Jalalifar H (2020) The impact of long-term curing period on the mechanical features of lime-geogrid treated soils. Geomech Geoeng 00:1–13. https://doi.org/10.1080/17486025.2020.1739753
5. Watanabe K, Nakajima S, Fujii K, Matsuura K (2020) Development of geosynthetic-reinforced soil embankment resistant to severe earthquakes and prolonged overflows due to tsunamis. Soils Found 60:1371–1386. https://doi.org/10.1016/j.sandf.2020.08.006
6. Gray DH, Ohashi H (1983) Mechanics of fiber reinforcement in sand. J Geotech Eng 109. https://doi.org/10.1061/(ASCE)0733-9410(1983)109:3(335)
7. Maher MH, Ho YC (1994) Mechanical properties of kaolinite/fiber soil composite. J Geotech Eng 120:1381–1393. https://doi.org/10.1061/(ASCE)0733-9410(1994)120:8(1381)
8. Consoli NC, Vendruscolo MA, Prietto PDM (2003) Behavior of plate load tests on soil layers improved with cement and fiber. J Geotech Geoenviron Eng 129:96–101. https://doi.org/10.1061/(asce)1090-0241(2003)129:1(96)
9. Yadav JS, Garg A, Tiwari SK (2019) Strength and ductility behaviour of rubberised cemented clayey soil. Proc Inst Civ Eng Gr Improv 1–38. https://doi.org/10.1680/jgrim.19.00017
10. Edil T, Bosscher P (1994) Engineering properties of tire chips and soil mixtures. Geotech Test J 17:453. https://doi.org/10.1520/gtj10306j
11. Rao GV, Dutta RK (2006) Compressibility and strength behaviour of sand-tyre chip mixtures. Geotech Geol Eng 24:711–724. https://doi.org/10.1007/s10706-004-4006-x
12. Mashiri MS, Vinod JS, Sheikh MN, Tsang HH (2015) Shear strength and dilatancy behaviour of sand-tyre chip mixtures. Soils Found 55:517–528. https://doi.org/10.1016/j.sandf.2015.04.004
13. Kong D, Deng M, Liu Y, Wang X (2018) The environmental study on consolidated undrained triaxial compression tests on lightweight soil mixed with rubber chips of scrap tires. Ekoloji 27:1503–1510
14. Edil TB (2008) A review of environmental impacts and environmental applications of shredded scrap tires. In: Proceedings of the International workshop on scrap tire derived geomaterials—opportunities and challenges (IW-TDGM 2007)
15. Banasiak L, Chiaro G, Palermo A, Granello G (2019) Recycling of end-of-life tyres in civil engineering applications: environmental implications. In: WasteMINZ 2019 conference, p 15
16. Mohajerani A, Burnett L, Smith JV, Markovski S, Rodwell G, Rahman MT, Kurmus H, Mirzababaei M, Arulrajah A, Horpibulsuk S, Maghool F (2020) Recycling waste rubber tyres in construction materials and associated environmental considerations: a review. Resour Conserv Recycl. https://doi.org/10.1016/j.resconrec.2020.104679
17. Veena U, James N (2022) Natural rubber latex for improving ductility characteristics of soil: a preliminary experimental investigation. Geotech Geol Eng. https://doi.org/10.1007/s10706-022-02162-1
18. IS:1498 (2007) Classification and identification of soils for general engineering purposes. Bureau of Indian Standards, New Delhi
19. IS:2720-Part 7 (1980) Methods of test for soils, determination of water content dry density relation using light compaction. Bureau of Indian Standards, New Delhi
20. Stark TD (2020) Postconstruction evaluation of fill compaction. J Leg Aff Disput Resolut Eng Constr 12:04520030. https://doi.org/10.1061/(ASCE)LA.1943-4170.0000415
21. Budhu M (2020) Soil mechanics and foundations. Wiley
22. Park SS (2011) Unconfined compressive strength and ductility of fiber-reinforced cemented sand. Constr Build Mater 25:1134–1138. https://doi.org/10.1016/j.conbuildmat.2010.07.017
23. Nagaraj HB (2016) Influence of gradation and proportion of sand on stress-strain behavior of clay-sand mixtures. Int J Geoeng. https://doi.org/10.1186/s40703-016-0033-8
24. IS:2720-Part 10 (1991) Methods of test for soils, determination of unconfined compressive strength. Bureau of Indian Standards, New Delhi

25. Bishop AW, Alpan I, Blight GH, Donald IB (1960) Factors controlling the strength of partly saturated cohesive soils. In: Proceedings of the ASCE research conference on shear strength of cohesive soils, pp 503–532
26. Bishop AW, Blight GE (1963) Some aspects of effective stress in saturated and partly saturated soils. Geotechnique 13. https://doi.org/10.1680/geot.1963.13.3.177
27. Fredlund DG, Morgenstern NR, Widger RA (1978) The shear strength of unsaturated soils. Can Geotech J 15:313–321. https://doi.org/10.1139/t78-029
28. Mitchell JK (1964) Shearing resistance of soils as a rate process. J Soil Mech Found Div 90:29–61. https://doi.org/10.1061/JSFEAQ.0000593
29. Eskisar T (2015) Influence of cement treatment on unconfined compressive strength and compressibility of lean clay with medium plasticity. Arab J Sci Eng 40:763–772. https://doi.org/10.1007/s13369-015-1579-z
30. Bekhiti M, Trouzine H, Rabehi M (2019) Influence of waste tire rubber fibers on swelling behavior, unconfined compressive strength and ductility of cement stabilized bentonite clay soil. Constr Build Mater 208:304–313. https://doi.org/10.1016/j.conbuildmat.2019.03.011
31. Bishop AW (1971) The influence of progressive failure on the choice of the method of stability analysis. Geotechnique 21:168–172. https://doi.org/10.1680/geot.1971.21.2.168

Study of Reinforced Soil Embankment Supported on Stone Column Improved Ground

Riddhi Gupta and Harish Madupuri

1 Introduction

Embankment construction on soft soil is a challenging task. To cater to the needs of industrial expansion, transportation plays a key role in the economic development of any country. Due to the development of railways and highways, which many times require construction at a higher grade level than the existing natural level on soft soil, their settlement and bearing requirements need to be satisfied to ensure safe and economic transportation. Various field and numerical studies have been conducted for stone column supported embankments [2, 7]. Based on available literature for embankments on soft soil, basal reinforcement has been extensively used in the construction of embankments on soft clays (Han and Gabber, [3, 10]).

It was found that a handful of studies have been observed on stone column supported geosynthetic reinforced embankments resting over soft soil [4]. A basal geosynthetic reinforcement can aid in resisting the earth pressure within the embankment and prevent lateral deformation of its foundation, thereby increasing bearing capacity and stability [9]. The application of geosynthetic reinforcement helps to transfer stresses from the soft soil to the columns [8]. The application of stone columns fulfills the combined benefit of stability and increased bearing capacity of embankment supported on the soft clay foundation [6].

In this paper, 2D plane strain modeling of stone columns with geosynthetic reinforced embankment is carried out following the gravel trench (G) and homogenization (H) approach. In the present parametric analysis, embankment stability

R. Gupta (✉)
Faculty of Technology and Engineering, Maharaja Sayajirao University of Baroda, Vadodara, India
e-mail: riddhigupta95@gmail.com

H. Madupuri
Sree Vidyanikethan Engineering College, Tirupati 517102, India

B. T. Jose et al. (eds.), *Proceedings of the Indian Geotechnical Conference 2022 Volume 4*, Lecture Notes in Civil Engineering 479,
https://doi.org/10.1007/978-981-97-1753-8_2

and deformation behavior were studied considering multiple factors such as basal reinforcement tensile strength, stone column material stiffness, spacing and length, soil strata thickness and shear strength parameters, and embankment height. The homogenization method consists of replacing the stone column and soft soil with an equivalent homogeneous soil with improved properties. The ratio of the amount of soft soil replaced by the stone column is called the area replacement ratio (as). By varying the area replacement ratio (as) and material parameter and without making a change in geometry, various combinations of different parameters can be analyzed at the design stage. Especially when the geometry is complex [5]. For the gravel trench method, Tan et al. (2008) based on available literature two different methods to convert the geometry of a single stone column in 2D plane strain from a three-dimensional model were studied. In the first method, soil permeability is changed while maintaining the column diameter constant while in the second method column diameter is changed by maintaining the soil permeability constant. After field validation, he found the second method to be more reliable than the first one. For the present parametric study, second approach is employed for analysis.

2 Numerical Modeling

2.1 Model Validation

In order to validate the modeling of the stone column supported embankment, the solution of Abusharar and Han [1] is selected. The dimension of geometry is provided as follows: the width of column is 0.8 m, the column length is 10 m, the c/c column spacing is 4.0 m, the embankment crest width is 20, the height of embankment is 5 m, and the angle of side slope is 2(H):1(V). The material parameters are shown in Table 1. The finite element software Plaxis 2D is used to validate the model. The location of the critical slip surface generated in both methods is observed and found reasonably similar. Figure 1 shows the critical failure pattern observed in cases with and without water table.

Table 1 Soil property considered in the present parametric study (after Abusharar and Han [1])

Parameter	Unit	Embankment fill	Clay	Sand	Stone column	Equivalent area
Thickness	M	5	10	2	10	10
Unsaturated unit weight	kN/m^3	18	16	18	17	16.2
Porosity	–	0.25	0.25	0.25	0.25	0.25
Cohesion	kPa	0	0	0	0	16
Friction angle	°	32	30	30	38	8.9

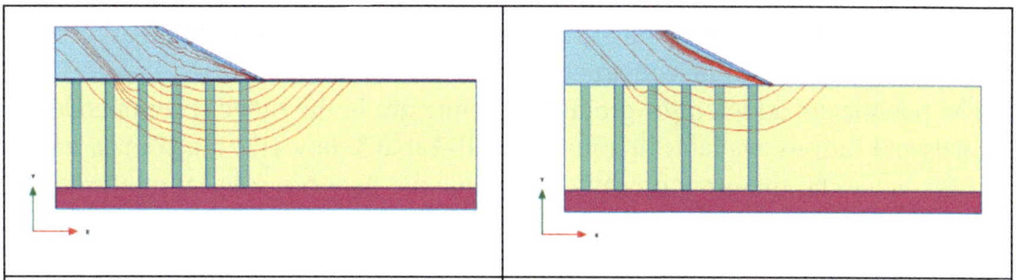

Case a) Individual column considering water table	Case b) Individual column without considering water table

Fig. 1 Embankment stability with gravel trench method (Case a) with considering water table and (Case b) without considering water table

Table 2 Factor of safety (after Abusharar and Han [1])

Ground water condition	Modified after Abusharar and Han [1]		Current FEM model	
	Individual column	Equivalent area method	Individual column	Equivalent area method
With water table	1.49	1.6	1.485	1.464
Without water table	1.53	1.69	1.531	1.48

Table 2 shows the factor of safety derived with the gravel trench method and homogenization method with the current FEM-based model and existing finite difference method-based software by Abusharar and Han [1]. The result derived from current model shows good agreement with that of model used for validation purpose for both homogenization and gravel trench method.

2.2 Present Parametric Study

For the present study, plane strain configuration is adopted for modeling stone column supported basal reinforced embankment. The governing parameters such as basal reinforcement tensile stiffness (J), stone column spacing (S), soil stratum thickness, soil shear strength parameter, and embankment height are considered for analysis.

In the study, the stone column termination is considered in two different soils. In Ground Model A, the stone column is terminated in soft soil. While in Ground Model B, the stone column is terminated in medium-dense sand. For Model A, soft clay layer depth is taken as 14.0 m, while for Model B, it is considered as 10.0 m. Thereafter, medium-dense sand is considered for both models. The stone column length and column diameter are kept constant for both the models. For each combination, modeling is carried out with gravel trench method and homogenization method. The embankment is constructed in 2.0 m lift. It is assumed that excess pore pressure is dissipated while construction.

The parameters adopted for ground modeling are in the range recommended by manuals and various available literature [2, 10]. Table 3 shows the material parameter used for modeling the ground profile. The uniaxial geogrid as basal reinforcement with tensile stiffness (J1) of 3000 (kN/m) at 5% strain is considered for analysis. The stone column spacing variation is kept at 1.5 m (S1) and 1.8 m (S2) c/c. Figure 2 shows the embankment geometry adopted for the analysis. One row of columns is extended from outer edge of the embankment to maintain the confinement. Here, (J0) indicates the absence of basal reinforcement.

Table 3 Soil properties considered in present parametric study

Material property	Embankment fill	Stone column	Soft clay	Medium-dense sand
Unit weight, kN/m^3	20	20	19.45	18.5
Friction angle, °	30	34	–	30
Undrained cohesion, kPa	5	–	25	–
Poison ratio	0.3	0.3	0.4	0.3
Modulus of elasticity, kPa	20,000	65,000	5500	22,000
Material behavior	Drained	Drained	Undrained	Drained

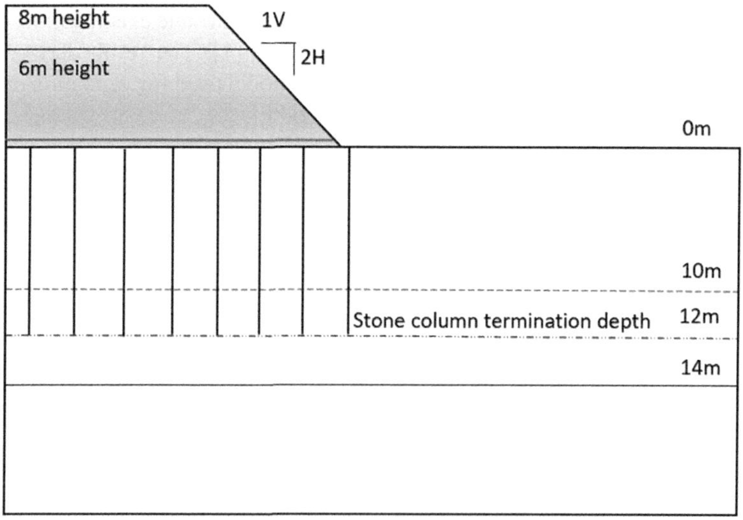

Fig. 2 Numerical model adopted for parametric study

3 Result and Discussion

The main objective of this present study is to understand the deformation and stability behavior of stone column supported embankments with basal reinforcement. The settlement profile along the embankment base, stability of embankment and horizontal soil deformation beneath the embankment toe are affected by basal reinforcement presence. Two ground models are studied with respect to stone column termination depth.

As shown in Figs. 3 and 4, uneven settlement was observed at the ground surface due to stiffness difference between the column and surrounding soil. The same is not reflected in Fig. 5 showing the base settlement with homogenization method. Upheaving deformations of soft soil foundation were noticed outside of the embankment. Maximum upheaval was observed in the ground with minimum area replacement ratio. The upheaval was observed at a distance from embankment edge. Settlement is measured at 0.1 m below from ground level.

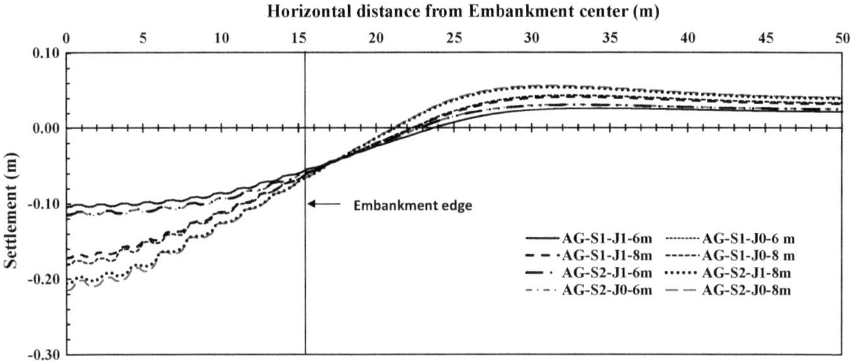

Fig. 3 Base settlement for Ground Model A

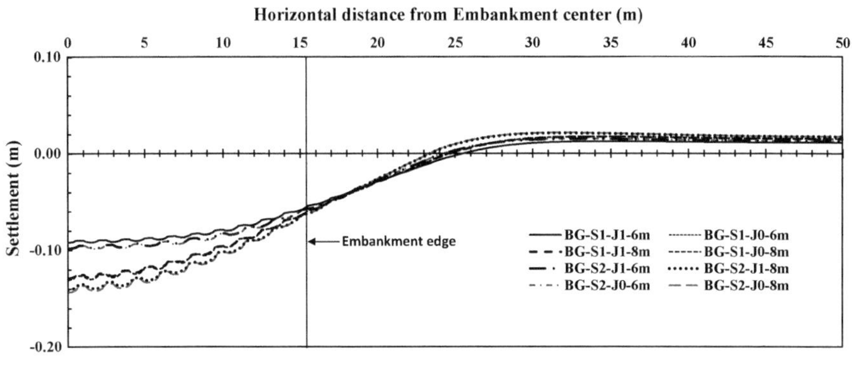

Fig. 4 Base settlement for Ground Model B

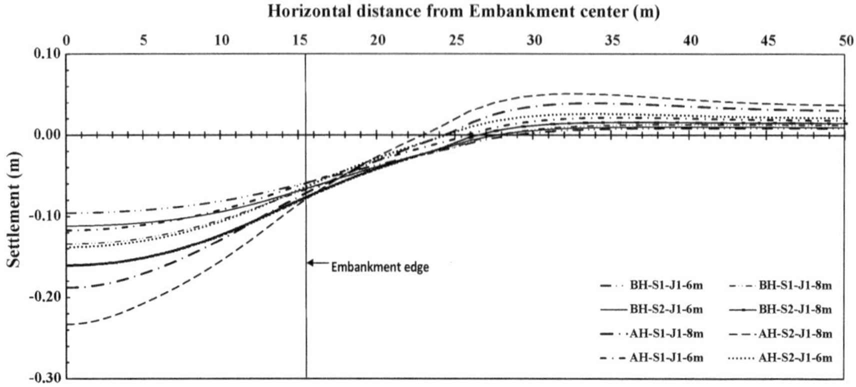

Fig. 5 Base settlement with homogenization method for Ground Models A and B

The lateral deformation of column at 1 m inside the embankment toe is shown in Figs. 6 and 7 for both the ground condition with stone column c/c spacing of 1.5 m and 1.8 m for embankment height of 6.0 and 8.0 m. Lateral deformation observed with stone column terminating in ground condition with Model A has significantly more deformation when compared to Ground Model B; i.e., stone column terminating in soft ground has significantly more deformation at top and bottom compared to column terminating in stiff soil with and without the basal reinforcement. Lateral deformation at top, middle, and bottom in stone column for a 6 m high embankment with columns terminating in medium-dense sand has 25, 36, and 51% variation compared to columns terminating in soft clay at respective levels for both the spacing condition. While for an 8 m high embankment, it is 33, 46, and 65% variation at top, middle, and bottom levels in stone column for both spacing. Nearly, similar behavior is observed in cases without basal reinforcement.

Stability analysis for embankment is conducted for ground condition with all the parametric combinations shown in Table 4. Figures 8 and 9 show the factor of safety for 6.0 and 8.0 m embankment height for all reinforced and unreinforced ground combinations with gravel trench and homogenization method. There is a 4–6% variation in the factor of safety in absence of basal reinforcement for Ground Model A. With Ground Model B, it is observed to be 9–12% in absence of basal reinforcement for both the column spacing.

From Figs. 8 and 9, it is clear that the gravel trench method and homogenization method predict reasonably accurate safety factor for Ground Model A, stone column terminating in soft clay, while the maximum error between both the methods is 10.67%, for Ground Model B, stone column terminating in medium-dense sand.

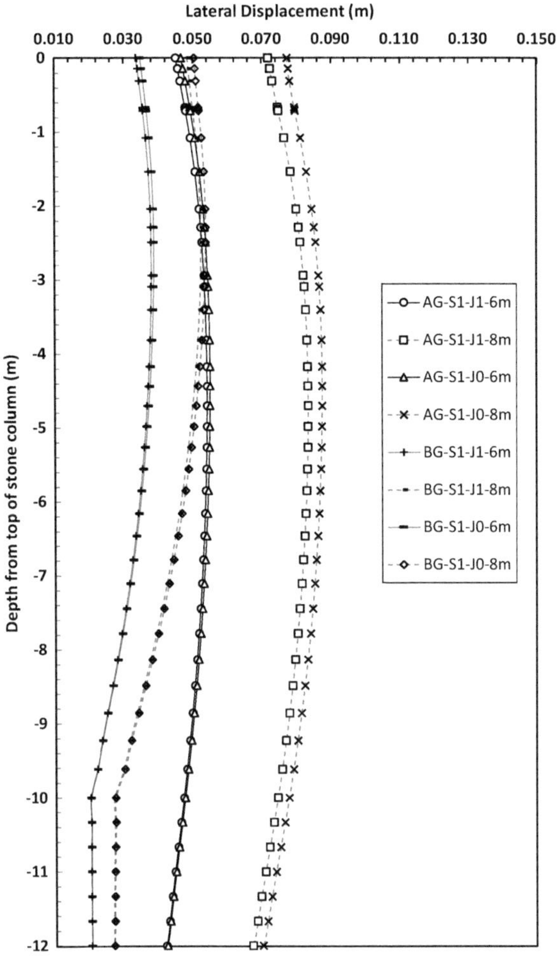

Fig. 6 Effect of basal reinforcement of lateral deformation of column near embankment toe for 1.5 m c/c stone column spacing with Case A and Case B ground condition

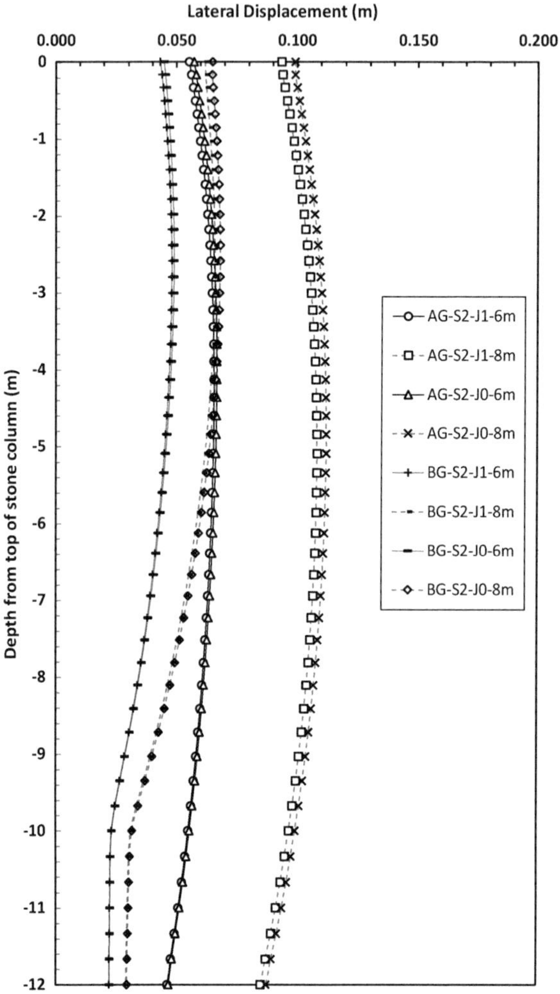

Fig. 7 Effect of basal reinforcement of lateral deformation of column near embankment toe for 1.8 m c/c stone column spacing with Case A and Case B ground condition

Table 4 Parametric study combination for stone column supported basal reinforced ground

Method of analysis	Combination	Height of embankment (m)	Method of analysis	Combination	Height of embankment (m)
Ground Model A-Gravel trench method	AG_6m	6	Ground Model B-Gravel trench method	BG_6m	6
	AG_8m	8		BG_8m	8
	AG-S1-J0_ 6m	6		BG-S1-J0_ 6m	6
	AG-S1-J0_ 8m	8		BG-S1-J0_ 8m	8
	AG-S1-J1_ 6m	6		BG-S1-J1_ 6m	6
	AG-S1-J1_ 8m	8		BG-S1-J1_ 8m	8
	AG-S2-J0_ 6m	6		BG-S2-J0_ 6m	6
	AG-S2-J0_ 8m	8		BG-S2-J0_ 8m	8
	AG-S2-J1_ 6m	6		BG-S2-J1_ 6m	6
	AG-S2-J1_ 8m	8		BG-S2-J1_ 8m	8
Ground Model A-Homogenization method	AH-S1-J0_ 6m	6	Ground Model B-Homogenization method	BH-S1-J0_ 6m	6
	AH-S1-J0_ 8m	8		BH-S1-J0_ 8m	8
	AH-S1-J1_ 6m	6		BH-S1-J1_ 6m	6
	AH-S1-J1_ 8m	8		BH-S1-J1_ 8m	8
	AH-S2-J0_ 6m	6		BH-S2-J0_ 6m	6
	AH-S2-J0_ 8m	8		BH-S2-J0_ 8m	8
	AH-S2-J1_ 6m	6		BH-S2-J1_ 6m	6
	AH-S2-J1_ 8m	8		BH-S2-J1_ 8m	8

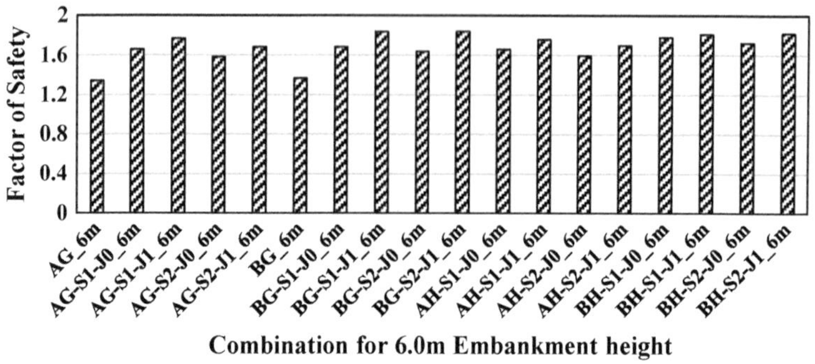

Fig. 8 Factor of safety with Ground Model A for different reinforcement and stone column spacing conditions

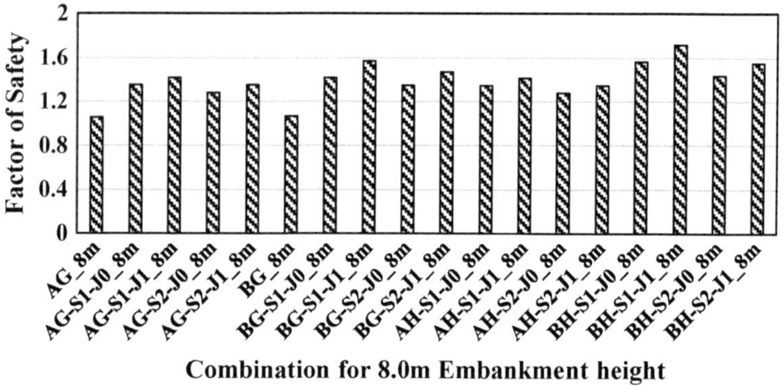

Fig. 9 Factor of safety with Ground Model B for different reinforcement and stone column spacing conditions

4 Conclusion

In this study, stability and deformation behavior of stone column supported geosynthetic reinforced embankment is analyzed with two different modeling methods and two different ground conditions. The effect of various parameters such as soil stratigraphy, stone column spacing, and geogrid stiffness was examined. From the analysis, the following conclusions can be drawn:

1. Basal reinforcement has noticeable influence on lateral soil deformation of the embankment and below the embankment ground for both stone columns terminating in soft soil and medium-dense sandy soil. The stability of the embankment is significantly influenced by uniaxial geogrid presence. For higher height embankment ($h/Hs > 0.57$), resting on soft clay basal has a noticeable influence on base settlement, specifically for stone columns terminating in soft soil.

2. Stability, base settlement, and lateral deformation significantly are affected more by stone columns spacing for stone columns terminating in soft soil rather than medium-dense sand.
3. Gravel trench method of stone column analysis has a similar output for base settlement and stability for stone column terminating in soft soil.
4. Lateral deformation is more significant for stone columns terminating in soft soil than in medium-dense sand. In absence of basal reinforcement, while measuring lateral ground deformation, there is 4.5% and 8% variation for 6 m and 8 m height, respectively, for Ground Model A with gravel trench method. Similarly, for Ground Model B, for 6 m and 8 m height, it is 3.7 and 6.5 and with gravel trench method. The maximum variation is observed at top of the stone columns for both ground models.
5. Homogenization method predicts a similar base settlement profile and factor of safety for stone columns terminating in soft soil when compared with gravel trench method, while there is a noticeable deviation in the condition when stone column terminates in medium-dense sand.

References

1. Abusharar SW, Han J (2011) Two-dimensional deep-seated slope stability analysis of embankments over stone column-improved soft clay. Eng Geol 120(1–4):103–110. https://doi.org/10.1016/j.enggeo.2011.04.002
2. Ambily AP, Gandhi SR (2012) Behavior of stone columns based on experimental and FEM analysis. https://doi.org/10.1061/ASCE1090-02412007133:4405
3. Bathurst RJ, Naftchali FM (2021) Geosynthetic reinforcement stiffness for analytical and numerical modelling of reinforced soil structures. Geotext Geomembr 49(4):921–940. https://doi.org/10.1016/j.geotexmem.2021.01.003
4. Deb K, Mohapatra SR (2013) Analysis of stone column-supported geosynthetic-reinforced embankments. Appl Math Model 37(5):2943–2960. https://doi.org/10.1016/j.apm.2012.07.002
5. Gupta RM, Gajjar H, Shukla J (2022) Slope stability analysis of stacker-reclaimer embankment under the influence of adjacent stockpile. Lect Notes Civ Eng 152:283–292. https://doi.org/10.1007/978-981-16-1831-4_26
6. Jayapal J, Rajagopal K (2020) Slope stability analysis of embankment resting on granular columns using FEM. Lect Notes Civ Eng 85:311–326. https://doi.org/10.1007/978-981-15-6086-6_25
7. Murugesan S, Rajagopal K (2006) Geosynthetic-encased stone columns: numerical evaluation. Geotext Geomembr 24(6):349–358. https://doi.org/10.1016/j.geotexmem.2006.05.001
8. Rowe RK, Li AL (2002) Geosynthetic-reinforced embankments over soft foundations
9. Yoo C, Kim SB (2009) Numerical modeling of geosynthetic-encased stone column-reinforced ground. Geosynth Int 16(3):116–126. https://doi.org/10.1680/gein.2009.16.3.116
10. Zheng G, Yu X, Zhou H, Wang S, Zhao J, He X, Yang X (2020) Stability analysis of stone column-supported and geosynthetic-reinforced embankments on soft ground. Geotext Geomembr 48(3):349–356. https://doi.org/10.1016/j.geotexmem.2019.12.006

Characterization of Soil–Jarosite Mixes for Geopolymer-Based Soil Stabilization

Surrender Singh, Abhishek Kumar, and T. G. Sitharam

1 Introduction

Jarosite is a waste byproduct from zinc (Zn) smelter produced during various hydrometallurgical processes, which are performed for the extraction of the metallic Zn from its ore. Jarosite is a hydrous sulfate of potassium and ferric iron with chemical formula $KFe_3(SO_4)_2(OH)_8$. Due to its hazardous nature, jarosite is mixed with 2% lime and 10% cement before its disposal into the tailings ponds (TP) [9]. Storage of jarosite in TPs adds no value to the mining industries, instead poses a serious threat to the environment and human health due to breach of tailings storage facilities. As a result, researchers have been investigating different alternatives to minimize the volume of jarosite to be disposed. Ray et al. [8] utilized jarosite as a partial replacement of cement in concrete mixtures and obtained satisfactory results in terms of mechanical strength and durability. On a similar note, Gared and Gaur [3] also employed jarosite as a substitute to cement in concrete mixes for its use in rigid pavement. According to them, with increase in the jarosite content, the mechanical strength of jarosite concrete mixtures was observed to increase, while the porosity and abrasion loss were found to decrease. Beena and Santhosh [1] utilized jarofix along with varying percentages of lime to enhance the geotechnical properties of soil to be used in road pavement. Most of the aforementioned studies employed jarofix for soil stabilization or as a cement substitute in concrete mixtures. Since jarofix is obtained by adding lime and cement to the jarosite, it is not an environmentally sustainable practice due to high carbon footprint associated with the production of lime and cement. Therefore, it is necessary to seek out alternate methods that not only detoxify jarosite but also stabilize the soil.

S. Singh (✉) · A. Kumar · T. G. Sitharam
Indian Institute of Technology, Guwahati, Assam 781039, India
e-mail: surendersingh@iitg.ac.in

B. T. Jose et al. (eds.), *Proceedings of the Indian Geotechnical Conference 2022 Volume 4*, Lecture Notes in Civil Engineering 479,
https://doi.org/10.1007/978-981-97-1753-8_3

Utilizing jarosite as a geopolymer binder for alkali-activated soil stabilization (geopolymerization) can be an alternative way to manage jarosite in an environmentally sustainable manner. In geopolymerization, aluminosilicate-rich material is combined with an alkali activator to create a new polymeric chain-like structure composed of aluminosilicate networks. To the best of authors' knowledge, no study has been carried out till date, which utilizes jarosite as a geopolymer binder for its use in road applications.

The main objective of this study is to characterize the jarosite for its use in alkali-activated soil stabilization. Various experiments are performed to examine the basic geotechnical properties such as particle size distribution, specific gravity, Atterberg limits, etc. of jarosite. Furthermore, an attempt has been made to evaluate the compaction characteristics of soil–jarosite (SJ) mixes. Two sets of SJ mixes, one with water and one with NaOH solution (activator), are prepared with varying jarosite content. Comparison is drawn between the compaction curves obtained for bare soil, jarosite, and SJ mixes. Finally, target dry density and optimum moisture content (OMC) values are suggested (for different SJ mixes) to be used in sample preparation required for the strength determination of geopolymerized soil.

2 Materials and Methods

2.1 Soil and Jarosite

Soil used in the current study was collected from northeastern state of Assam, whereas jarosite was procured from Hindustan Zinc Limited (HZL) smelter at Debari, Rajasthan. Basic characterization of soil and jarosite is done following the experiments specified by Bureau of Indian Standards (BIS). Specific gravity (G) of soil and jarosite is determined by density bottle method [5], whereas particle size analysis is done using dry sieve analysis coupled with hydrometer analysis [6]. Liquid limit (LL) of soil and jarosite is measured by Casagrande method, whereas plastic limit (PL) is determined by rolling the sample into a thread of 3 mm as per IS 2720-5 [6]. The water holding capacity of soil and jarosite is measured by immersing a known amount of oven-dried sample in distilled water for 48 h. Further, the wet weight of sample is measured after being gravity drained for 2 h through Whatman grade-1 filter paper. The water holding capacity was then calculated (in percentage) as ratio of wet weight to initial dry weight of the sample [7]. In addition, microstructure analysis is carried out in a field emission scanning electron microscope (FESEM; manufacturer–Zeiss, model–Gemini) which has a Schottky-type field emitter system imaging at very high resolution (0.8 nm at 15 kV; 1.4 nm at 1 kV).

2.2 Alkali Activator

In addition to water, sodium hydroxide (NaOH) solution (alkali activator) is used to prepare the SJ mixes. NaOH being a strong base activates the geopolymerization process by dissolving the silica and alumina present in the material. The concentration of NaOH solution in the present study is kept at 5 M based on the previous research done on alkali-activated soil stabilization [11].

2.3 Preparation of SJ Mixes

In the present work, jarosite content in the SJ mix was fixed at 15 and 30%. Two sets of mixes were prepared (one with distilled water and other with NaOH solution). Oven-dried soil and jarosite are mixed together thoroughly for 5 min to obtain a homogeneous mix. Thereafter, the dry mix thus prepared is blended with distilled water/NaOH solution roughly for 10 min to devoid the formation of any lumps.

2.4 Standard Mini Compaction Test

The compaction characteristics of SJ mixes are determined utilizing standard mini compaction test apparatus [10]. Mini compaction test not only involves 1/10 volume of soil as required for standard proctor test but can be performed in less time and effort [10]. The various parts of mini compaction test apparatus are shown in Fig. 1. The test apparatus consists of a brass mold attached with a base plate, removable collar, and a steel drop hammer with guide frame (see Fig. 1). The mold has an internal diameter of 3.81 cm and height of 10 cm leading to an overall volume of 114 cc. Weight of drop hammer is 1 kg for light compaction test with a height of fall of 16 cm.

Test Procedure. About 200 g dry sample of soil, jarosite, and SJ mix is taken for mini compaction test. Soil and jarosite are prepared with distilled water whereas SJ mixes are blended with both distilled water as well as alkali activator solution. The wet mix is then compacted in three layers in the mold by giving 36 blows with 1 kg hammer. Once the compaction of sample is done, weight of mold with compacted soil and water content is measured to calculate the dry density at respective trials. At least six compaction tests were conducted for each mix to obtain an accurate compaction curve. The results obtained from different compaction tests are discussed in Sect. 3.

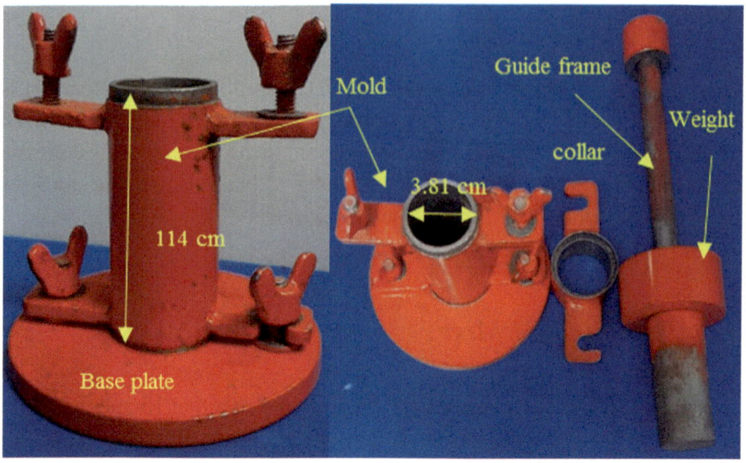

Fig. 1 Mini compaction test apparatus [10]

3 Results and Discussion

3.1 Physical Properties of Soil and Jarosite

Table 1 summarizes the various physical properties of soil and jarosite. The soil has reddish brown physical appearance whereas jarosite is pale yellowish tinge (see Fig. 2). In comparison to soil, the G of jarosite was found to be on higher side. The G of jarosite is measured as 3.01 in comparison to 2.64 for soil. The higher G of jarosite can be attributed to the presence of heavy metals in jarosite.

Figure 3 shows the particle size distribution curve obtained from the sieve analysis coupled with hydrometer analysis. It can be observed from Fig. 3 that jarosite has greater finer fraction (particle size < 75 μ) than the red soil. The sand, silt, and clay

Table 1 Physical properties of soil and jarosite

Properties	Soil	Jarosite
Specific gravity	2.64	3.01
Clay content (< 0.002 mm)	19.05	26.60
Silt content (0.075–0.002) mm	51.03	72.80
Sand fraction (4.75–0.075) mm	29.81	1.12
Liquid limit	38.51	66.84
Plastic limit	15.14	50.13
Plasticity index	23.37	16.71
ISCS classification	CI	MH
Water holding capacity (%)	190.10	140.72

Fig. 2 Physical appearance of soil and jarosite

content in soil was found to be 29.81%, 51.03%, and 19.05%, respectively. On the other hand, the sand, silt, and clay content clay content in jarosite was noted as 1.12%, 72.80%, and 26.60%, respectively. Based upon the grain size analysis, jarosite has a potential to be used as a filler material for various geotechnical applications.

LL of soil was determined to be 38.51% as opposed to 66.4% for jarosite. Similarly, PL of soil and jarosite was found to be 15.14% and 50.13%, respectively. As per Indian Standard Soil Classification System (ISCS) [4], soil was classified as clay with intermediate plasticity (CI), while jarosite was categorized as silt of high plasticity (MH).

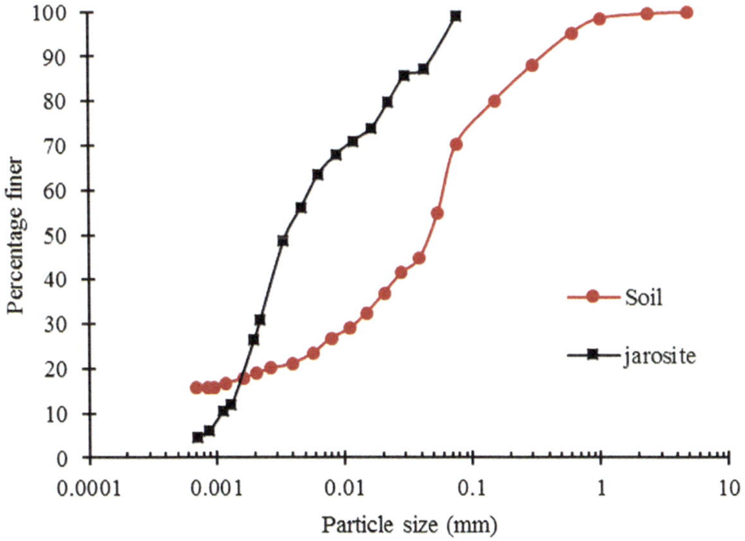

Fig. 3 Particle-size distribution curve for soil and jarosite

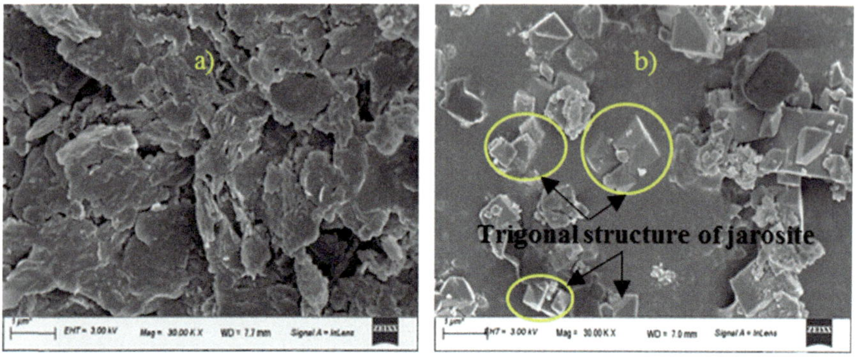

Fig. 4 FESEM images showing the microstructure of **a** soil and **b** jarosite particle

The water holding capacity of jarosite was observed to be higher (190.10%) as compared to soil (140.72%). This might be due to the fact that jarosite contains a much higher finer fraction (particle size < 75 μ) than soil, which increases its specific surface area and consequently its capacity to hold water.

3.2 Microstructure of Soil and Jarosite

Figure 4 depicts the microstructure of soil and jarosite obtained from FESEM analysis. It can be seen from Fig. 4a that soil particles have irregular structure without any particular shape. On the other hand, jarosite particles have trigonal crystal structure as evident from Fig. 4b. Similar observations regarding the crystalline structure of jarosite have been also reported by Cruells and Roca [2] in their study.

3.3 Compaction Characteristics of Soil, Jarosite, and SJ Mixes

The MDD and OMC of different materials obtained from mini compaction test are summarized in Table 2. The MDD of jarosite (1.24 g/cc) is determined to be significantly lower as opposed to soil (1.78 g/cc), while the OMC is noticed to be higher than soil. The lower MDD of jarosite can be attributed to its smaller particle size as compared to soil, contributing to its increased air void content. On the other hand, the proportions of sand, silt, and clay in soil are such that the voids left by coarser fractions are filled by finer fractions (see Fig. 5), hence increasing the MDD. Since jarosite has a higher water holding capacity than soil, a greater quantity of water is required for lubrication, resulting in a higher OMC compared to soil.

Table 2 Maximum dry density (MDD) and optimum moisture content (OMC) of mixes

Material	MDD (g/cc)	OMC (%)
Bare soil	1.78	16.5
Jarosite	1.24	45.5
Soil + jarosite (15%), with water	1.71	19.8
Soil + jarosite (30%), with water	1.69	20.5
Soil + jarosite (15%), with NaOH solution	1.60	23.8
Soil + jarosite (30%), with NaOH solution	1.48	27.8

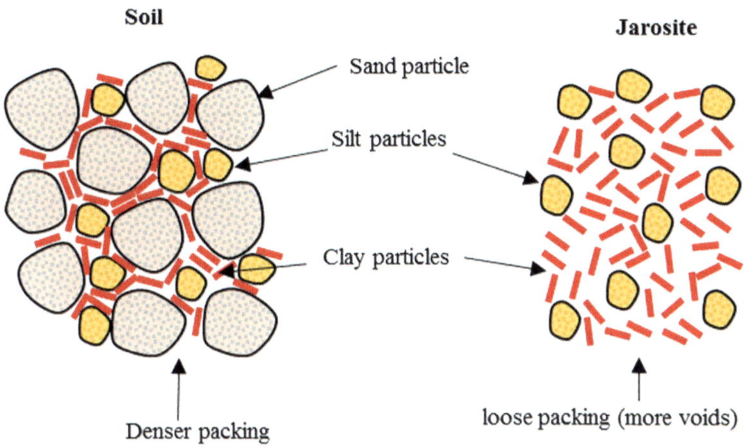

Fig. 5 Particles arrangement in soil and jarosite

Figure 6 illustrates the compaction curves obtained for soil, jarosite, and SJ mixes. In case of SJ mixes, with increase in jarosite content, a decrease in MDD and increase in OMC are observed. For instance, the MDD of SJ mix (prepared with water) with 15% jarosite content decreased by 3.9% in comparison to bare soil. On the other hand, a 20% increase in OMC is observed with the addition of 15% jarosite to the bare soil. Similarly, for SJ mix with 30% jarosite content, a 5% decrease in MDD and 24.24% increase in OMC are observed as compared to bare soil (see Fig. 6). Since jarosite contains a greater proportion of finer particles than soil, air voids existing in soil should be filled with jarosite particles. However, in this case, the amount of jarosite added (i.e., 15% and 30%) is substantially higher than the volume of air voids present in the bare soil, resulting in the reduction of MDD values. The higher OMC of SJ mixes (in compared to bare soil) can be attributed to the high-water holding capacity of jarosite (i.e., 190.12%).

Furthermore, the MDD of SJ mixes blended with alkali activator, i.e., NaOH solution was observed to be significantly lower than SJ mixes prepared with water (see Fig. 6). For example, MDD of SJ mix (with 15% jarosite content) prepared with NaOH solution decreased by 6.43% in comparison to SJ mix blended with water.

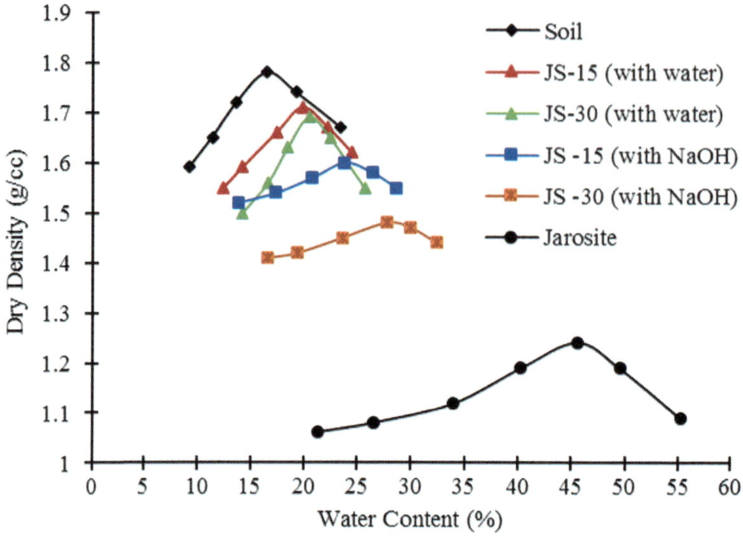

Fig. 6 Compaction curve for soil, jarosite, and SJ mixes

On a similar note, a 12.42% decrease in MDD was observed for SJ mix (with 30% jarosite content) prepared with NaOH solution as compared to that prepared with water. This is due to fact that a portion of soil and jarosite particles in SJ mix is replaced by Na_2O solids present in NaOH solution (see Fig. 7), reducing the MDD of SJ mixes. Furthermore, the OMC of SJ mixes prepared with NaOH solution was found to be higher than that prepared with water. Because NaOH solution has a higher viscosity than distilled water, a larger amount of solution is required to lubricate soil and jarosite particles, increasing the OMC values (see Fig. 7).

In conclusion, the addition of jarosite content and NaOH solution decreased the MDD of SJ mixes while increasing the OMC.

4 Conclusions

The main objective of the present study is to characterize the jarosite for its potential use as a geopolymer binder for soil stabilization. Basic physical properties of jarosite including particle size analysis, consistency limits, microstructure analysis, etc. are initially investigated. Thereafter, compaction characteristics of SJ mixes prepared with the addition of 15% and 30% jarosite to soil are examined with an objective to find the target dry density and OMC required for the strength determination of geopolymerized soil. The following conclusions are drawn from the study:

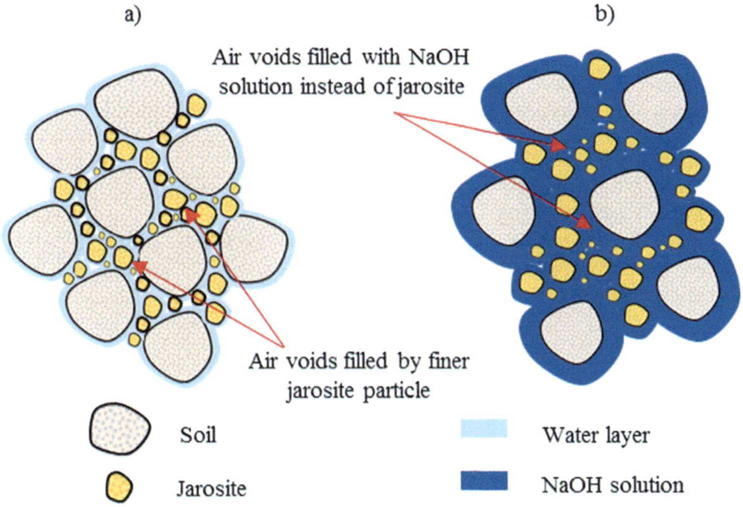

Fig. 7 Mechanism of particles lubrication in SJ mix prepared with a) distilled water and b) NaOH solution

- The finer fraction (particle size < 75 μ) in jarosite is found to be higher than soil. Jarosite is classified as silty soil with high plasticity (MH), whereas soil is categorized as clay of intermediate plasticity as per ISCS system.
- MDD of jarosite is observed to be significantly lower than the soil; however, the OMC is found to be higher than bare soil.
- A decrease in MDD and increase in OMC is noted for SJ mixes with increase in the jarosite content.
- In contrast to water, SJ mixes blended with NaOH solution are found to have lower MDD and higher OMC.

References

1. Beena KS, Santhosh G (2021) Transport geotechnical properties of lateritic soil stabilized with Jarofix and lime. Environ Dev Sustain 23(2):1601–1622
2. Cruells M, Roca A (2022) Jarosites: formation, structure, reactivity and environmental. Metals 12(5)
3. Gared O, Gaur A (2020) Feasibility study of Jarosite as cement replacement in rigid pavement. Mater Today: Proc 44:4337–4341
4. IS 1498 (1970) Classification and identification of soils for general engineering purposes, Bureau of Indian Standards, New Delhi
5. IS 2720-Part 3 (1) (1980) Methods of test for soils—determination of specific gravity. Bureau of Indian Standards, New Delhi
6. IS 2720-Part 5 (1985) Methods of test for soils—determination of liquid limit and plastic limit. Bureau of Indian Standards, New Delhi

7. Patwa D, Chandra A, Ravi K, Sreedeep S (2021) Influence of biochar particle size fractions on thermal and mechanical properties of biochar-amended soil. J Mater Civ Eng 33(9):04021236
8. Ray S, Daudi L, Yadav H, Ransinchung GD (2020) Utilization of jarosite waste for the development of sustainable concrete by reducing the cement content. J Clean Prod 272:122546
9. Seyer S, Chen TT, Dutrizac JE (2001) Jarofix: addressing iron disposal in the zinc industry. JOM 53(12):32–35
10. Sridharan A, Sivapullaiah PV (2005) Mini compaction test apparatus for fine grained soils. Geotech Test J 28(3):240–246
11. Yaghoubi M, Arulrajah A, Disfani MM, Horpibulsuk S, Bo MW, Darmawan S (2018) Effects of industrial by-product based geopolymers on the strength development of a soft soil. Soils Found 58(3):716–728

An Experimental Investigation and Numerical Modeling of Glass Fiber-Reinforced Expansive Subgrade Soil With Alkaline Stabilizer

Mazhar Syed, Sai Kiran Chukka, and Anasua GuhaRay

1 Introduction

The twin character of expansive soil, especially in desiccation cracking, impacts the pavement structure's long-term sealing properties. With active montmorillonite and smectite groups, the periodic changes in soil-moisture balance lead to severe distresses on subgrade layers upon the load application [1]. Lime and Portland cement-based binders can effectively control the soil volumetric stability by forming cementitious products through the pozzolanic-hydration mechanism [2]. Besides, these traditional stabilizers' annual carbon production contributes 7–8% globally. Using low energy stabilizer with residual end products such as alkaline binder is an alternative solution for efficiently strengthening the unstable expansive soil. AB is a long mechanism of aluminosilicate-polymeric sodium compounds when dry precursors are activated [3]. AB reduces the cost of disposing of solid waste in landfills. It also provides better workability and mechanical strength than traditional binders, emitting 75% less CO_2. Although AB-soil's peak shear and compressive resistance improve significantly, it still has a low tensile and penetrating strength. Numerous research has demonstrated that fiber reinforcing effectively regulates the volume and surface area of linear cracks on the subgrade layer.

The current study focuses on utilizing eco-friendly waste by-products in efficient and cost-effective applications to prolong the subgrade pavement structure's service life. Comprehensive research is needed to estimate the impact of uncertainties on stabilized soil, particularly subgrade performance, by including fiber reinforcement and industrial-based low-carbon precursors in an alkaline soil mixture. A detailed model study is needed to predict the subgrade stability behavior under the influence of vehicular load at fiber-mixed expansive soil. This study uses geopolymerization

M. Syed · S. K. Chukka · A. GuhaRay (✉)
BITS-Pilani Hyderabad Campus, Secunderabad, Telangana 500078, India
e-mail: guharay@hyderabad.bits-pilani.ac.in

B. T. Jose et al. (eds.), *Proceedings of the Indian Geotechnical Conference 2022 Volume 4*, Lecture Notes in Civil Engineering 479,
https://doi.org/10.1007/978-981-97-1753-8_4

Table 1 Characteristics of the soil and other raw materials

Properties	Soil	FA	SBA	Properties	GF
pH	8.5	9.2	10.5	Diameter (μm)	33
Specific gravity	2.6	1.8	2.5	Specific gravity	0.88
Swelling index (%)	96	–	–	Tension resistance (MPa)	342
Liquid limit (%)	64	21	–	Elastic modulus (MPa)	3960
Dry density (g/cc)	1.5	–	–	Flaming point (°C)	205
Moisture content (%)	22	–	–	Warming point (°C)	610
Soaked CBR	1.96	–	–	Coefficient of friction	0.20

and glass fiber reinforcement at different fly ash-bagasse ash doses in the alkaline combination to try and strengthen the expansive subgrade soil. The study examines GF subgrade soils stabilized with lime and AB microstructural and geomechanical behavior.

2 Material Characteristics

2.1 *Expansive Soil*

In this research, expansive soil was taken from the Kazipet region of Telangana, India. The collected soil (excavated around 12 cm below the surface) was categorized as high plastic clay (CH) with 70% fine particles as per the Indian standard soil classification system. Low calcium-fly ash (FA) and sugarcane bagasse ash (SBA) were obtained from the Ramagundam Power Plant and Medak Sugar Industry as a dry precursor in the alkaline binder. Hychem Chemicals Ltd. in Hyderabad provided hydrated lime and glass fiber (GF). The current investigation used GF with a circular radius of 17 μm under a static height of 1.2 cm. A previous study found that GF exhibits a stronger soil-fiber matrix, good interfacial bridging, and a greater inclination to control the desiccating cracks compared to other fiber lengths [4, 5]. Table 1 shows the specifications of various raw materials.

2.2 *Alkali Stabilizer (AS)*

An alkaline binder was produced by combining sodium hydroxide (NaOH) in sodium silicate (Na_2SiO_3) solution with varying content of FA (100, 90, 80, and 70%) and SBA (0, 10, 20, and 30%). With 0.4 water-to-solid ratios (amount of water used to prepare the binder), the mass ratio of NaOH: FA: SBA: Na_2SiO_3 was maintained at 10.57:360:40:129.43 [6, 7]. The Na_2SiO_3 comprises 56% H_2O, 28% SiO_2, and

15% Na_2O. Both chemicals (NaOH and Na_2SiO_3) were supplied by Organic-Chem Laboratories, Hyderabad, in pellet and solution form.

2.3 Sample Preparation

Different fiber dosages (0–0.5%) were randomly added into oven-dried untreated soil with a 0.1% variance. Before fiber inclusion, expansive soil was thoroughly mixed with 6% hydrated lime and 6% AS (total soil mass) with 0.4 w/s in the geopolymeric compounds. Different quantities of fly ash-bagasse ash material-based alkaline soil mixture were compacted into three layers separately in containers of (950 × 480 × 150) mm^3 size by covering with moist jute bags for 28 days.

3 Results and Discussions

3.1 X-Ray Diffraction (XRD)

The crystalline profile was investigated through XRD tests by generating geopolymeric products in fiber-reinforced alkaline soil using a RIGAKU Ultra IV diffractometer with CuK rays with operating 2θ up to 80° at 2 s/step. The XRD peak for untreated soil, lime, and alkaline-treated soil was exhibited in Fig. 1 at 28 days of curing. Raw soil attains the crystalline peaks of montmorillonite (M) $2\theta = 18.9°$, 26.9°, 35.9°, and 61°), which predominantly influence the soil swell behavior [8, 9]. Besides, a small amount of muscovite (Ms = 26.1°, 26°, and 60.4°) was also observed. The highest peaks exhibited the existence of quartz ($Q = 27.3°$, 36.1°, and 61°), which was strong crystallinity in clay structure. The profile of montmorillonite (M) and muscovite (Ms) showed dramatic variations (reduction in diffraction peak) in alkaline and lime-treated soil. As expected, the quartz (Q) diffraction peak in both binder mixed soil remained unaltered. The geopolymerization reaction in alkaline soil can cause the development of Mullite (Mu = 18.20°, 19.85°, 34.25°, 47.2°, 50.9°, and 54.4°) and calcium silicate hydrate (CSH). In addition, two new cementitious compounds, calcium alumina hydrate (CAH) and kaolinite (K), were identified in lime soil.

3.2 Stereomicroscope-Scanning Electron Microscope

Figure 2a–f showed the soil surface images from a stereomicroscope (using an Sx27 Olympus device. The most petite measurement that could be marked under this micromachine is 20 μm) and SEM analysis (using FEI-Thermo Scientific Apreo

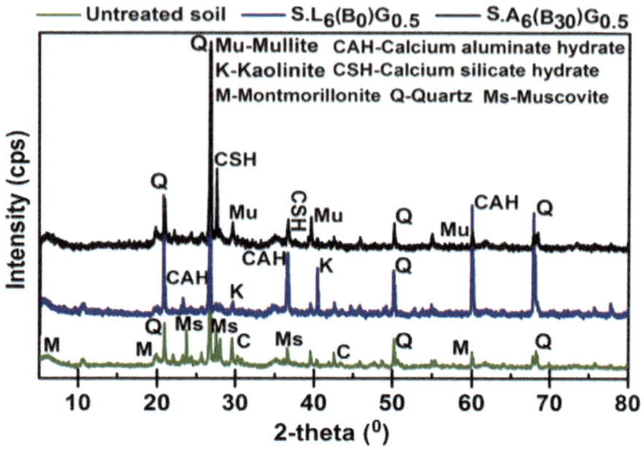

Fig. 1 XRD mineralogical peaks of untreated, GF-reinforced lime and alkaline mixed soil

device) at various magnifications. The untreated soil (Fig. 2a) demonstrated the existence of iron and illite in the region of light yellow and brown color. Similarly, Fig. 2b shows the crumbly-type morphological structure corresponding to the clayey smectite and montmorillonite groups. In addition, significant layer soil cracks were seen, which decreased the volumetric stability during saturation. Glass fiber inclusion in lime-stabilized soil was shown in Fig. 2e and d, acting as a bridge surface [10, 11]. The discrete GF reinforcement in the alkaline soil combination could act as a spatial trench framework that interlocks the particles through contact roughness, resulting in higher deformation energy before failure (Fig. 2e, f).

3.3 California Bearing Ratio (CBR)

CBR penetration resistance tests on fiber-mixed subgrade soil treated in lime and AS mixture were carried out according to the ASTM D-1883. At respective MDD-OMC values, different fiber doses in lime and AS soil specimens were compacted in a cylindrical CBR mold of 17.5 cm under a constant diameter of 15 cm, respectively. A static compactor with 56 blows was tested at a predefined strain rate of 1.25 mm/min under a loading frame with a 50-mm plunger after being submerged in water for 96 h with five layers of soil-fiber samples.

Figure 3 showed soaked CBR penetration tests for fiber-reinforced lime and alkaline soil at varying FA-SBA doses. It was also noted that the soaked CBR values of the AS soil mixture increased with an increase in SBA-GF dosages. The graph showed that adding lime and 100% fly ash-based (zero bagasse ash) alkaline mixtures (B0) to GF-reinforced soil raised the penetration value from 2 to 5% and 7.1%, respectively. The pozzolanic reactions that catalyzed the flocculation and geopolymerization processes during the soaking phase might be attributed to the drastic improvement in the strength-bearing ratio. The combined addition of GF and SBA to the

Fig. 2 Soil surface morphology. **a, b** Untreated soil, **c, d** lime-stabilized GF-soil, **e, f** alkaline-treated GF-reinforced soil

geopolymerized soil composite was beneficial as it effectively improves the confinement bonding and stretching resistance during load penetration [12, 13]. Therefore, it became difficult for the cementitious structure to form clayey particles between the GF matrix to alter the bonding position from one surface area to another, thus strengthening the penetration resistance during static load application. Additionally, GF could effectively disperse in the soil specimen and support the compacting stress with a greater coupling effect [14]. It was also discovered that between 0.2 and 0.4% fiber doses, the influence of CBR on lime and AS soils rose significantly, regardless of bagasse ash proportions. As a result, the rapid changes in soil-fiber interlocking morphology and frictional density had the potential to improve penetration resistance.

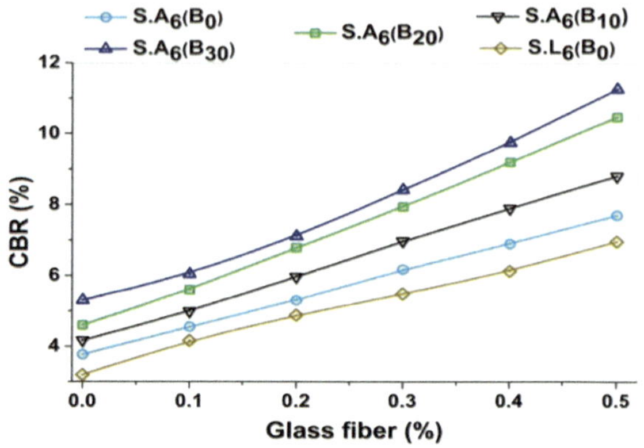

Fig. 3 Variations of CBR of lime and alkaline stabilizer mixed GF-soil at varying FA-SBA

3.4 *Indirect Tensile Strength (ITS)*

A Marshal stability machine with a 12.5 mm loading strip attached to the load frame and a 50.5 mm/min strain rate was used to conduct the ITS test. Lime and alkaline soils were compacted in 100-mm-diameter by 80-mm-thick cylindrical molds. The variation in ITS of both lime and alkaline stabilized subgrade soil with varying GF doses was depicted in Fig. 4a. In comparison with unreinforced lime and alkaline soil specimens at various fly ash-bagasse ash contents, the data shown in Fig. 4b showed a significant decline in the ITS ratio. The combination of SBA-FA-GF dosages in the geopolymerized soil composite played a critical part in improving soil stiffness resisting potential, correlating to trends in tensile strength and ITS ratio values by imparting the compressive shear stresses to the fibers. [10, 15].

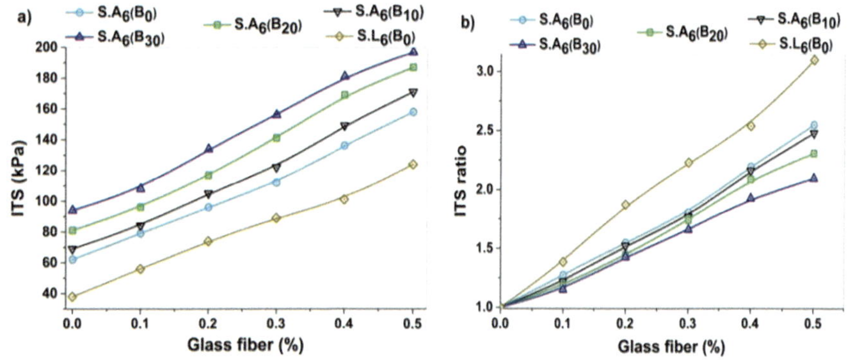

Fig. 4 Variations of ITS and ITS ratio of lime and alkaline stabilizer mixed GF-reinforced soil at different fly ash-bagasse ash dosages

4 Finite Element Analysis

The finite element mechanism is a practical network-based computational method effective in countering difficult geotechnical problems. In this method, the basic approach is to divide the continuous and complex model into a finite number of elements such that stresses and strains are calculated computationally. There was a less number of FEM studies applied for expansive soils in pavement subgrade application. The present study addresses the deformation behavior of a subgrade under the application of wheel load due to static conditions using PLAXIS 2D geotechnical FE software.

4.1 FE Model Inputs

The model was analyzed for single-wheel load due to symmetry under both wheels, and the plain strain condition was assumed for analysis. The wheel load was applied at the center of the span of 3 m. The section depth was 1.5 m, in which the top 0.3 m thickness layer was considered subgrade resting on the dense soil. The total thickness of pavement was about 1 m, in which the top 0.7 m thick base and sub-base layers were applied as overburden load, and the equivalent load of 11.06 kPa was placed uniformly over the subgrade layer. IRC:37 [16] standard load of 80 kN was acting due to single axle and load distributed equally on either side of wheels resulting in a value of 40 kN. The load was assumed to be distributed 1H:2 V below the ground, under a pressure of 40 $kN/m^2/m$ on the subgrade's surface. Figure 5 presented the typical model with subgrade, dense layer, wheel load pressure, and surcharge loads on the subgrade.

The analysis was investigated for unstabilized and chemically stabilized subgrade soils in PLAXIS; the material characteristics used in the model were presented in Table 2. Demonstrated values in the Table 2 were calculated in the laboratory using standard proctor compaction, unconfined compressive strength, and 1D consolidation tests per ASTM standards. The boundary conditions defined in the model by extremes

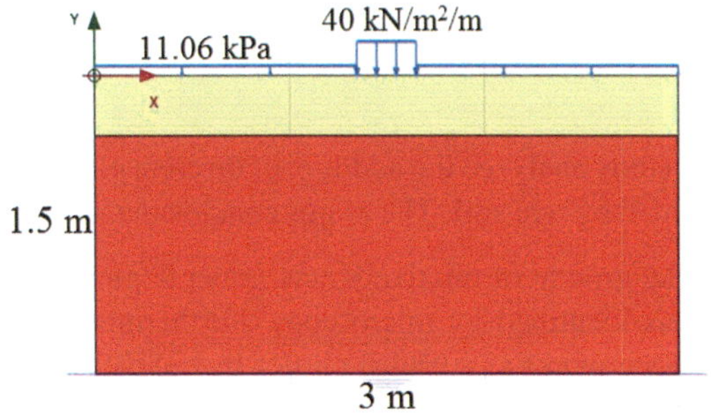

Fig. 5 FE model of pavement section with single-wheel load and surcharge load

Table 2 Parameters used for FE modeling

Identification	Untreated soil	Lime-GF-soil	Alkaline-GF-soil	Dense soil	Unit
γ_d	15.80	17.80	18.90	21.04	kN/m^3
γ_{sat}	16.80	19.40	21.11	22.60	kN/m^3
e_{int}	1.09	0.88	0.65	0.75	–
E	5010	21,500	34,000	47,000	kN/m^2
μ	0.3	0.32	0.33	0.35	–
c	85	288	466	550	kN/m^2
φ	4	12	18	35	–

of X-min and X-max were horizontally restrained, and Y at the surface allowed horizontal and vertical deformations. The bottom was completely restrained. To get an accurate solution, fine elemental meshing was used in the analysis.

4.2 FE Analysis Results

The outcomes of FE analysis were compared in view of vertical deformations below the wheel load with a change of subgrade treatment with GF-lime and GF-AS. Figure 6a–c shows the results of vertical deformations in subgrade soil under a wheel load. The deformation decreased with the treatment of soils. The vertical deformations for untreated subgrade soils were about 0.85 and 0.29 mm with GF-lime-treated and 0.22 mm with GF-AS-treated soils. Less vertical settlements were observed as compared to GF-lime and untreated soils.

Similarly, the overall factor of safety of subgrade was estimated by using shear strength reduction in PLAXIS FE analysis. The GF-AS-treated soils exhibited higher safety factors than GF-lime-treated and untreated soils. Figure 7 presents the factor of safety variation in the subgrade with the addition of GF-lime and GF-AS. The alkaline-treated soils exhibited around five times higher than untreated soils and nearly two higher when compared with GF-lime treated.

5 Conclusions

This study analyzed the addition of discrete glass fibers in lime and alkaline stabilizer mixed subgrade soil. The major conclusions are as follows from this analysis:

- The growth of relatively new mineralogical phases in the lime and alkaline soil was identified by microscopy. Furthermore, the GF-AB-soil's physical surface shape works as a bridge cluster by tightly encircling the soil-fiber matrix.

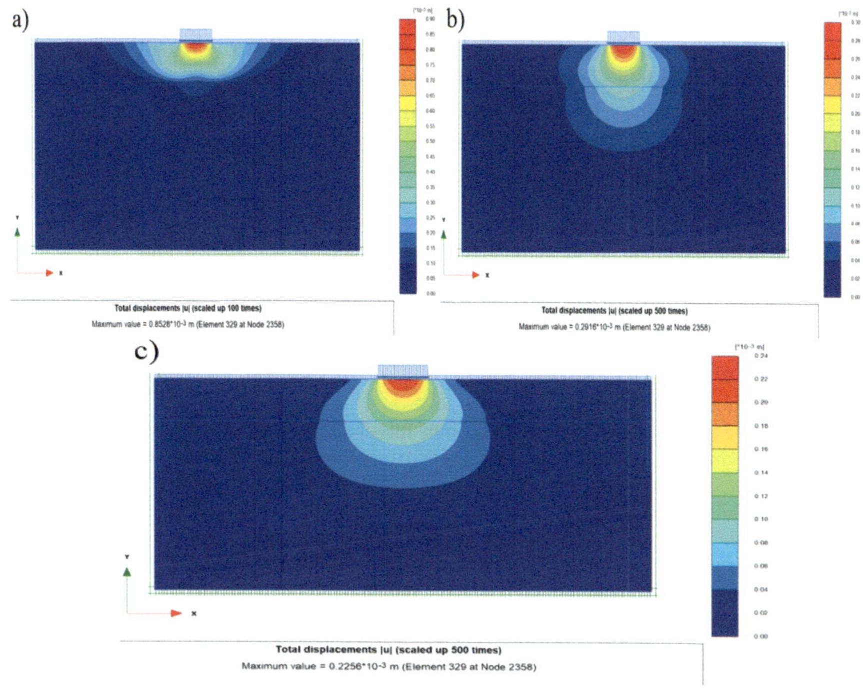

Fig. 6 Variation of vertical settlements in subgrade soils. **a–c** Untreated, GF-lime-treated, and GF-AS-treated subgrade soils

Fig. 7 Variation of factor of safety with untreated, GF-lime-treated, and GF-AS subgrade soils

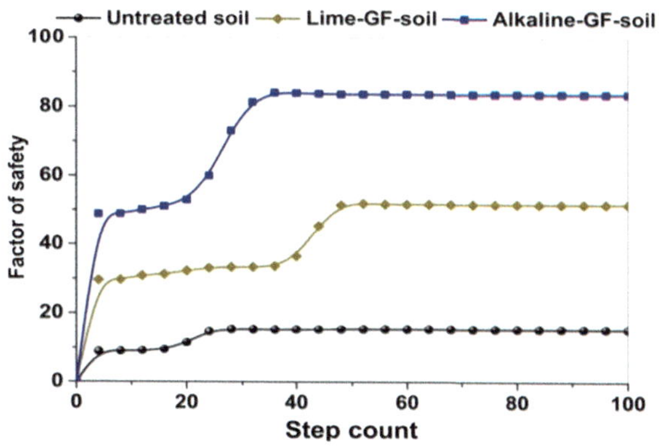

- At different doses of fly ash-bagasse ash and fibers, GF-alkaline-soil achieved higher penetration stiffness resistance than lime-GF-soil due to the significant high fluctuation of interfacial density and frictional bonding. Furthermore, compared to untreated soil, the CBR of alkaline soil is enhanced by 66% for 30% bagasse replacement and 39% for lime soil.

- Through active cementitious bonding and the formation of interfacial roughness within the fiber-clay matrix, GF-AB-soil effectively controls tensile cracking by converting brittle behavior into ductile behavior.
- The outcomes of the FE analysis reveal that GF-AS-treated subgrade soils have exhibited lower deformation and higher safety factors compared to GF-lime-treated and untreated subgrade soils.

References

1. Ackroyd LW, Husain R (1986) Residual and lacustrine black cotton soils of north-east Nigeria. Géotechnique 36:113–118
2. Bell FG (1996) Lime stabilization of clay minerals and soils. Eng Geol 42:223–237
3. Mazhar S, Guharay A (2020) Stabilization of expansive clay by fibre-reinforced alkali-activated binder: an experimental investigation and prediction modelling. Int J Geotech Eng 15:1–17
4. Syed M, Guharay A (2020) Effect of chemically treated fiber reinforcement on expansive soil subgrade stabilized with alkali activated binder. Constr Build Mater J 1–17
5. Syed M, GuhaRay A, Garg A (2021) Performance evaluation of lime, cement and alkali-activated binder in fiber-reinforced expansive subgrade soil: a comparative study. J Test Eval 51
6. Gupta S, GuhaRay A, Kar A, Komaravolu VP (2018) Performance of alkali-activated binder-treated jute geotextile as reinforcement for subgrade stabilization. Int J Geotech Eng 1–15
7. Kar A (2013) Characterizations of concretes with alkali-activated binder and correlating their properties from micro-to-specimen level. Arkamitra Kar Dissertation submitted to the Benjamin M. Statler College of Engineering and Mineral in partial fulfillment of Doctor. West Virginia University
8. Sivapullaiah PV, Sankara G, Allam MM (2010) Mineralogical changes and geotechnical properties of an expansive soil interacted with caustic solution. Environ Earth Sci 60:1189–1199. https://doi.org/10.1007/s12665-009-0260-9
9. Miao S, Wei C, Huang X et al (2017) Stabilization of highly expansive black cotton soils by means of geopolymerization. J Mater Civ Eng 29:04017170
10. Tang C, Shi B, Gao W et al (2007) Strength and mechanical behavior of short polypropylene fiber reinforced and cement stabilized clayey soil. Geotext Geomembr 25:194–202
11. Mazhar S, GuhaRay A (2020) Effect of fiber reinforcement on mechanical behaviour of alkali activated binder treated expansive soil: a reliability-based approach. Int J Geomech 20(12):04020225-1-14. https://doi.org/10.1061/(ASCE)GM.1943-5622.0001871
12. Sudhakaran SP, Sharma AK, Kolathayar S (2018) Soil stabilization using bottom ash and areca fiber: experimental investigations and reliability analysis. J Mater Civ Eng 30:1–10
13. Syed M, GuhaRay A, Goel D (2021) Strength characterization of fiber reinforced expansive subgrade soil stabilized with alkali activated binder. Road Mater Pavement Des 20:1–25
14. Mazhoud B, Collet F, Pretot S, Lanos C (2017) Mechanical properties of hemp-clay and hemp stabilized clay composites. Constr Build Mater 155:1126–1137
15. Pourakbar S, Asadi A, Cristelo N, Fasihnikoutalab MH (2016) Application of alkali-activated agro-waste reinforced with Wollastonite fibers in soil stabilization. J Mater Civ Eng 29:04016206
16. IRC:37-2018 (2018) Guidelines for the design of flexible pavements, 4th Revision. IRC 37-2018, New Delhi, India, pp 1–94

Stabilization of Kuttanad Soil Using Terrasil Nanochemical and GGBS

Aleena Mathew and Nirmal John Joy

1 Introduction

Kuttanad is one of the globally important agroheritage areas which contain low-lying region of 0.6–2.2 m below sea level. Many geotechnical failures necessitate studies on engineering properties. The characteristics of Kuttanad soil's permeability and compressibility have been identified through a variety of studies. According to reports, this soil is unsuitable for construction of foundations, embankments and unlined canals so some stabilization techniques should be incorporated to modify the properties [1]. Recently, the use of nanocompounds is increasing in various applications; in the case of soft soil also, additions have to be incorporated to improve the properties so that road pavement or embankment or structures can be constructed safely. Nanoparticles are small objects which behave as a whole unit, and they come under a size range of 1–100 nm. Nanoparticles behave differently physically and chemically than conventional materials do.

1.1 Soil Interaction With Nanoparticles

The terrasil chemical is evolving as a new substance for soil stability. Terrasil contains 100% organosilane atoms, which are water soluble, prevent ultraviolet and heat; they are used as a highly reactive soil improvement technique which makes the soil waterproof. Nobody has a clear idea about the fundamental mechanisms of how soil behavior changes when nanoparticle is added to the soil. But some researchers made an attempt to do the same.

A. Mathew (✉) · N. J. Joy
Saintgits College of Engineering, Kottayam, Kerala, India
e-mail: aleenamathew565@gmail.com

B. T. Jose et al. (eds.), *Proceedings of the Indian Geotechnical Conference 2022 Volume 4*, Lecture Notes in Civil Engineering 479,
https://doi.org/10.1007/978-981-97-1753-8_5

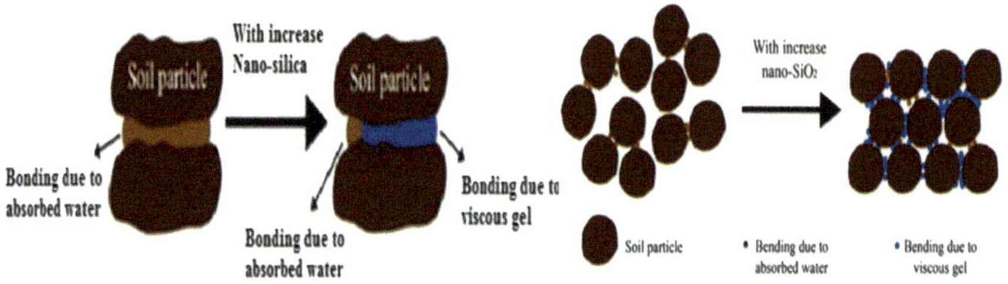

Fig. 1 Behavior of soil with addition of nanoparticles [2]

Normally, soil particles are bonded due to the adsorbed water molecules in between the soil particles. When nanomaterial is added to the soil, a viscous gel will be created in between the soil particles. The silica nanoparticles present create a viscous gel-like substance through absorption double-layer water. Due to this viscous gel, the particles get tightly bonded with each other. When compared with untreated soil, nanosoil matrix has stronger bonding due to the presence of viscous gel in between the soil particles. In Fig. 1, it is observed that as the silica nanoparticle is added to the soil, the space between the clay particles reduces and thus improves the interparticle bonding and friction [2].

2 Materials Used

2.1 Kuttand Soil

The samples for the testing program were acquired from Kerala's Kunnamkeri region, close to the Kuttanad area. The samples of disturbed and undisturbed soil were gathered from 1 m below the surface of the earth. The laboratory tests were conducted according to IS 2720–1965 and determined the geotechnical properties of untreated soil samples. Table 1 shows the test results.

2.2 Terrasil

Terrasil nanochemical used as the stabilizer was acquired from Zydex Industry. According to the technical information offered by Zydex Industries Pvt. Ltd., the recommended dosage of terrasil is between 0.5 and 1 kg per cubic meter of soil in order to achieve higher UCS values. To find the ideal dosage that corresponds to the maximum strength value, terrasil percentage is varied in the current work between 0.02 and 0.05% of soil weight. After mixing terrasil into soil in amounts of 0.02, 0.03, 0.04 and 0.05%, a total of 4 soil combinations were created. The terrasil agent

Table 1 Geotechnical properties of soil samples

Sl. No.	Property	Values
1	Specific gravity	2.49
2	*Atterberg's limits*	
	wl-Liquid limit (%)	93
	wp-Plastic limit (%)	69
	Ip-Plasticity index (%)	24
3	IS grain size analysis and soil classification	High compressible Silt-MH
4	*Engineering properties*	
	Density g/cc	1.207
	Water content (%)	90
5	Unconfined compressive strength (kpa)	10
6	Cohesion (kN/m^2)	5.5
7	Angle of internal friction	6.2°
8	Modulus of elasticity (E)	2357 kN/m^2

is added in a predetermined dosage along with the necessary amount of water to create the initial nanochemical solution. Additionally, the soil mixtures are made by evenly mixing the nano-14 chemical solution after spraying it on soil. For curing, the mixture was sealed against the air.

2.3 GGBS

GGBS was collected from India Mart. It is a by-product material made during the production of iron. It is primarily made of silicate, alumina and lime. The ingredients used for the iron production process determine the slag's chemical composition. 40% Calcium oxide, 35% Silica, 13% Alumina and 8% Magnesia make up a typical chemical composition.

3 Experimental Investigation

3.1 Optimum Dosage of GGBS

Sample Preparation. The GGBS was mixed with various proportions like 6, 9 and 12% by its dry weight of soil to obtain the optimum percentage of GGBS. For each percentage, samples were prepared and properly sealed to prevent the moisture to come in contact. To determine the optimum value, SPT and UCS tests were conducted.

Laboratory Tests. To determine the optimum percentage, different experimental investigations were conducted like UCS and SPT.

Unconfined Compressive Strength Test. The mixed soil samples were filled in a mold, and it was extruded and kept for 14 days of curing. After the curing period, UCS test was carried out, and it was inferred that the soil samples mixed with 9% GGBS to the dry weight of soil showed a better result when compared to 6 and 12% GGBS mix.

The relationship between UCS and GGBS percent after 14 days of curing is depicted in Fig. 2. From the results, it was concluded that 9% GGBS addition shows a better result of 15 kPa when compared to zero percent GGBS addition.

Standard Proctor Test. Samples for the compaction test were made by combining 3 kg of dry soil with various GGBS additions (6, 9 and 12%); in total, three samples were prepared. These samples were kept for 28 days of curing. After curing, standard proctor test was done. Figure 3 shows the results obtained. Soil samples with 9% GGBS addition show a better result with a MDD of 1.416 g/cc and the OMC value of 21%. As the GGBS percentage increases, the MDD increases and OMC decreases up to a particular limit of 9%, and beyond that, the MDD got dropped down and OMC value got raised.

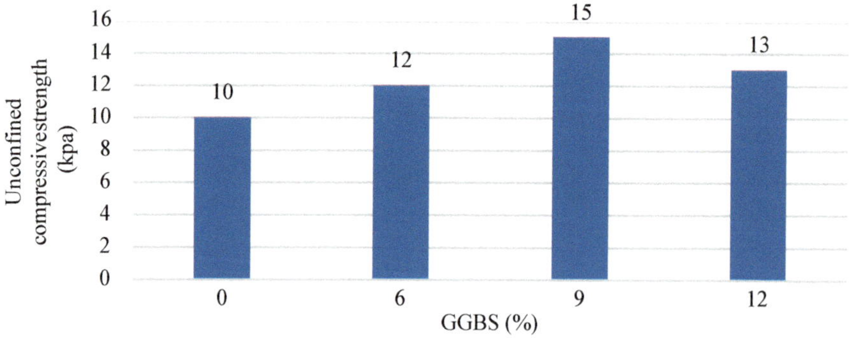

Fig. 2 UCS test results of 6, 9 and 12% GGBS addition

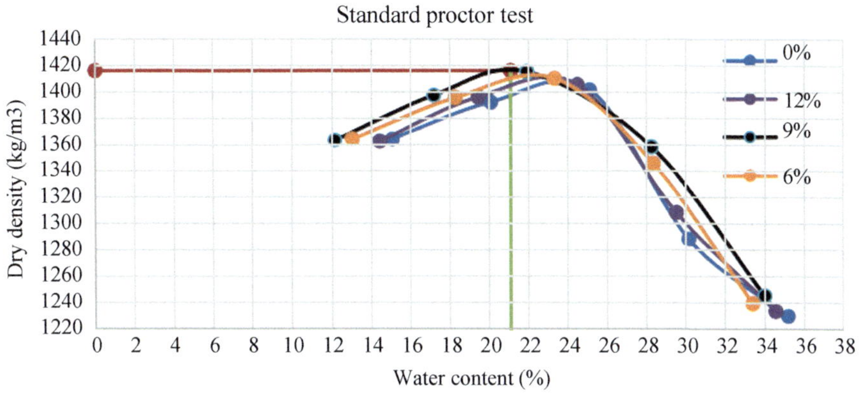

Fig. 3 Standard proctor test results

3.2 *Optimum Dosage of Terrasil*

Sample Preparation. Terrasil dosage varies in the current work and ranges from 0.02 to 0.05% of the weight of the dry soil. Initially, the ideal percentage of GGBS was mixed to the soil, later terrasil was mixed with the OMC and it was inputted to the soil-GGBS mixture. Later water was incorporated to the mix to attain field condition. Figure 4 shows the soil-GGBS-terrasil sample mix prepared with different dosage of terrasil (0.02, 0.03, 0.04 and 0.05%) for conducting the laboratory test. Following a seven-day curing period, laboratory tests were conducted on these samples to determine the ideal terrasil percentage and the level of soil improvement.

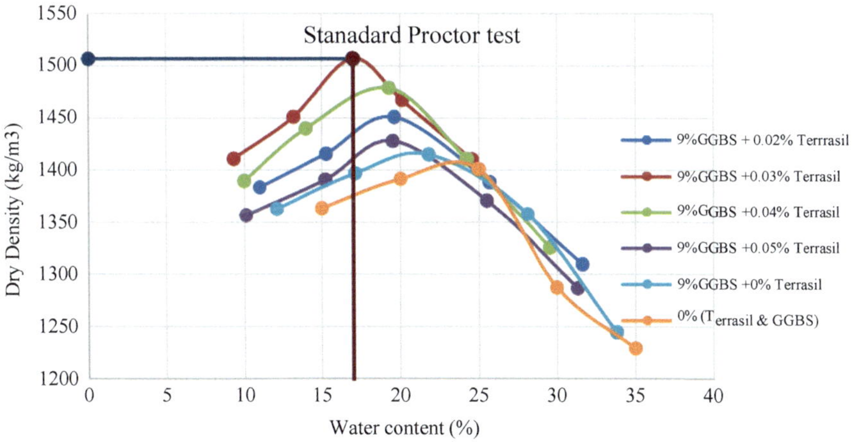

Fig. 4 Standard Proctor test results of terrasil and GGBS treated soil

Laboratory Test. To establish the optimum terrasil dosage, various laboratory tests including SPT, UCS, triaxial and Atterberg limits tests were carried out.

Standard Proctor Test. This test was done for each terrasil dosage to identify the compaction characteristics of the soil. 3 kg of soil samples were taken for each dosage of terrasil; in total, four samples of different dosages like 0.02, 0.03, 0.04 and 0.05% were taken for the test. Figure 4 shows the results obtained after conducting standard proctor test. Soil mixed with 9% GGBS and 0.03% terrasil shows a peak value of 1.507 g/cc and OMC value reduced to 17%. From the graph, it is visible that as the terrasil dosage increases, the maximum dry density increases up to a particular limit of 0.03% terrasil addition and beyond that the value decreases. So the optimum percentage of terrasil addition was observed as 0.03%.

Atterbergs Limits Test. To ascertain how soil behaves plastically, the Atterberg limits test was performed. Using Casagrande's liquid limit apparatus, the test was conducted. Table 2 shows the Atterberg limits results of nanochemical treated soil. From the test results, it was inferred that as the terrasil dosage increases, liquid limit and plastic limit values decrease. The trend of plasticity index values is in such a way that initially it starts decreasing up to a particular dosage of 0.03% terrasil and beyond that it increases. Thus, it is inferred that the plastic behavior of soil starts decreasing up to the optimum value.

Unconfined Compressive Strength Test. UCS of each dosage soil samples was tested. Figure 5 shows the UCS test results of treated soil. From the graph, it is clear that initially the UCS starts increasing up to a particular limit of 0.03% after that it starts decreasing. For virgin soil, the UCS value was 10 kPa, whereas for treated soil, the value increases up to 40 kPa, whereas for 0.04 and 0.05% terrasil, the value reduces to 25 and 20 kPa. Thus, it was concluded that only up to optimum terrasil addition it indicates an improvement in the unconfined compressive strength value by 30 kPa beyond that the value starts decreasing.

Triaxial Test. It was conducted in order to determine the shear strength characteristics. Three soil samples were taken, and different confining pressure of 50, 75 and 100 kN/m^2 was applied. Figure 6 shows the triaxial test results of terrasil treated soil. Trials 1, 3 and 2 correspond to the Mohr circle of 50, 75 and 100 kN//m^2 confining pressures, respectively. The equation of Mohr envelope is given as $y = 0.1586x + 20.2$, and the value of cohesion and angle of internal friction is obtained as 20.2 and 9.01 kN/m^2.

Table 2 Atterberg limits test results of terrasil and GGBS treated soil

Soil type	Liquid limit (%)	Plastic limit (%)	Plasticity index (%)
Soil + 0% (GGBS, Terrasil)	93	69	24
Soil + 9% GGBS + 0.02% Terrasil	86	66	20
Soil + 9% GGBS + 0.03% Terrasil	83	65	18
Soil + 9% GGBS + 0.04% Terrasil	81	62	19
Soil + 9% GGBS + 0.05% Terrasil	80	59	21

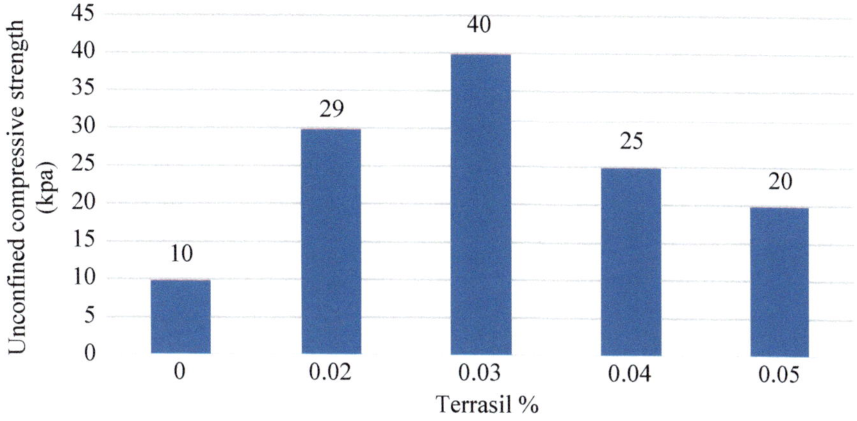

Fig. 5 UCS test of GGBS and terrasil treated soil

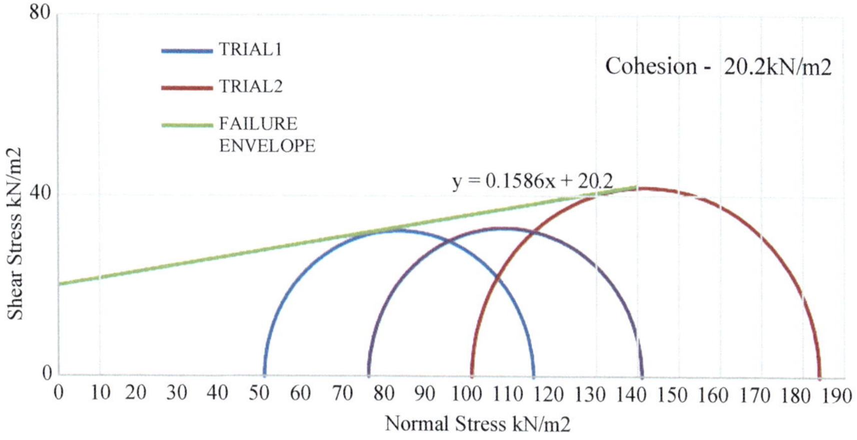

Fig. 6 Triaxial test results

4 Numerical Analysis

The numerical modeling was done using the two-dimensional geotechnical finite element analysis PLAXIS 2D. The lateral extents of the model were taken as follows $x_{min} = -50$ m, $x_{max} = 50$ m, $y_{min} = -50$ m, $y_{max} = 50$ m. An embankment rests here on both treated and untreated soil. The treated and untreated soil is modeled up to a depth of 10 m below the surface, and a subsoil depth of 20 m is provided below that.

Table 3 Material properties of soil interface

Parameters	Embankment	Untreated soil	Treated soil	Subsoil
Young's modulus, E (kPa)	20,000	2357	6964	50,000
Saturated unit weight, γ_{sat} (kN/m^3)	18	12.51	14.49	21
Unsaturated unit weight γ_{unsat} (kN/m^3)	16	14.32	18.05	17
Angle of internal friction, Φ (°)	31	6.65	9.01	35
Cohesion, c (kN/m^2)	5	5.5	20.2	1

Table 4 Material property of plate element [3]

Plate type	Type	EA (kN/m)	EI (kN/m)	d (m)	υ
Steel	Plastic	5.72E + 0.7	2040	0.02	0.3

4.1 Material Properties

Table 3 shows the material properties of soil interfaces inputted in PLAXIS 2D. Mohr–Coulomb is used as the material model for all soil interfaces. Different line loads of 9, 12, 13.5 and 15 kN/m^2 were applied along the top of the embankment. These line loads were applied to the embankment using a plate element. Table 4 exhibits the material properties of plate element applied.

4.2 Modeling and Meshing

The above-mentioned geometry and material properties were inputted in PLAXIS 2D, and the embankment resting on soil was modeled. Figure 7 shows the 2D model and the meshed of soil generated based on the above properties. In this analysis, medium-size mesh is used.

4.3 Analysis

Settlement. Line loads of 9, 12, 13.5 and 15 kN/m^2 were applied on the top of the embankment for both treated and untreated soil. Figure 7 indicates the settlement behavior of both treated and virgin soil. From the graph, it is clear that for untreated soil, the displacement value is greater than 0.5 m, whereas for terrasil-GGBS treated soil, the value is less than 0.15 m. From the graph, we can see that as the load increases, there are huge settlement increases in the case of untreated soil, whereas for treated soil, the settlement increases slightly as the load increases. Figures 7 and 8 show the software results obtained for treated and untreated soil.

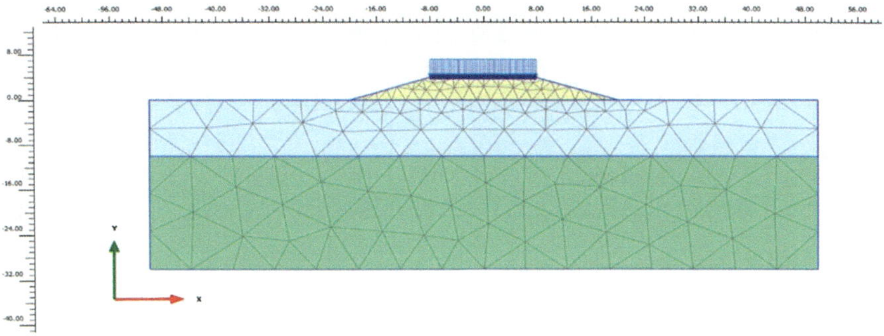

Fig. 7 Meshed structure of 2D model generated in PLAXIS 2D

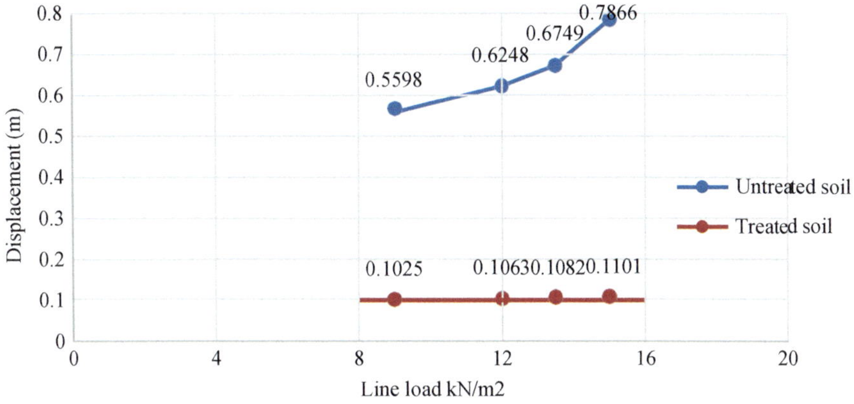

Fig. 8 Load versus displacement graph for treated and untreated soil

Factor of Safety. It is expressed as the safety margin of the slopes. Factor of safety indicates the reliableness for geotechnical structures such as embankment dams and soil slopes. FOS of embankment resting on treated and untreated soil was analyzed.

Figure 9 shows the FOS versus load graph of treated and untreated soil under different loading cases. From the graph, it is clear that the FOS of treated soil is greater than 1.5; therefore, it is safe, whereas for untreated soil, it is less than 1.5 but greater than 1; therefore, it is stable but not in a safe condition.

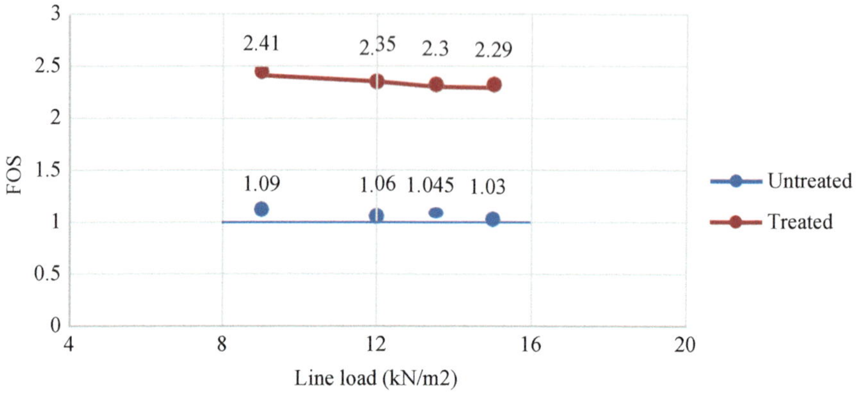

Fig. 9 FOS versus load graph for treated and untreated soil

5 Conclusions

The studies looked at how terrasil nanochemical and GGBS affected the characteristics of Kuttanad soil. The economic aspects of terrasil were also analyzed using the data collected from Zydex industry. For 20 kg terrasil nanochemical, it costs 19,800 ₹; since we are using only a small quantity of product, the overall cost will be very low, and also since this product does not produce any harmful effect, it is considered as an economic and environmental friendly stabilizing material. An experimental investigation was done and identified the properties of Kuttand soil. Later, ideal dosage of terrasil and GGBS was added to the virgin soil and conducted some laboratory tests to identify the improvement in the properties. Numerical analysis of treated and untreated soil was done and analyzed the settlement and FOS parameters. The findings of the current study led to the conclusions below.

1. The optimum percentage of GGBS and terrasil was obtained as 9% and 0.03%, respectively, by conducting different laboratory tests like SPT, Atterberg limits test and UCS test.
2. When 9% GGBS was mixed with soil, the MDD value increased by 0.4% and OMC reduced by 12%, similarly soil mix of 9% GGBS addition and 0.03% terrasil addition improved the compaction characteristics by 7% and reduced the OMC value by 29%.
3. The plasticity index value was reduced from 24 to 21% by the addition of GGBS and Terrasil; it states that the soil becomes less plastic by the addition of optimum percentage of terrasil (0.03%) and GGBS (9%). Thus, it is established that terrasil increases soil stiffness.
4. The shear strength parameters obtained from the triaxial test like the c and ϕ values have increased by the addition of optimum dosage of terrasil and GGBS. Thus, the shear strength of treated soil increases.
5. The UCS values of GGBS-terrasil treated soil increase from 10 to 40 kPa.

6. From the PLAXIS 2D analysis, it was observed that the FOS of untreated soil was less than 1.5, whereas for treated soil, it was in the range of 2.29–2.41 which was safe when compared to untreated soil. For untreated soil, the value is greater than 1; therefore, it is stable but not in safe condition.
7. The settlement behavior of treated soil has been reduced by about 86%. There is not much settlement observed even if the load is increased.

References

1. Meeravali K, Ruben N, Rangaswamy K (2020) Stabilization of soft-clay using nanomaterial: terrasil. Mater Today: Proc 27:1030–1037
2. Krishnan J, Shukla S (2019) The behaviour of soil stabilised with nanoparticles: an extensive review of the present status and its applications. Arab J Geosci 12(14):1–25
3. Saing Z (2017) Vertical deformation of lime treated base (LTB) model of laterite soil using numerical analysis. Int J Civ Eng Technol 8(5):758–764
4. Al-Khafaji R, Jafer H, Dulaimi A, Atherton W, Weida Z (2017) Soft soil stabilisation using ground granulated blast furnace slag. In: The 3rd BUiD doctoral research conference at British University in Dubai
5. Babu S, Sh J (2016) Effect of nano materials on properties of soft soil. Int J Sci Res 5(8):634–637
6. Biswas T, Sarkar D (2021) Stabilization of subgrade soil using nano-chemicals and fly ash. In: Proceedings of the Indian geotechnical conference. Springer, Singapore, pp 65–74
7. Dayalan J, Dayalan J (2016) Comparative study on stabilization of soil with ground granulated blast furnace slag (GGBS) and fly ash. Int Res J Eng Technol 3(5)
8. Hussain SA (2016) Soil stabilization using nano-materials for rural roads–a case study. Int J Innov Res Sci Eng Technol 5
9. Karumanchi M, Kodi R (2018) Compressibility and permeability characteristics of nano chemical treated Kuttanad soft clay. J Emerg Technol Innov Res. ISSN 2349-5162
10. Shirahatti PN, Kallimani PM, Rajashekhar MS (2016) Application of geo-textile and terrasil chemical to reduce permeability of soil. Int J Sci Res Dev (IJSRD) 4(6)
11. Thomas ANSU, Tripathi RK, Yadu LK, Roy S (2016) Soil stabilisation using terrasil. Int. J. Earth Sci. Eng 9:1049–1052
12. Yadu L, Tripathi RK (2013) Stabilization of soft soil with granulated blast furnace slag and fly ash. Int J Res Eng Technol 2(2):115–119
13. IS 2720 Part 3 (1980) Methods of test for soils–determination of specific gravity. Bureau of Indian Standards, New Delhi
14. IS: 2720-Part 4 (1985) Methods of test for soils: grain size analysis
15. IS 2720 Part 5 (1985) Determination of liquid limit and plastic limit
16. IS 2720-Part 8 (1983) Methods of test for soils—determination of water content—dry density relation. Bureau of Indian Standards, New Delhi

Stabilization of Black Cotton Soil Using Red Mud, Phosphogypsum and Lime

Sanjeet Sahoo, Bithika Ghosh, Janarul Shaikh, Krutideepan Kar, and Subhakant Pati

1 Introduction

Expansive soil with variation of water content shows undesirable behavior for construction. Expansive soil is present all around the globe. India also has large area of expansive soil known as black cotton soil (BCS). It covers around 0.8 million square kilometers which are 20% of total land area [1]. Around 4.71 million tons of red mud is produced every year in India which is 6.25% of world's total generation [2]. In general, for minimizing the effect of swelling and shrinkage behavior of BCS, red mud along with some additives (lime, cement, fly ash and phosphogypsum) is added. Several investigations were conducted by many researchers all over the world to reduce the expansive behavior of the soil.

Mane and Rajashekhar [3] studied the properties of BCS with addition of red mud and sodium silicate. They observed that 30% replacement of red mud along with Sodium silicate gave better result in the values of California bearing ratio (CBR Value). Singh and Vasaikar [4] found that the addition of lime with BCS increased the CBR value and decreased the maximum dry density (MDD). Anusha et al. [5] conducted a study to stabilize BCS with phosphogypsum and lime. In their study, they noted a significant improvement in unconfined compressive strength (UCS). Sridevi et al. [6] investigated on BCS to improve its properties with the addition of red mud and lime. They found that its UCS value as well as compaction characteristics increases significantly. Degirmenci et al. [7] carried out several tests on expansive soil with additives like phosphogypsum, cement and fly ash. He observed that the optimum moisture content (OMC) decreases and maximum dry density increases

S. Sahoo (✉) · J. Shaikh · K. Kar · S. Pati
C V Raman Global University, Bhubaneswar, Odisha 752054, India
e-mail: sanjeet476@gmail.com

B. Ghosh
Odisha University of Technology and Research, Bhubaneswar, Odisha 751003, India

B. T. Jose et al. (eds.), *Proceedings of the Indian Geotechnical Conference 2022 Volume 4*, Lecture Notes in Civil Engineering 479,
https://doi.org/10.1007/978-981-97-1753-8_6

with increase in phosphogypsum quantity. However, there is a dearth of study which investigated the use of phosphogypsum as an additive to stabilize the BCS along with the considerable quantity of red mud.

An effort has been made in the current study to utilize red mud a largely available industrial waste for stabilizing highly expansive black cotton soil. In addition, the study investigated the effectiveness of phosphogypsum and lime as the admixture to enhance various properties in different mixes of BCS and red mud. In this study, various laboratory tests have been conducted to measure their OMC, MDD, UCS and CBR value to explore the possibilities of improving various properties of black cotton soil.

2 Material and Methodology

2.1 Materials

In this present study, black cotton soil and red mud were used as the main soil materials, whereas lime and phosphogypsum were considered as the admixture for investigating their effects on various geotechnical and engineering properties of different mixes. Black cotton soil was collected from Banapur of Khordha district located in state of Odisha. Red mud was collected from NALCO located in Koraput district of Odisha. Phosphogypsum and lime were procured from local market nearby the university where the study was carried out.

2.2 Basic Characterizations

The geotechnical properties of black cotton soil and red mud were evaluated in the laboratory as per the guidelines mentioned in the corresponding parts of IS code: 2720. The results obtained are summarized in Table 1.

2.3 Sample Preparation

A total of twelve different combinations of black cotton soil, red mud, phosphogypsum and lime were considered for the study. Three mixes were first prepared by blending 60%, 70% and 80% of black cotton soil with 40%, 30% and 20% of red mud respectively. Six samples were then prepared by adding 1 and 2% of phosphogypsum to each of the three mixes. This range of percentage of adding phosphogypsum was decided for its economic utilization in enhancing the soil strength based on previous studies. Similarly, another six samples were prepared by adding 2.5 and 5% of lime.

Table 1 Geotechnical properties of black cotton soil and red mud

Properties	Black cotton soil	Red mud
Specific gravity	2.816	2.61
Liquid limit (%)	59.29	27.55
Plastic limit (%)	35.56	24.95
Plasticity index (%)	23.73	2.6
Shrinkage limit (%)	23.77	16.65
OMC (%)	23.6	22.7
MDD (g/cc)	1.62	1.97
Unconfined compressive strength (cohesion kN m^{-2})	48.78	98.013
Unsoaked CBR (%)	3.43	5.64
Free swell index (%)	27.7	NA
Percentage of finer fraction (less than 0.075 mm)	95.12	64.29

Note NA—Not Applicable

Such range of percentage of lime was incorporated to the soil mixes for its utilization in improving suitable properties according to several past researches [4]. Table 2 presents all of these twelve samples along with the designations indicating their various mix proportions. Test specimens were finally prepared for evaluating UCS and CBR values from all these sample mixes at their respective OMC (refer to Fig. 1) and MDD (refer to Fig. 2) by adding suitable quantity of water.

Table 2 Proportions of various sample mixes used in the study

Mix designation	Percentage (%) of				Sample No.
	Black cotton soil	Red mud	Phosphogypsum	Lime	
B60R40	60	40	1	–	1
	60	40	2	–	2
	60	40	–	2.5	3
	60	40	–	5	4
B70R30	70	30	1	–	5
	70	30	2	–	6
	70	30	–	2.5	7
	70	30	–	5	8
B80R20	80	20	1	–	9
	80	20	2	–	10
	80	20	–	2.5	11
	80	20	–	5	12

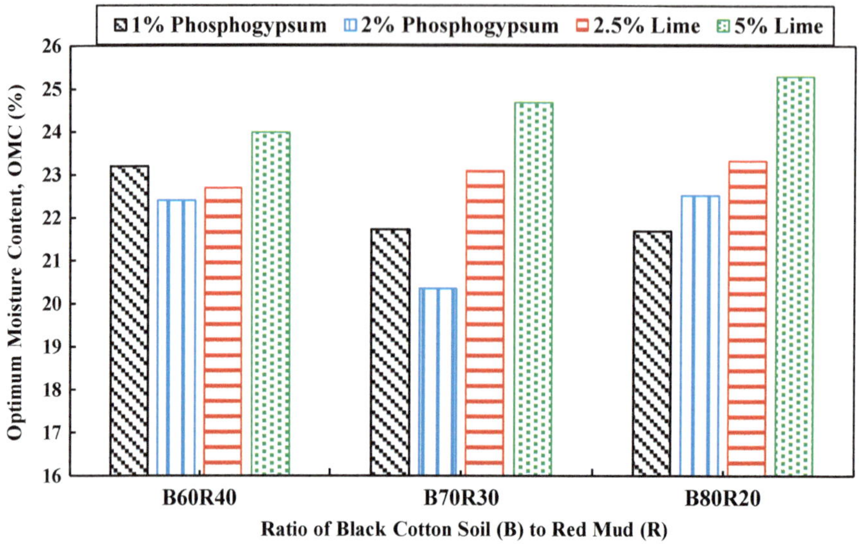

Fig. 1 Variation of optimum moisture content (OMC) in different combinations of black cotton soil, red mud, phosphogypsum and lime

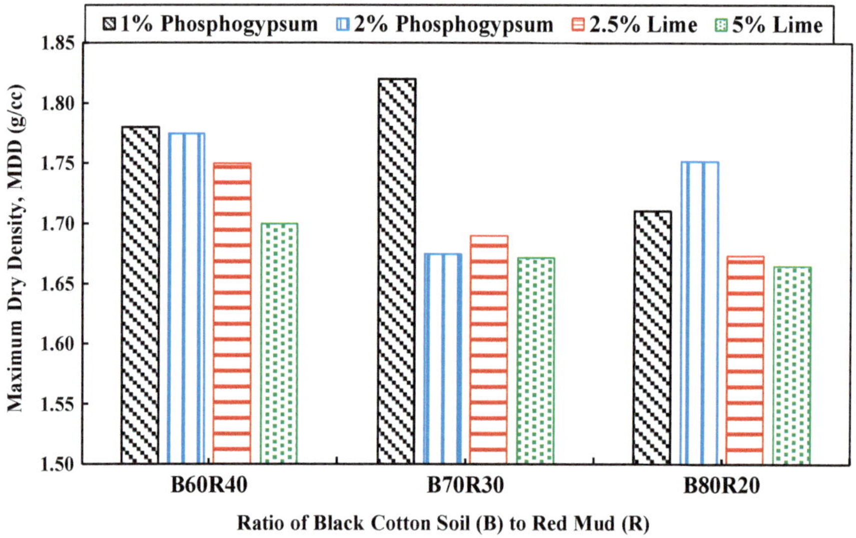

Fig. 2 Variation of maximum dry density (MDD) in different combinations of black cotton soil, red mud, phosphogypsum and lime

3 Result and Discussion

This present study investigated the variation of optimum moisture content (OMC), maximum dry density (MDD), unconfined compressive strength (UCS) and California bearing ratio (CBR) in twelve different combinations of black cotton soil, red mud, phosphogypsum and lime.

3.1 *Variation in Optimum Moisture Content*

Optimum moisture contents (OMCs) were determined in the laboratory by conducting standard proctor tests on three various mixes of 60%, 70% and 80% of black cotton soil with 40%, 30% and 20% of red mud, respectively. Four tests were performed on each mix twice by adding 1 and 2% of phosphogypsum and twice by adding 2.5 and 5% of lime. Figure 1 showcases all these test results of OMCs which range from 20.3 to 25.3% in terms of bar charts. The figure shows the higher values of OMCs in the case of lime addition than in the case of addition of phosphogypsum. The least value of OMC was noted in mix of 70% black cotton soil and 30% red mud when 2% of phosphogypsum was added, whereas the highest value of OMC was observed in the mix of 80% black cotton soil and 20% red mud when 5% of lime was added. For a specific mix, OMC increases with the increase in lime, whereas it decreases with increase in percentage of phosphogypsum. Degirmenci et al. [7] observed the similar trends in the results of OMC from their investigation.

3.2 *Variation in Maximum Dry Density*

The standard proctor test was performed in the laboratory for evaluating the maximum dry densities (MDDs) of twelve different combinations of black cotton soil, red mud, phosphogypsum and lime. Three mixes were first prepared by blending 60%, 70% and 80% of black cotton soil with 40%, 30% and 20% of red mud, respectively. A total of twelve samples were then prepared including six samples by incorporating 1 and 2% of phosphogypsum and six samples by adding 2.5 and 5% of lime to each of three mixes. Figure 2 illustrates the observed results of MDDs of all these samples with the help of bar diagrams. The figure indicates higher values of MDDs when phosphogypsum was added and lower values of MDDs when lime was incorporated. It also portrays that the values of MDDs remain in between 1.665 g/cc and 1.820 g/cc. The minimum MDD was found when 5% of lime was incorporated to the mix of 80% black cotton soil and 20% red mud. The maximum MDD was noticed when 1% of phosphogypsum was added to the mix of 70% black cotton soil and 30% red mud. Similar observations were found in several previous studies [4, 7]. However, the red mud was not considered as a component material in their study.

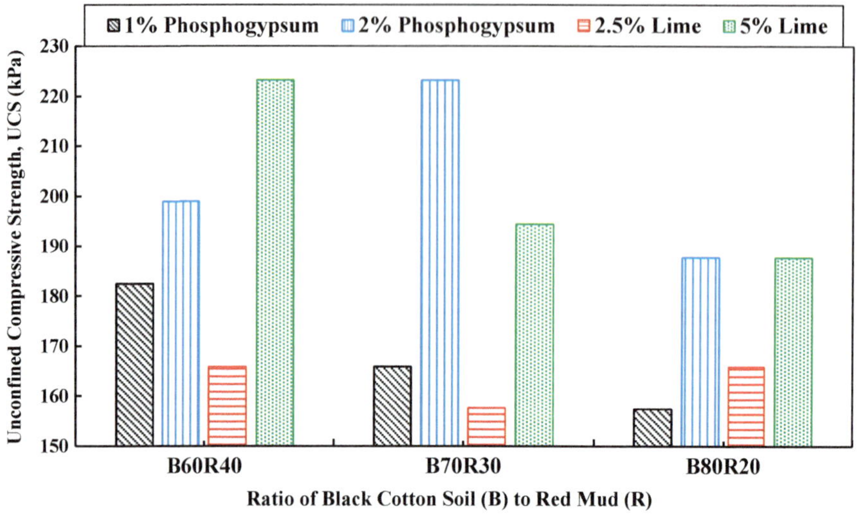

Fig. 3 Variation of unconfined compressive strength (UCS) in different combinations of black cotton soil, red mud, phosphogypsum and lime

3.3 Variation in Unconfined Compressive Strength

Unconfined compressive strength tests were carried out in the laboratory on twelve disturbed and remolded cylindrical specimens which were prepared by following the way explained in the above Sects. 3.1 and 3.2. Figure 3 demonstrates the experimental observations of unconfined compressive strengths (UCSs) for all the soil specimens in terms of bar graphs. The UCS values vary in a range from 157.5 to 223.5 kPa. The figure indicates the higher values of UCSs when both phosphogypsum and lime were added in higher percentage whereas the lower values of UCSs were observed in the case of lower percentage of both phosphogypsum and lime. In this study, the unconfined compressive strength reaches its peak when 5% of lime was added to the mix of 60% black cotton soil and 40% red mud. The UCS value was found to be the least when 1% phosphogypsum was incorporated to the mix of 80% black cotton soil and 20% red mud. Anusha et al. [5] found the similar results in their research study where red mud was not utilized unlike the present study.

3.4 Variation in California Bearing Ratio

Values of California bearing ratios (CBRs) were determined in the laboratory by conducting CBR tests on twelve disturbed and remolded samples which were prepared based on the above procedures described in Sects. 3.1 and 3.2. All the CBR values obtained from the experiment were depicted in forms of bar diagrams

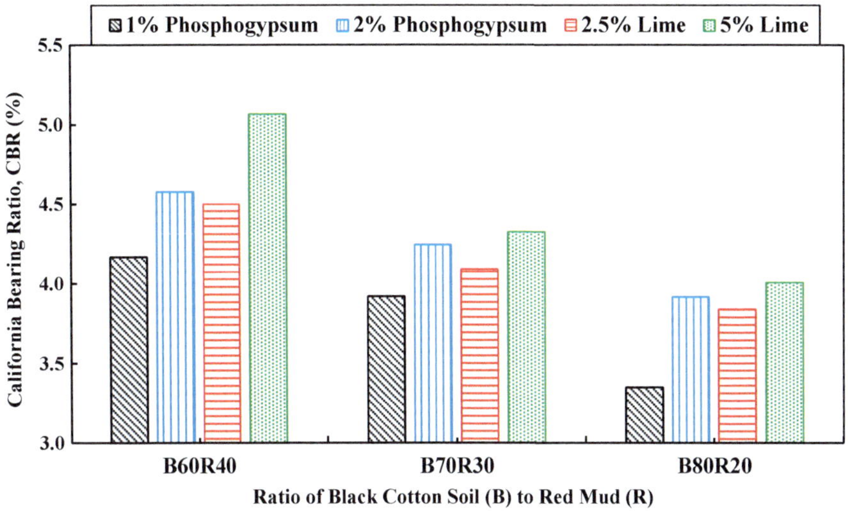

Fig. 4 Variation of California bearing ratio (CBR) in different combinations of black cotton soil, red mud, phosphogypsum and lime

in Fig. 4. The figure clearly shows that the higher CBR values were observed when 60% black cotton soil was mixed with 40% red mud and lower CBR values were obtained when 80% black cotton soil and 20% red mud were mixed. For a particular addition of either phosphogypsum or lime to a particular mix, CBR value improves with the increase in red mud to the mix. In this study, the CBR value was found to be the maximum when 2% lime was added to the blend of 60% black cotton soil and 40% red mud. Minimum CBR value was noted in the case of mix of 80% black cotton soil and 20% red mud when 1% phosphogypsum was added. [4] noticed the similar trends in their observations.

4 Conclusion

The following conclusions can be derived from this current study:

- Addition of lime influences the optimum moisture content of the mixtures of black cotton soil and red mud significantly than the addition of phosphogypsum. For a particular mix, the optimum moisture content rises with the increase in lime, whereas it falls with the rise in percentage of phosphogypsum.
- Incorporation of phosphogypsum improves the maximum dry density of various blends of black cotton soil and red mud better than the addition of lime. The maximum dry density was noticed to be highest when 1% of phosphogypsum was incorporated to the mix of 70% black cotton soil and 30% red mud.

- Relatively higher amount of both lime and phosphogypsum improves unconfined compressive strength of various mixes of black cotton soil and red mud. The unconfined compressive strength was found to reach its peak when 5% of lime was added to the mix of 60% black cotton soil and 40% red mud.
- For a particular addition of either lime or phosphogypsum, enhancement in CBR value is found if the higher quantity of red mud is mixed with black cotton soil. The CBR value was noted to be the maximum when 2% lime was mixed to the blend of 60% black cotton soil and 40% red mud.
- The mixture of 60% black cotton soil and 40% red mud performs the best to enhance the soil properties in terms of UCS and CBR values.

Acknowledgements The authors would like to acknowledge the Indian Institute of Technology Bhubaneswar for providing laboratory facilities to conduct the research work reported in this paper.

References

1. Mehta KS, Sonecha RI, Daxini PD, Ratanpara PB (2014) Analysis of engineering properties of black cotton soil & stabilization using by lime. Int J Eng Res Appl 4(5): 25–32 (Version 3)
2. Paramguru RK, Rath PC, Misra VN (2005) Trends in red mud utilization—a review. Miner Process Extr Metall Rev 26(1):1–29
3. Mane N, Rajashekhar MS (2017) Stabilization of black cotton soil by using red mud and sodium silicate. Int Res J Eng Technol 4(07)
4. Singh S, Vasaikar HB (2015) Stabilization of black cotton soil using lime. Int J Sci Res 4(5)
5. Anusha HM, Bariker P, Viswanath B (2019) Analysis of Strength properties of lime stabilized black cotton soil with phosphogypsum. In: Proceedings of the Indian geotechnical conference
6. Sridevi G, Sahoo S, Sen S (2016) Stabilization of expansive soil with red mud and lime. In: Indian geotechnical conference
7. Degirmencia N, Okucub A, Turabib A (2007) Application of phosphogypsum in soil stabilization. Build Environ 42:3393–3398, Elsevier
8. IS: 2720 (Part 5)—1970, "Methods of Test for Soils–Part 5: Determination of Liquid and Plastic Limit", BIS, New Delhi, India
9. IS: 2720 (Part 7)—1980, "Methods of Test for Soils—Part 7: Determination of Water Content—Dry Density Relation Using Light Compaction", BIS, New Delhi, India (reaffirmed in 1987)
10. IS: 2720 (Part 10)—1973, "Determination of Unconfined Compressive Strength", BIS, New Delhi
11. I.S: 2720 (Part XVI)-1965: "Indian standard for laboratory determination of CBR", Bureau of Indian Standards Publications, New Delhi
12. I.S: 2720 (Part III/Section 2)-1980: "Indian standard for determination of specific gravity (fine, medium and coarse grained soil)", Bureau of Indian Standards Publications, New Delhi

A Study on Recycled Plastic Pins as Micropile in Stabilization of Cohesive Slope

Vinu Sunny and Nirmali Borthakur

1 Introduction

Different approaches are used for stabilization of natural or man-made slopes. Some methods aim to increase the shear strength of soil to enhance its properties and factor of safety of the slope. The frequently used mechanical and chemical stabilization techniques include compaction, mixing, dewatering, lime, cement, fly ash, etc. Methods such as pre-stressed anchors, rock bolts, piles, micropiles, soil nailing, geosynthetic reinforcement, retaining walls are used for structural support to increase the stability of the slope. The applicability of micropiles for slope stabilization projects is increasing due to some advantages such as simple construction process, requires small working area, can be installed easily in zones with restricted equipment admittance, requires less excavation in comparison with earth retaining systems, can be installed in all types of soil where installation of driven piles or drilled shafts may be difficult and costly, produce less noise and vibration during installation and improves aesthetics of the area [1]. Large-scale physical model tests were performed on different pattern of micropiles installed on earthen slopes to create an experimental database related to the spacing and micropile batter angle to enhance the prediction of limit loads for micropiles in slope stabilization works [2, 3]. Numerical and experimental study was performed on micropiles utilized to strengthen tall railway embankments [4]. Case studies are also available related to the application of micropiles in stabilization of slopes in different types of soils such as loose deposits of sandy soil, dense sand, stiff clay soil, $c - \phi$ soil, karstic limestone deposit, layered soil, hard

V. Sunny (✉) · N. Borthakur
Civil Engineering Department, National Institute of Technology Silchar, Silchar, Assam 788010, India
e-mail: vinusunny47@gmail.com

N. Borthakur
e-mail: nirmali@civil.nits.ac.in

B. T. Jose et al. (eds.), *Proceedings of the Indian Geotechnical Conference 2022 Volume 4*, Lecture Notes in Civil Engineering 479,
https://doi.org/10.1007/978-981-97-1753-8_7

rock. These include location of the site and its general description, problem statement, approved remedial measures, technology adopted for boring/drilling holes, choice of gravity grouted/pressure grouted micropiles, its design (diameter, length, spacing, number, etc.) and layout, instruments used for micropiling work, load testing procedures adopted for different types of loads, etc. High-strength micropile groups were used for the restoration of Cannon Place (1868), an ancient monumental structure in London [5]. Micropile and anchor system was applied to meet up environmental and technical challenges in the construction of Banjo Bill rock catchment basins situated on SR 89A in Oak Creek Canyon near Sedona, Arizona [6]. Micropiling was used for retrofitting of slope of an aqueduct in Manchester, UK, as the retrofitting work was permitted with no direct loading and minimum vibration on the conduit or slope [7]. Micropiles were installed successfully for slope stabilization in inaccessible areas and under adverse climatic conditions, one in a ropeway station in high alpine mountain range in Europe and another in East–West gas pipeline over the mountain region of Western Ghats near Mumbai in India [8]. The reconstruction of Oak Hill Cemetery and restoration of a tall wall in Port Deposit, MD, were done with A-frame micropiles [9]. Micropiles are also installed as a remedial measure to mitigate landslides. Some examples are: The Blue Trail Landslide project located on US Highway between Alpine Junction and Hoback Junction, Wyoming; the SUM-271 project located on Interstate 271 northbound in Richfield Township, Summit County, Ohio; the Littleville landslide project located on a section of US Route 43, Alabama [1].

A detailed design procedure of micropiles for soil slope stabilization is available [1], but the grout-to-ground ultimate bond strength value required to calculate the axial force to be resisted by the micropile below the potential slip surface is not available for soft cohesive soil. This type of deep-seated soft cohesive soil having high liquid limit (> 50%) and plasticity index (> 20%) is available in many parts of North-East India. So, in this paper, an experimental and numerical approach has been adopted to study the load-settlement behavior of cohesive slope reinforced with A-pattern micropiles constructed with recycled plastic pin (RPP). Recycled plastic pins inserted through the embankment upstream face offer extra resistance that restrains the further movement of slip surface and thus increases the safety factor. However, experimental study is very limited on the potential use of RPPs in stabilization of cohesive slope.

2 Materials Used in the Study and Properties

Materials used for this research work are recycled plastic pins and clayey soil.

Table 1 Material properties of RPP

Property	Value	ASTM Standard
Density	0.93 gm/cm^3	ASTM D792
Tensile strength	16 MPa	ASTM D638
Shear strength	14.6 MPa	ASTM D638
Compressive strength	19.6 MPa	ASTM D695
Continuous service temp	180 °C	N/A
Thermal coefficient of expansion	0.00001/°C	ASTM D696
Flexural modulus	685 MPa	ASTM D790
Modulus of elasticity	3000 kPa	N/A

2.1 Recycled Plastic Pin

Recycled plastic pin (RPP) is a lightweight material derived from the used plastic and waste materials like fly ash, sawdust and polymer [10]. It is chemically and biologically less reactive compared to some other underpinning elements. The use of RPPs in geotechnical applications can reduce the volume of waste plastic that is generally used in landfilling work and also can increase its demand [11]. The different constituents of RPP are: high-density polyethylene (HDPE) (55–70%); low-density polyethylene (LDPE) (5–10%); polystyrene (PS) (2–10%); polypropylene (PP) (2–7%); polyethylene-terepththalate (PET) (1–5%); and additives like sawdust, fly ash, etc. (0–5%) [12]. Modulus of elasticity of plastic lumber can be increased sufficiently by adding glass and wood fiber as additives [13]. The material properties of RPP used for this experimental study are given in Table 1 (as supplied by the vendor).

2.2 Soil

The soil used in the study is cohesive soil collected from the fields situated near National Institute of Technology Silchar campus, Assam. The physical properties of the soil were determined in the laboratory as per Bureau of Indian Standards code of practice and given in Table 2.

Table 2 Properties of soil

Properties of soil	Value
Gravel (%)	0
Fine sand (%)	12
Silt + clay (%)	88
Plastic limit (%)	26.5
Liquid limit (%)	55
Plasticity index (%)	28.5
Maximum dry density, MDD (gm/cc)	1.422
Optimum moisture content, OMC (%)	23
Unconfined compressive strength, UCS (kPa)	49.64
Specific gravity	2.54
Soil classification	CH

3 Methodology for Experimental Investigation

3.1 Preparation of Test Tank

Small-scale model tests were planned to perform in a steel tank of size 1 m × 1 m × 1 m. Cohesive soil was collected from nearby fields and were crushed into fine powdered form and mixed properly with optimum moisture content. A polythene sheet was used to cover the steel tank to minimize the frictional drag between tank and soil. Soil was filled inside the tank by compacting with a hammer. The soil slope is made with an angle of 30° as shown in Fig. 1.

Fig. 1 Preparation of test tank

3.2 Construction of Micropile with Recycled Plastic Pins

Micropiles of two different diameters (*d*) 22 and 30 mm are as shown in Fig. 2, and two different lengths (*l*) 20*d* and 30*d* were selected for testing. It was constructed along a single row 1 × 4 numbers having A-Pile configuration making inclinations (θ) of 0°, 5°, 10° and 15° with respect to vertical axis of loading. Each group was constructed with three different spacings (*s*) of 3*d*, 4.5*d* and 6*d*. At the specific locations of the micropiles, holes were made with hand auger to the required depth. Cement (43 grade) with water–cement ratio of 0.45 was used as grout. A total of forty-nine tests were conducted as shown in Table 3.

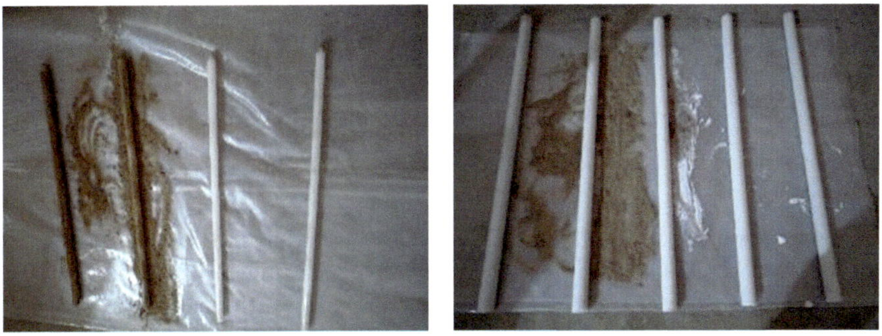

Fig. 2 Recycled plastic pins used as micropile of diameter 22 and 30 mm

Table 3 Summary of experimental program

Diameter (*d*) (mm)	Length/dia. (*l*/*d*) of MP	Number (*n*) of MP	Spacing/dia. (*s*/*d*) of MP	Inclination of MP with vertical (θ)	Total tests
22	20	4	3, 4.5, 6	0°, 5°, 10°, 15°	12
	30	4	3, 4.5, 6	0°, 5°, 10°, 15°	12
30	20	4	3, 4.5, 6	0°, 5°, 10°, 15°	12
	30	4	3, 4.5, 6	0°, 5°, 10°, 15°	12
Test conducted on soil without micropile					01
Total number of tests conducted					49

3.3 Experimental Setup and Procedure

A steel plate was used as micropile cap. Small holes having diameter smaller than the micropile diameter were made on the steel plate at the specific spacing. The top portion of the micropile was tapered to smaller diameter so that it can be inserted into the hole and placed in a fixed position. Two dial gauges were used to measure vertical displacement and placed at two corners of the cap. Pressure gauge and hydraulic jack were used to apply load on the steel plate and one proving ring was used to measure the applied load. The picture of experimental setup during a test is shown in Fig. 3. The load corresponding to settlement having 10% of micropile diameter used was assumed as the failure load for the embankment [14]. The maximum load corresponding to the maximum settlement obtained from different micropile configurations was observed and applied for base soil without having micropiles.

4 Numerical Analysis

Finite element analysis is a numerical technique used in geotechnical engineering for finding approximate solutions to actual problems as it can handle various types of materials, geometries and boundaries and can show the distribution of stresses and displacements. However, to validate the results, it should be supplemented by experimental investigation. Generally, more accurate results are obtained by finer meshes, but it increases the analysis time. In the study, Plaxis-3D software was used to model the embedded micropile. The micropile was demonstrated by embedded beam elements that were connected with the cap. The slope stability was analyzed with a plain strain model using 15-noded elements. The standard fixities were used to define boundary conditions. The Mohr–Coulomb material model was used for undrained condition. Table 4 shows the different properties of soil and steel plate. Figure 4. shows the generated numerical model in Plaxis-3D software.

Fig. 3 Experimental setup

Table 4 Parameters assigned for numerical calculation in Plaxis-3D

Parameters	Soil	Steel plate
Material model	Mohr–Coulomb	Elastic, isotropic
Drainage type	Undrained	Non-porous
Unit weight above phreatic line (γ_{unsat})	16.5 kN/m^3	78 kN/m^3
Unit weight below phreatic line (γ_{sat})	18 kN/m^3	–
Young's modulus (E)	1.25×10^3 kPa	200×10^6 kPa
Poison's ratio (ν)	0.45	0.3
Cohesion (c)	24.8 kPa	–
Friction Angle (Φ)	0°	–

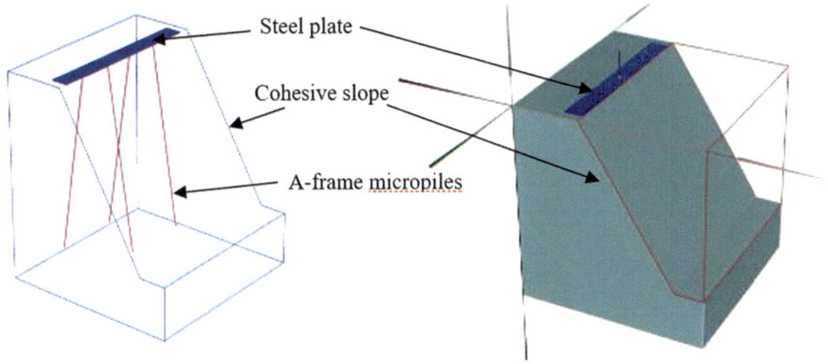

Fig. 4 Numerical model created in Plaxis-3D software

5 Results and Discussion

5.1 Results from Experimental Analysis

The load corresponding to the settlement equal to 10% of micropile diameter was assumed as the failure load for the embankment reinforced with micropile [14]. The diameter of micropiles used was known, so load carrying capacity was recorded directly during the experiment from the proving ring reading corresponding to the settlement of the 10% of micropile diameter. So, load-settlement graphs were not plotted, and the experimental results are tabulated in Tables 5 and 6 for two different *l*/*d* ratios 20 and 30, respectively.

Variation of load carrying capacity of A-frame micropile group along with the degree of inclination of micropile with vertical is plotted for 22 mm diameter micropile group for two different *l*/*d* ratios of 20 and 30 and shown in Fig. 5a, b, respectively.

Table 5 Load carrying capacity of micropile group for l/d ratio 20

Diameter (d), mm	Inclination of micropile with vertical (θ), °	Load carrying capacity, kN		
		$s/d = 3$	$s/d = 4.5$	$s/d = 6$
22	0	0.928	0.9915	1.104
	5	1.25	1.307	1.424
	10	1.2765	1.442	1.567
	15	1.4125	1.553	1.69
30	0	1.202	1.3055	1.3785
	5	1.3765	1.5405	1.595
	10	1.533	1.67	1.756
	15	1.657	1.8115	1.8665

Table 6 Load carrying capacity of micropile group for l/d ratio 30

Diameter (d), mm	Inclination of micropile with vertical (θ), °	Load carrying capacity, kN		
		$s/d = 3$	$s/d = 4.5$	$s/d = 6$
22	0	1.127	1.157	1.236
	5	1.449	1.4725	1.556
	10	1.4755	1.6075	1.699
	15	1.6115	1.7185	1.822
30	0	1.401	1.471	1.5105
	5	1.5755	1.706	1.727
	10	1.732	1.8355	1.888
	15	1.856	1.977	1.9985

Variation of load carrying capacity of A-frame micropile group along with the degree of inclination of micropile with vertical is plotted for 30 mm diameter micropile group for two different l/d ratios of 20 and 30 and shown in Fig. 5c, d, respectively.

It is observed from Tables 5 and 6 and Fig. 5a–d that, at the same l/d ratio, the load carrying capacity increases with the increase in diameter of micropiles. Similarly, for a particular diameter of micropile, capacity is more at larger l/d ratio. For the same diameter of the micropile, same l/d ratio and s/d ratio, the capacity of the group increases with the degree of inclination. Likewise, the capacity of a group increases along with the increase in s/d ratio if other parameters remain constant. However, optimum value could not be found from this small-scale model test.

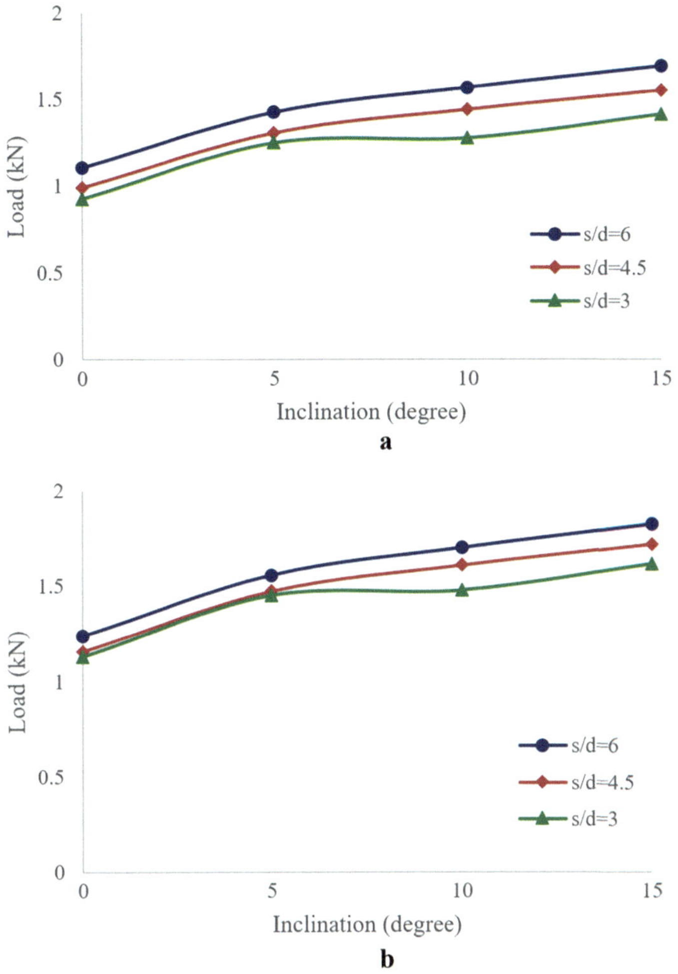

Fig. 5 Load carrying capacity versus angle of inclination of micropile with vertical for $d = 22$ mm and $l/d = 20$; **b** load carrying capacity versus angle of inclination of micropile with vertical for $d = 22$ mm and $l/d = 30$; **c** load carrying capacity versus angle of inclination of micropile with vertical for $d = 30$ mm and $l/d = 20$; **d** load carrying capacity versus angle of inclination of micropile with vertical for $d = 30$ mm and $l/d = 30$

5.2 *Results from Numerical Analysis*

Generally, the factor of safety value is assumed to be 2 in the design of micropile [1]. So, safe load was calculated from the failure load (ultimate load) obtained from the experimental investigation. This safe load carrying capacity obtained from the experimental analysis was applied in numerical models constructed with different variables using Plaxis-3D software. The settlement under this load was determined

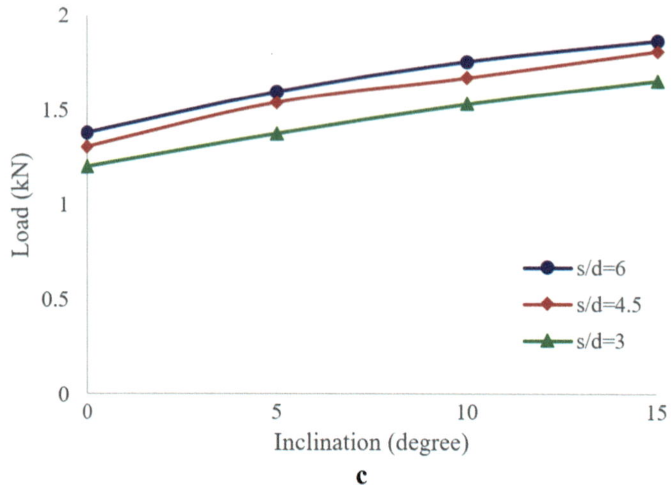

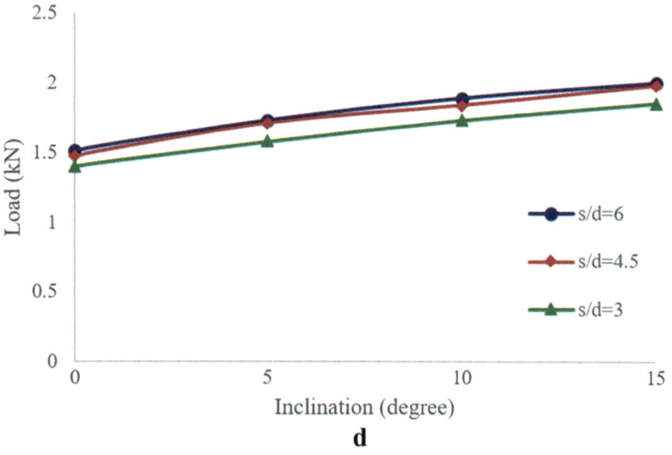

Fig. 5 (continued)

and compared with the values obtained in the respective experiment. As for example, the generation of mesh and displacement (settlement) for slope without micropile and for 22 mm diameter micropile group with *l*/*d* ratio 20, *s*/*d* ratio 4.5 and inclination of micropile of 15° with vertical are shown in Fig. 6a, b, respectively. The results of numerical analysis are tabulated in Tables 7 and 8 for two different *l*/*d* ratios 20 and 30, respectively.

Variation settlement of A-frame micropile group along with the degree of inclination of micropile with vertical id plotted for 22 mm diameter micropile group for two different *l*/*d* ratios of 20 and 30 and shown in Fig. 7a, b, respectively.

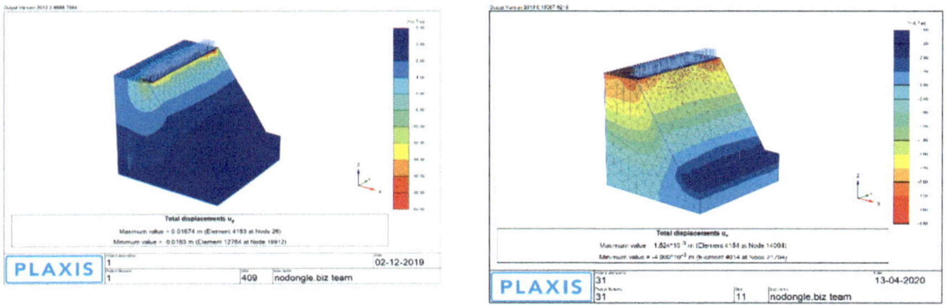

a b

Fig. 6 Displacement of slope without micropiles; **b** displacement of slope with diameter micropile 22 mm, *l*/*d* ratio 20, *s*/*d* ratio 4.5 and inclination with vertical 15°

Table 7 Settlement of micropile group for *l*/*d* ratio 20

Diameter (*d*), mm	Inclination of micropile with vertical (θ), °	Settlement, mm		
		$s/d = 3$	$s/d = 4.5$	$s/d = 6$
22	0	6.629	5.052	4.891
	5	5.358	5.008	4.866
	10	5.358	4.911	4.862
	15	5.161	4.9	4.858
30	0	4.828	4.721	4.686
	5	4.826	4.72	4.638
	10	4.8	4.712	4.555
	15	4.751	4.705	4.552

Table 8 Settlement of micropile group for *l*/*d* ratio 30

Diameter (*d*), mm	Inclination of micropile with vertical (θ), °	Settlement, mm		
		$s/d = 3$	$s/d = 4.5$	$s/d = 6$
22	0	4.49	4.161	4.031
	5	4.483	4.157	4.031
	10	4.18	4.108	4.012
	15	4.164	4.067	4.004
30	0	4	3.827	3.583
	5	3.937	3.803	3.552
	10	3.868	3.749	3.499
	15	3.85	3.713	3.242

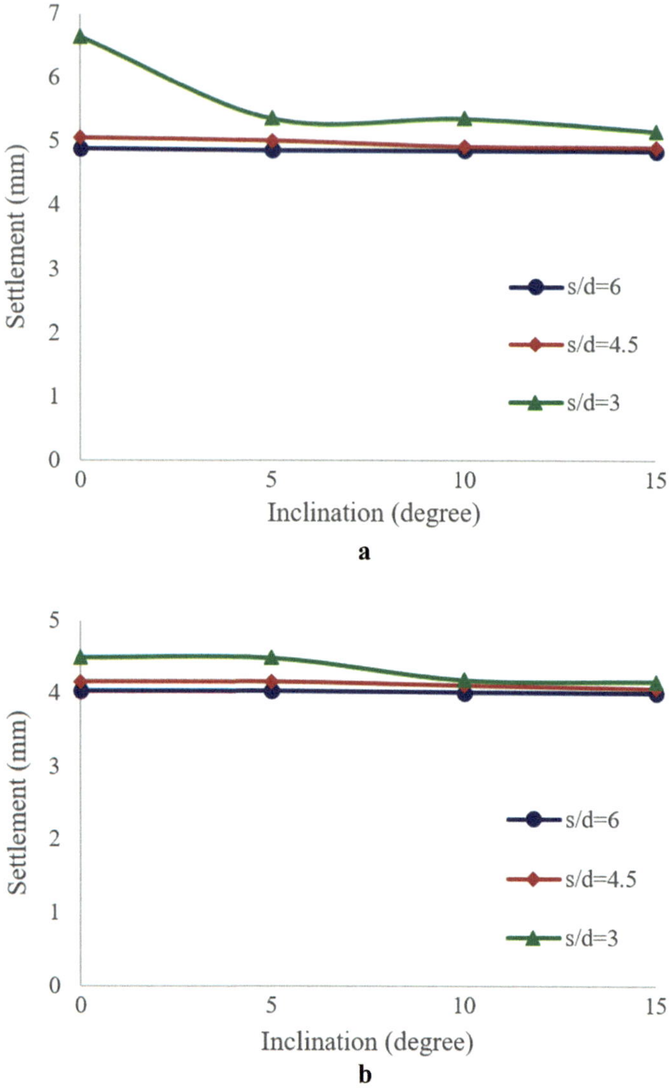

Fig. 7 Settlement plotted against angle of inclination of micropile with vertical for $d = 22$ mm and $l/d = 20$; **b** settlement plotted against angle of inclination of micropile with vertical for $d = 22$ mm and $l/d = 30$; **c** settlement plotted against angle of inclination of micropile with vertical for $d = 30$ mm and $l/d = 20$; **d** settlement plotted against angle of inclination of micropile with vertical for $d = 30$ mm and $l/d = 30$

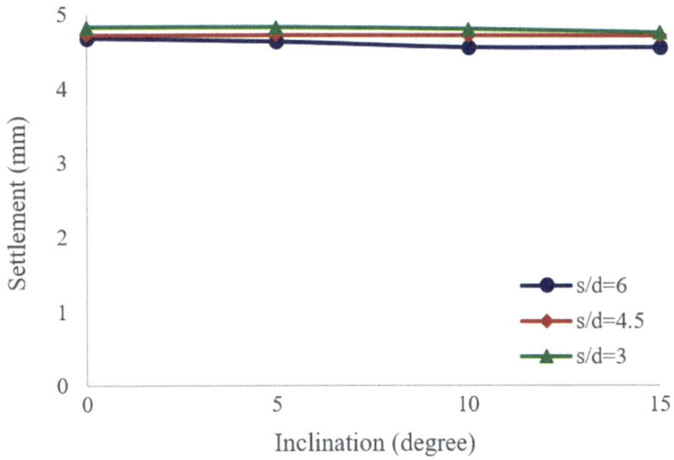

c

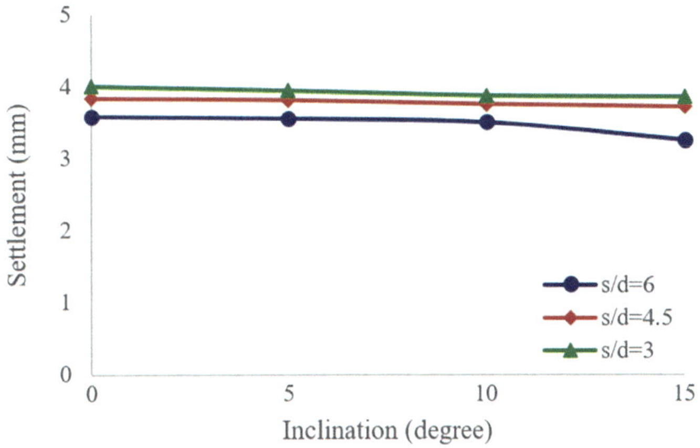

d

Fig. 7 (continued)

Variation settlement of A-frame micropile group along with the degree of inclination of micropile with vertical is plotted for 30 mm diameter micropile group for two different *l*/*d* ratios of 20 and 30 and shown in Fig. 7c, d, respectively.

It is observed that the A-frame micropile group with diameter 30 mm, length 30*d*, spacing 6*d*, and placed at an inclination of 15° with vertical shows least group settlement. This may be due to the less overlapping of stress zone inside the group due to the large spacing.

6 Conclusions

This paper presents the experimental results of load tests conducted on cohesive model slope with unconfined compressive strength value around 50 kPa reinforced with A-frame micropile groups. Slope was made inside a steel tank of dimensions 1 m × 1 m × 1 m. Half the load obtained from experimental investigation corresponding to settlement of 10% of micropile diameter was applied to the numerical models made with Plaxis-3D software, and the settlement resulted from numerical analysis was cross-validated. Based on the experimental and numerical analysis, the following conclusions are drawn.

The load carrying capacity of a A-frame micropile group was observed to be higher for higher diameter micropile. For an illustration, capacity of vertical ($\theta = 0°$) micropile group with l/d ratio 20 and s/d ratio 4.5 increases to 31.6%, when the diameter is increased from 22 to 30 mm. The load carrying capacity of a micropile group was observed to be higher for higher length of the micropile. For an example, the capacity of vertical micropile group with diameter 22 mm and spacing 3*d* increases to 21.4%, when the length is increased from 20 to 30*d*. The load carrying capacity of a micropile group increases with the spacing also. As an instance, capacity of micropile group with diameter 22 mm, l/d ratio 20 and placed at an angle 15° with vertical (θ) increases to 20%, when the spacing increases from 3*d* to 6*d*. The load carrying capacity of a micropile group increases with the increase in the inclination angle with vertical (θ). For an illustration, capacity of micropile group placed vertically ($\theta = 0°$) with l/d ratio of 20 and s/d ratio of 4.5 increases to 56.6%, when the inclination is increased from 0° to 15°. Settlement of a micropile group under the ultimate load in experimental investigation is less compared to that of numerical analysis. This may be due to the resistance offered by lateral confining action of steel tank. For an illustration, the settlement of vertical micropile group of 22 mm diameter with l/d ratio 20 and s/d ratio 3, is obtained 6.629 mm (instead of 2.2 mm) under the same load of 0.928 kN in the numerical model.

References

1. Sabatini PJ, Tanyu B, Armour T, Groneck P, Keeley J (2005) Micropile design and construction reference manual. FHWA-NHI 05(039). US Department of transportation: McLean
2. Boeckmann AZ (2006) Load transfer in micropiles for slope stabilization from tests of large -scale physical models (Doctoral dissertation, University of Missouri—Columbia)
3. Bozok O (2009) Response of micropiles in earth slopes from large-scale physical model tests (Doctoral dissertation, University of Missouri—Columbia
4. Esmaeili M, Nik MG, Khayyer F (2013) Experimental and numerical study of micropiles to reinforce high railway embankments. Int J Geomech 13:729–744
5. Fraser J, Taylor T, John HT (2009) High capacity micropile groups for the cannon place redevelopment in London. In: Proceedings of the 9th international workshop on micropiles, London, England
6. Groneck PB, Armour T (2009) Banjo Bill rock catchment basins. In: Contemporary topics in deep foundations, pp 398–405

7. Preece D (2010) Slope stabilization for the Thirlmere Aqueduct at Nab Scar. In: Proceedings of the 10th International workshop on micropiles, Washington, DC
8. Brandner A (2010) Successful slope stabilization under severe conditions. In: Proceedings of the 10th International workshop on micropiles, Washington, DC
9. Englert C (2011) Use of micropiles for slope stabilization, Case Histories—East Coast. In: Proceedings of the 36th annual conference on deep foundations. Boston
10. Chen CW, Salim H, Bowders J, Loehr E, Owen J (2007) Creep behavior of recycled plastic lumber in slope stabilization applications. J Mater Civ Eng 19(2):130–138
11. Loehr JE, Fennessey TW, Bowders JJ (2007) Stabilization of surficial slides using recycled plastic reinforcement. Transp Res Rec J Transp Res Board 2(1989):79–87
12. McLaren MG (1995) Recycled plastic lumber and shapes design and specification. In: Proceedings 13th structures congress, ASCE, vol 1, pp 819–833
13. Breslin VT, Senturk U, Berndt CC (1998) Long-term engineering properties of recycled plastic lumber in pier construction. Resour Conserv Recycl 23:243–258
14. IS 2911-Part 4: Indian Standard Code of Practice for Design and Construction of Pile Foundations: Load Test on Piles., Bureau of Indian Standards, New Delhi (1985)

Effect of Fly Ash and Synthetic Fibers on the Desiccation Cracking of Expansive Clay

Uma Chaduvula, Purvi Agrawal, Rashi Desai, and Swati Bharti

1 Introduction

Black cotton soil is one of the major regional soil deposits in India, covering an area of about 3.0 lakhs km^2 [1]. In India, the area covered by expansive soils is nearly 22.2% of the total land area. The desiccation of clay and swelling pressure lead to structural distresses. Desiccation cracking is a common problem found in fine-grained soils, especially expansive soils. Quantification of cracking properties like crack intensity, crack dimensions, crack initiation, and crack pattern is an important aspect to study behavior of cracked soil. Desiccation alters the properties of soil adversely, viz. the seepage, consolidation, and compressibility, and shear strength characteristics of the soil [2–5].

One of the most important pillars of Sustainable development goals in to use locally available material and use of industrial waste. Waste containment systems use clay-rich soils having low hydraulic conductivity as landfill liners and covers. Since expansive clays have a high clay content and low hydraulic conductivity (in the order of 1×10^{-9} m/s), they can potentially be used as an impervious barrier material in landfill lining systems. However, due to their shrink-swell nature, they have a tendency of severe desiccation cracking leading to distress in impervious barriers of landfill lining systems. In recent times, the concept of reinforcing the soil by including randomly distributed fibers is being investigated. One of the main advantages of randomly distributed fibers over conventional geosynthetic reinforcement is the absence of potential plane of weakness [6, 7]. However, more recently there has been an increasing interest in studying the behavior of fiber-reinforced soil to increase the desiccation cracking resistance [8–13]. It is found that geofibers reduce the soil cracking. This study experimentally investigates the use of fly ash

U. Chaduvula (✉) · P. Agrawal · R. Desai · S. Bharti
Pandit Deendayal Energy University, Gandhinagar, India
e-mail: Uma.chaduvula@sot.pdpu.ac.in

B. T. Jose et al. (eds.), *Proceedings of the Indian Geotechnical Conference 2022 Volume 4*, Lecture Notes in Civil Engineering 479,
https://doi.org/10.1007/978-981-97-1753-8_8

Table 1 Particle size analysis of soil used

Particle	Value (%)
Gravel	2
Coarse sand	5
Medium sand	8
Fine sand	9
Silt + clay	76

and fiber-reinforced expansive clay as a barrier material with an emphasis on desiccation cracking. The objective of this study is to identify the potential of randomly distributed discrete fiber reinforcement in mitigating desiccation cracking of barrier systems.

2 Materials

2.1 *Soil*

In the current study, expansive clay collected from a construction site near the city of Surat, Gujarat, was used. When the soil's Atterberg Limits were examined, it was discovered that the Liquid Limit and Plasticity Index were 52 and 22, respectively, according to ASTM D 4318-93 (1994) and ASTM D 427-93 (1994). The Free Swell Index for it was found to be 50. As a result, the black cotton soil utilized in the study was categorized as high plasticity clay under the Unified Soil Classification System. The grain size distribution of the soil used is provided in Table 1.

2.2 *Fly Ash*

The fly ash used in the present study was collected from nearby thermal power plant in Gandhinagar, Gujarat. The properties of fly ash are provided in Table 2.

2.3 *Fibers*

Polyester (PET) fibers commercially available as Recron3s, Reliance Industries Ltd., were used in this study. The specific gravity of the PP fibers is higher than water (Table 3).

Table 2 Physical properties of fly ash used

Parameters	Test result
SiO_2	51.40%
Al_2O_3	18.74%
Fe_2O_3	13.89%
CaO	4.35%
MgO	1.12%
Na_2O	0.11%
SO_3	0.65%
LOI	7.28%
Specific gravity	2.24
Bulk density	1545 kg/m^3
Porosity	20.74%
Water absorption	7.62%

Table 3 Properties of polyester fiber used

Properties	Value
Specific gravity	1.35
Effective diameter (μm)	40
Linear density (tex)	0.331
Breaking force for single filament (mN)	112.72
Tenacity (mN/tex)	340.56
Strain at break (%)	45.53
Fusion point (°C)	165

2.4 Sample Preparation

The soil sample was broken into smaller pieces using a compaction machine which was further broken down into fine clay using the Jaw Crusher machine. A mixture of fine dry clay, fly ash, and fibers with varying amounts was prepared as shown in Table 4, to which water was gradually added to make a slurry and mixed for approximately 10 min. All specimens were prepared near soil's liquid limit. To obtain a specimen thickness of 15 mm, the produced slurry was put onto a glass petri dish with a 100 mm diameter. To get rid of the air bubbles, the soil slurry container was gently pounded on a wooden platform. To ensure a homogeneous drying surface, the interior surface of the glass containers utilized in the study was free of any scratches, protrusions, or depressions.

Table 4 Soil samples with varying amounts of additive

S. No.	Experiment	Soil (%)	Fly ash (FA) (%)	Fiber content (%)
1	Unreinforced (UR)	100	0	0
2	UR + 7% FA	93	7	0
3	0.25% FR + 7% FA	93	7	0.25
4	0.50 % FR + 7% FA	93	7	0.5
5	0.75% FR + 7% FA	93	7	0.75
6	UR + 10% FA	90	10	0
7	0.25% FR + 10% FA	90	10	0.25
8	0.50% FR + 10% FA	90	10	0.5
9	0.75% FR + 10% FA	90	10	0.75

2.5 *Test Setup and Procedure*

The test setup mainly consists of a soil sample, adjustable stand, two Halogen lamps each of 500 W, a digital camera, and a 100 mm diameter dish for keeping the soil sample as shown in the schematic setup in Fig. 1. The main objective of this setup is to measure the crack width, length, and area of the soil sample in presence of proper heat and light with and without synthetic fibers.

The sample was kept under the observation of the halogen lamps of 500 W and a digital camera for 8 h at a height of 50 cm. The camera has captured the picture

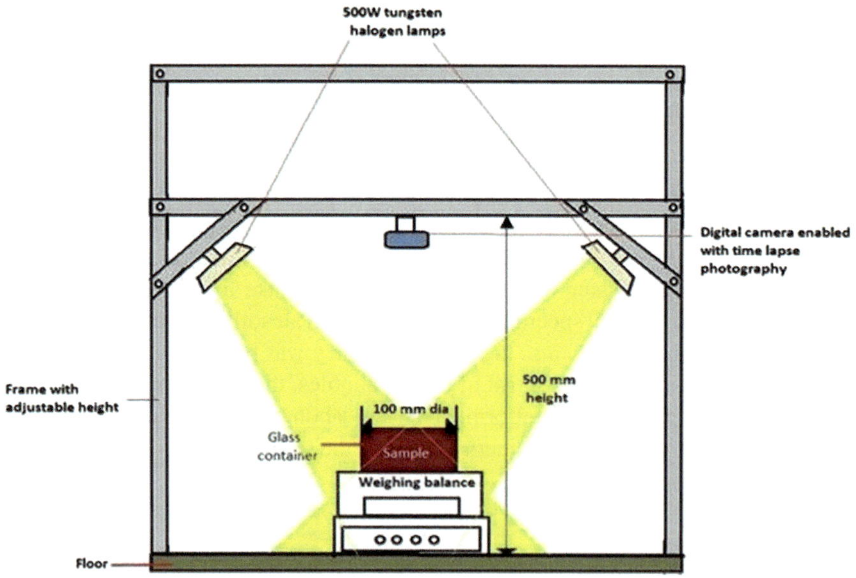

Fig. 1 Schematic diagram of the test setup

of the specimen in every one hour. The weight for each specimen was individually weighed to see how much moisture was being lost and recorded.

Digital image analysis was used to analyze the images captured during the drying process. Studies have been ongoing for the past several years to determine the exact dimensions of cracks. Image analysis is the best method out of all those that are accessible. It is a relatively quicker procedure and measuring cracks and shrinking takes less time. For picture processing and quantification of surface cracks and shrinkage in the specimens, a Java-based software ImageJ, Version 1.50a, was used to measure the crack features during all stages of tests through image analysis [14, 15].

3 Results and Discussion

In order to expedite the drying process, halogen lamps were used during the study. From the lab experiment, a few pictures were chosen to demonstrate how the expanded clay behaved as it dried. The performance of each of the chosen improvement methods was investigated using the information gathered throughout the image analysis process. Figure 2 shows the cracking pattern of the soil specimen during time interval of every 2 h. It is clear that specimens lacking fiber exhibited cracks that were longer and wider but less numerous. With increasing fiber content, cracks become more prevalent. A higher fiber content, however, causes cracks to be narrower and shorter.

At the end of the experiments, the crack width, crack length, crack area, and final crack intensity factor (CIF) for all the soil specimens, were determined using digital image analysis as given in Table 5.

As observed from the table, the CIF is highest for the unreinforced sample, whereas it is lowest for the sample whose content for soil, fly ash, and fiber is 90%, 10%, and 0.75%, respectively. With the increase of additives, the crack parameters are decreasing which shows that the additives were effective for the reduction of cracks.

3.1 Crack Width

Crack width is obtained by measuring the shortest distance between the boundaries of the crack in the captured images using the distance measurement tool in ImageJ software. The addition of fly ash and fibers reduces the crack width of the soil. The variation of crack width with varying fiber contents is shown in Fig. 3. The crack width is increasing in case of 10% fly ash and 0.5% fiber content but is then decreased at drastic rate.

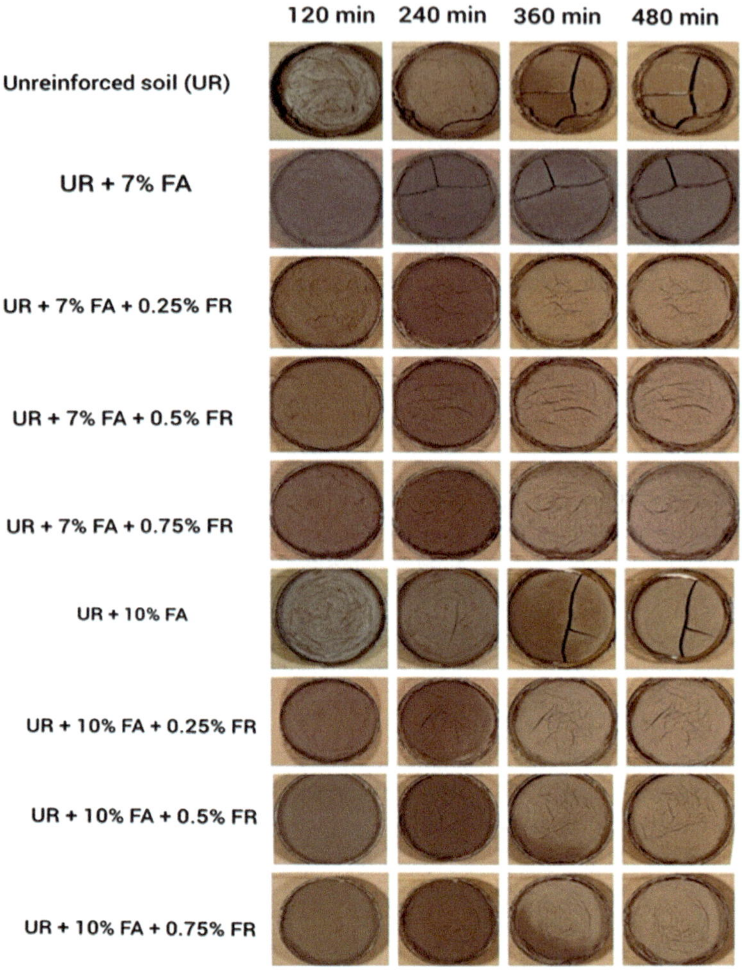

Fig. 2 Desiccation test results for every two hours

3.2 *Crack Length*

Crack length is the length of the crack which changes by varying the content of fly ash and fibers as shown in Fig. 4.

Table 5 Crack parameters for all soil samples

Experiment	Crack width (mm)	Crack length (mm)	Crack area	CIF %
UR	4.2	31.8	284.6	3.6
UR + 7%FA	2.9	39.8	275.5	3.5
0.25% FR + 7% FA	0.6	15.4	100.0	1.2
0.50% FR + 7% FA	0.6	11.7	73.3	0.9
0.75% FR + 7% FA	0.4	9.1	65.5	0.8
UR + 10% FA	4.1	56.0	82.2	1.0
0.25% FR + 10% FA	0.5	16.4	27.4	0.3
0.50% FR + 10% FA	0.5	4.5	12.3	0.1
0.75% FR + 10% FA	0.3	1.7	7.2	0.09

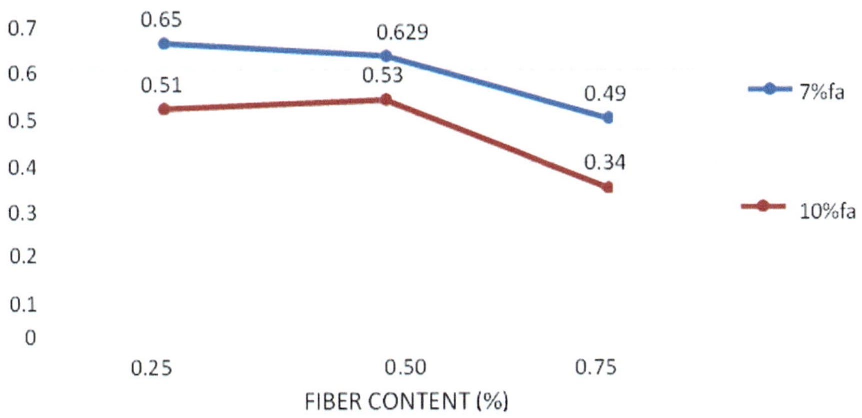

Fig. 3 Variation of crack width with fiber content

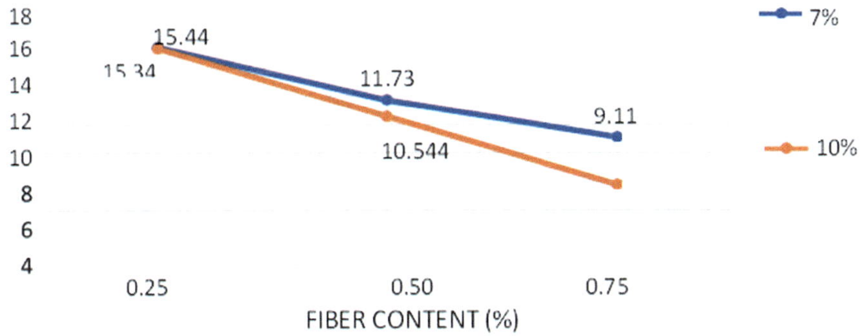

Fig. 4 Variation of crack length with fiber content

4 Conclusion

Over the past several decades, numerical researchers have given close attention to the cracking behavior and mechanisms in clayey soils connected to desiccation. According to the results of the experiments conducted for this study, inclusion of fly ash and fiber decreased cracking. The largest crack width was found in the unreinforced soil, which was reduced by the addition of fly ash and further reduced by the addition of fibers.

In unreinforced specimens, large cracks developed that extended all the way to the glass dish's bottom. When fiber was added, surface cracks developed because the polyester fibers had a propensity to hold the cracks from enlarging. Fibers altered the soil's characteristics, causing a decrease in tensile strength and swelling pressure. The utilization of the image processing technique demonstrated its efficacy in quantifying crack features and providing meaningful data. This investigation resulted in a substantial 89.9% reduction in cracks across various fiber lengths and contents. The incorporation of fibers led to a decrease in crack width, crack spacing, and cell area, indicating an improvement in crack-related characteristics.

References

1. Subbarao GV, Siddartha D, Muralikrishna T et al (2011) Industrial wastes in soil improvement. Int Sch Res Network, ISRN Civ Eng 2011:1–5. https://doi.org/10.5402/2011/138149
2. Tang C, Shi B, Liu C et al (2008) Influencing factors of geometrical structure of surface shrinkage cracks in clayey soils. Eng Geol 101:204–217. https://doi.org/10.1016/j.enggeo.2008.05.005
3. Tang C, Shi B, Cui YJ, et al (2012) Desiccation cracking behavior of polypropylene fiber—reinforced clayey soil. Can Geotech J 49:1088–1101. https://doi.org/10.1139/T2012-067
4. Costa S, Kodikara J, Shannon B (2013) Salient factors controlling desiccation cracking of clay in laboratory experiments. Geotechnique 63:18–29
5. Chaduvula U, Viswanadham BVS, Kodikara J (2016) A study on desiccation cracking behavior of polyester fiber-reinforced expansive clay. Appl Clay Sci 142:1–10. https://doi.org/10.1016/j.clay.2017.02.008
6. Maher MH, Ho YC (1994) Mechanical properties of kaolinite/fiber soil composite. J Geotech Eng 120:1381–1393
7. Maher MH, Gray DH (1990) Static response of sands reinforced with randomly distributed fibers. J Geotech Eng ASCE 116:1661–1677
8. Harianto T, Hayashi S, Du Y-J, Suetsugu D (2008) Effects of fiber additives on the desiccation crack behavior of the compacted Akaboku soil as a material for landfill cover barrier. Water Air Soil Pollut 194:141–149. https://doi.org/10.1007/s11270-008-9703-2
9. Wahab RMA Al, El-Kedrah MA (1995) Using fibers to reduce tension cracks and shrink/swell in a compacted clay. In: Acar YB, Daniel DE (eds) Geoenvironment 2000: characterization, containment, remediation, and performance in environmental geotechnics, vol 46. ASCE Geotechnical Special Publication. pp 791–805
10. Rifai SM, Miller CJ (2009) Theoretical Assessment of increased tensile strength 1857–1862
11. Ziegler S, Leshchinsky D, Ling HI (1998) Effect of short polymeric fibers on crack development in clays. Soils Found 38:247–253
12. Indu Priya M, Chaduvula U, Viswanadham BVS (2019) Studies on desiccation cracking behavior of geofiber reinforced clay

13. Chaduvula U, Viswanadham BVS, Kodikara J (2016) Desiccation cracking behaviour of geofiber-reinforced expansive clay. In: Geo-Chicago 2016. Geotechnical Special Publication 271. pp 368–377
14. Wojnar L (1999) Image analysis—applications in materials engineering, 1st edn. CRC press LLC, Taylor and Francis group, Newyork, London
15. Abràmoff MD, Magalhães PJ, Ram SJ (2014) Image processing with ImageJ. Biophotonics Int 11:36–42

Performance Assessment of Waste Plastic Inclusions on the Strength of Expansive Clays Blended with Vitrified Polish Waste and Lime

D. Appanna, L. N. V. N. Varaprasad, B. Varaprasada Rao, and R. DayakarBabu

1 Introduction

Expansive are the soils which have tendency to decrease in the volume in the absence of water and increase the volume in the presence of water. This expansive behavior in the soil is due to presence of clay minerals in expansive clay mostly of swelling clay minerals like montmorillonite and kaolinite, illite in some cases. Due to presence of high amount of clay minerals in expansive clays causes cracks in buildings, damage to the building foundations and to underground infrastructures. So it is necessary to find the problems with the expansive soils and prevent damage of structures. It is, therefore, important to use new techniques by utilizing the locally generated waste materials that are produced from industrial and agricultural byproducts for improving the properties of problematic soil. Due to low cost and convenience of locally available waste materials, shows more interest in soil stabilization. The stabilization with pozzolana and especially with silica rich admixtures can be explained with the formation of gelation process when mixed with the cementing agents like lime, etc. The main importance of this paper is to find the performance of new industrial waste

D. Appanna (✉)
Department of Civil Engineering, Lenora College of Engineering, JNTUK, Rampachodavaram 533288, India
e-mail: appannanani78@gmail.com

L. N. V. N. Varaprasad
Department of Civil Engineering, UCEK, JNTUK, Kakinada 533003, India

B. Varaprasada Rao
School of Civil Engineering, REVA University, Bangalore 560064, India

R. DayakarBabu
Freelance Academician and Consultant, Kakinada 533003, India

B. T. Jose et al. (eds.), *Proceedings of the Indian Geotechnical Conference 2022 Volume 4*, Lecture Notes in Civil Engineering 479,
https://doi.org/10.1007/978-981-97-1753-8_9

material vitrified polish waste (VPW), different waste plastics, and also a binder lime.

Stabilization using waste materials with and without a binder like lime, cement, etc., is one of them. Umar et al. [1] stated that the metakaolin can be used as stabilizing agent for increase in the lateritic soil properties.

Kolovos et al. [2] reported that stabilization of expansive soils with metakaolin is to be helpful for improving the properties of expansive soils and can be used as a structural material. According to Binici [3], about 30% of daily production in the ceramic industry goes to waste are some of the prominent waste materials which have been successfully utilized for stabilization of expansive soil.

Koyuncu et al. [4] reported that the potential of using ceramic tile dust as an addition up to 40% to investigate its effect on swelling potential and swelling pressure of Na –bentonite and the results showed that for the addition of 40% ceramic tile dust waste, swelling pressure decreased by 86% and swelling potential decreased by 57%. Gaeta Rani et al. [5] investigated the effect of tile waste on expansive soil. Different percentages of tile waste (0, 10, 20, and 30%) were added to find the variation in its strength. From the laboratory test results, optimum amount of tile waste was determined as 20% and it can be used for increase in the strength of expansive soil sub-grade of flexible pavement with a substantial save in the construction cost.

From the available literature, it is found that limited research has been done to study the use of waste ceramic dust on different strength properties of clayey soil. Therefore, the present study is undertaken to investigate the performance of waste plastic inclusions (WPI) on the strength of expansive clays blended with vitrified polish waste and lime.

2 Methodology

Expansive clay was taken from Amalapuram region at a depth of 2 m below the ground level. The soil is dark gray to black in color. The properties of the expansive clay are assessed based on relevant I.S. Code provisions, which are given in Table 1.

2.1 Vitrified Polish Waste (VPW)

VPW is a waste produced from ceramic tile industry with low porosity. A lot of waste is produced during the process of formation and transportation of ceramic tiles. These vitrified polish waste mainly consisting of CaO and silica. Vitrified polish waste was collected from a local industry RAK Ceramics, Samarlakot, Kakinada district, Andhra Pradesh (Table 2).

Table 1 Properties of expansive clay

S. No.	Property	Value
1	Specific gravity	2.68
2	Differential free swell index (%)	101
3	Liquid limit (%)	68.6
4	Plastic limit (%)	25.2
5	Plasticity index (%)	43.4
6	Sand size particles (%)	12
7	Silt and clay size particles (%)	88
8	IS soil classification	CH
9	Max. dry density (g/cc)	1.45
10	Optimum moisture content (%)	27.9
11	CBR—soaked (%)	1.5
12	CBR—unsoaked (%)	3.3
13	Unconfined compressive strength (kN/m^2)	82

Table 2 Properties of vitrified polish waste

S. No.	Property	Value
1	Specific gravity	2.46
2	Optimum moisture content	12.5
3	Maximum dry density (g/cc)	1.58

2.2 Lime

The lime used for the present study was taken from local market in Kakinada, Andhra Pradesh state, for the purpose of stabilizing the weak soil.

2.3 Waste Plastics

Pieces of polythene bags which may be cut into standard dimensions of waste plastic inclusions (2 mm wide and 50 mm long) are cut into discrete fibers and used in the project. The usage of waste plastics is as the soil reinforcement and was taken from locally available factory.

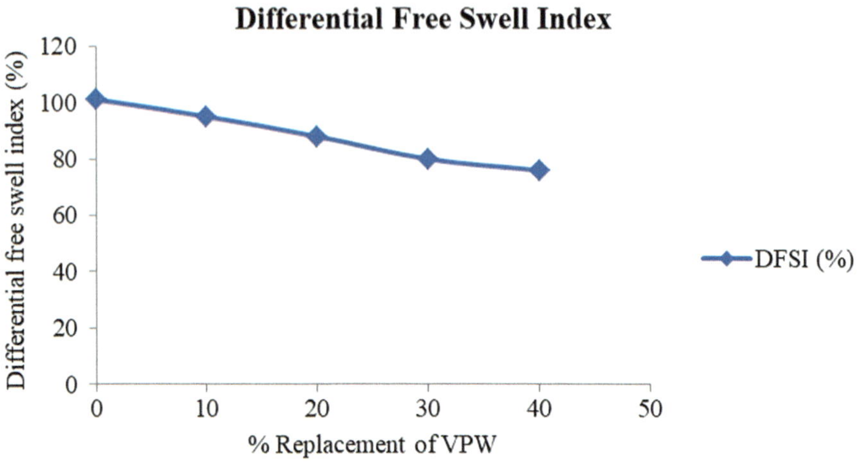

Fig. 1 Variation of DFS with % replacement of VPW

3 Results and Discussions

Different laboratory experiments were performed to evaluate the optimum combination of vitrified polish waste and lime and further treated with waste plastics, and the experimental results are discussed below.

3.1 Influence of Vitrified Polish Waste and Lime on the Swelling Property of Expansive Clay

The influence of VPW with varying percentages 0, 10, 20, 30, and 40% on swelling property is presented in Fig. 1, and to the optimum percentage of VPW, the lime was added with percentages varied from 0 to 6%. All these materials showed a decrease in DFSI value with increase in % of material as replacement. Among varying percentages of VPW, 40% replacement of VPW shows decrease in DFSI about 24.7% and further decrement was more when lime is added.

3.2 Influence of VPW and Lime on the Compaction Properties of Expansive Clay

The compaction properties are determined using Standard Proctor Compaction test in accordance with IS: 2720, Part VII [6]. The variation in compaction properties for the replacement of VPW is presented in Fig. 2, and further lime was added

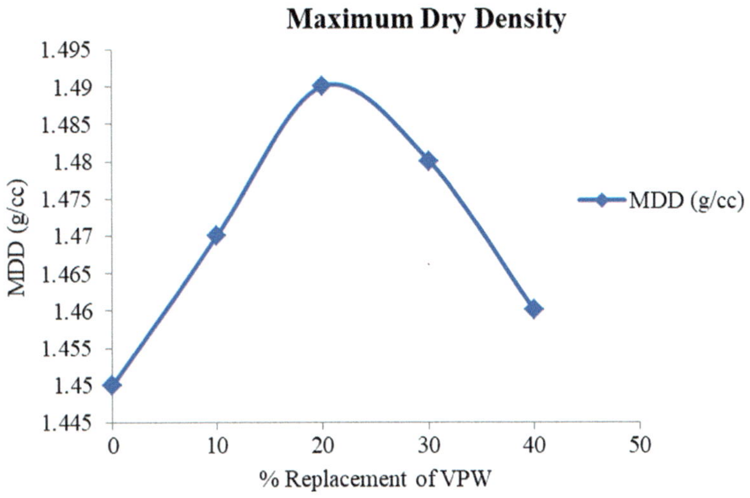

Fig. 2 Variation of MDD with % replacement of VPW

to the optimum percentage of VPW and the variation is presented in Fig. 6. From the graphs, it is found that the addition of VPW leads to an improvement in the compaction parameters and further addition of 6% lime to the 20% VPW, there is an improvement in expansive soil about 6.71%.

3.3 Influence of VPW and Lime on the CBR Values of Expansive Clay

The CBR test was performed on expansive clay according to IS 2720-Part XVI [7]. From Fig. 3, it can be seen that there is an increment in CBR value of soil treated with VPW individually. It is noticed that soil treated with 20% VPW has shown more effect on expansive clay. Further lime was added to expansive clay with 20% VPW as replacement and the results are clearly presented in Fig. 7. From the graphs, it is found that with the addition of lime, CBR value of expansive clay for soaked and unsoaked has been improved by 191.5 and 109%, respectively, for 6% addition of lime.

3.4 Influence of VPW and Lime on the UCS Values of Expansive Clay

The unconfined compressive strength test was conducted for all the samples in accordance with IS 2720-Part X [8]. The changes observed in UCS values with treatment

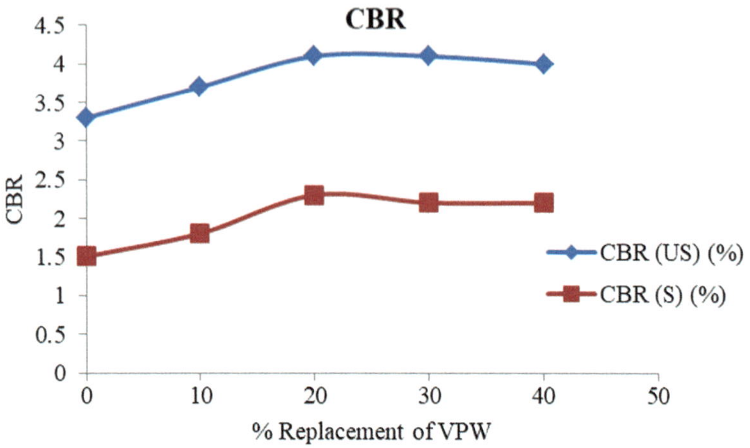

Fig. 3 Variation of CBR with % replacement of VPW

of VPW are shown in Fig. 4. From Fig. 4, it is noticed that UCS value of expansive clay is increased up to 20% replacement of VPW individually and increment is about 20.7%. Further, with the addition of lime with different percentages varied from 0 to 6% to the 20% VPW and expansive clay. It has been further increment by 73.7% when 6% lime is added (Figs. 3 and 4).

It can be inferred from the graphs that there is a gradual increase in soil properties with percentage replacement of VPW. The above results indicated that a prominent improvement is found with 20% replacement of expansive clay with VPW. Further investigated the optimum content of lime content with 20% VPW as replacement. From the experimental results, the following graphs are drawn (Figs. 5, 6, 7 and 8).

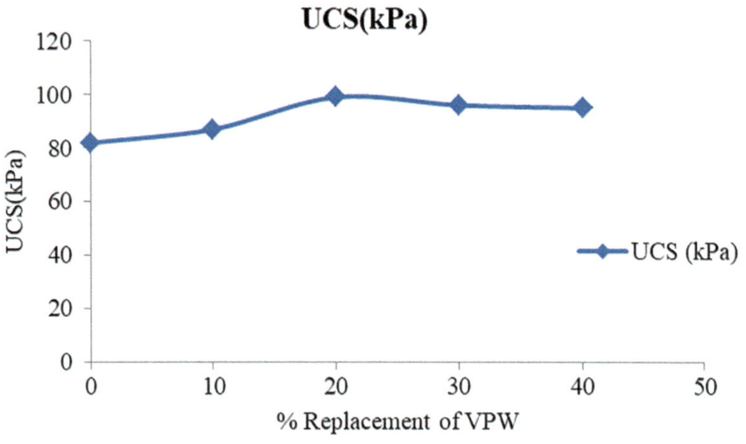

Fig. 4 Variation of UCS with % replacement of VPW

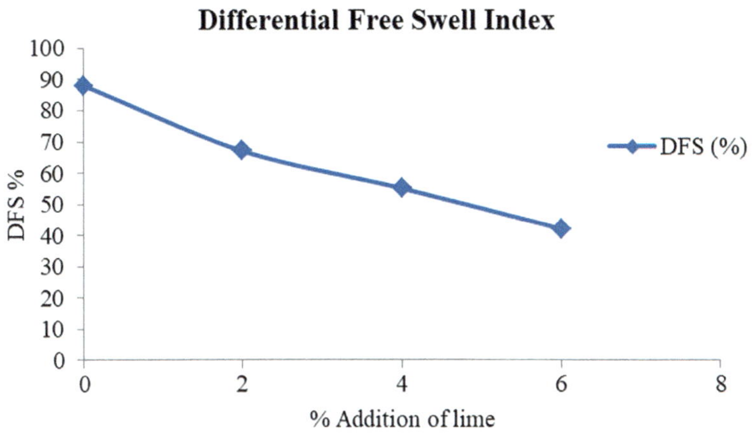

Fig. 5 Variation of DFS with % addition of lime at 20% VPW as replacement

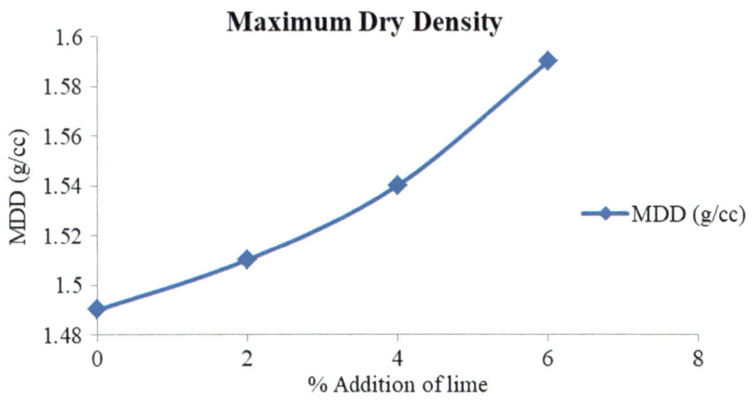

Fig. 6 Variation of MDD with % addition of lime at 20% VPW as replacement

It can be inferred from the graphs that there is a gradual increase in soil properties with percentage replacement of VPW and percentage addition of lime. From the above results, the optimum content of lime with 20% VPW as replacement of expansive soil is 6%. The addition of lime showed a constant increase in the soil properties and the trend also showed the same, and hence, the percentage lime was limited to 6% keeping in view of economic criteria and further waste plastics were used to enhance the strength of expansive soil to achieve a sustainable solution.

Further different waste plastic inclusions were added to the VPW-treated expansive soil with an optimum percentage of lime, i.e., 6% and the studies were done (Figs. 9 and 10).

Finally, from the above discussions and from the test results variations, the optimum percentage of waste plastics with 6% addition of lime and 20% replacement of VPW to the expansive clay is 1.0%. The findings indicated that the addition of

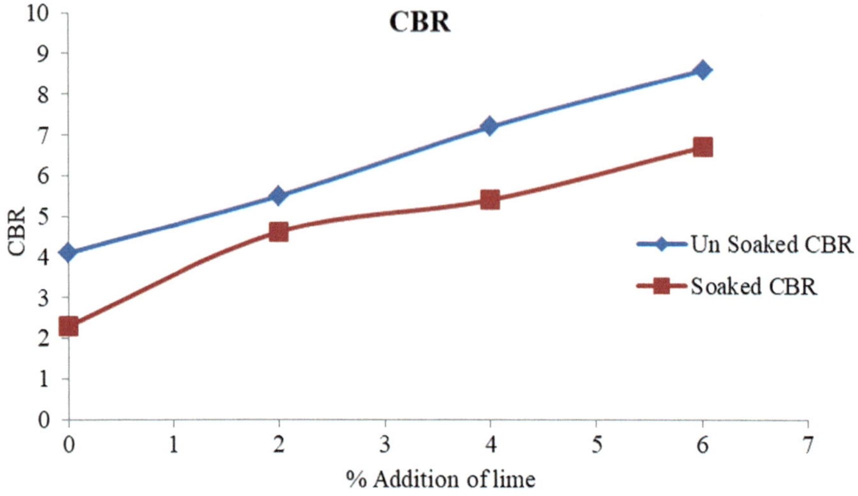

Fig. 7 Variation of CBR with % addition of lime at 20% VPW as replacement

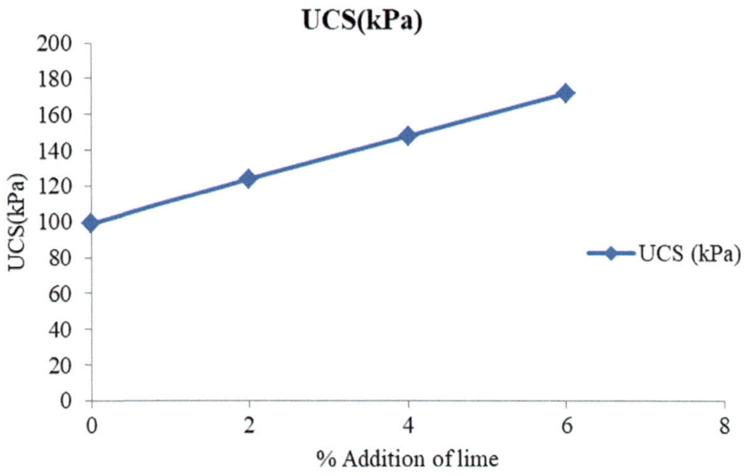

Fig. 8 Variation of UCS with % addition of lime at 20% VPW as replacement

VPW to the expansive clay showed major effect on compaction and strength characteristics. Further added with lime and waste plastics, there is more enhancement in strength characteristics. This may be due to CSH gel formation, chemical reactions between pozzolanic material and lime, and further the waste plastic inclusions added interparticular discrete reinforcement and hence the further enhancement in the strength.

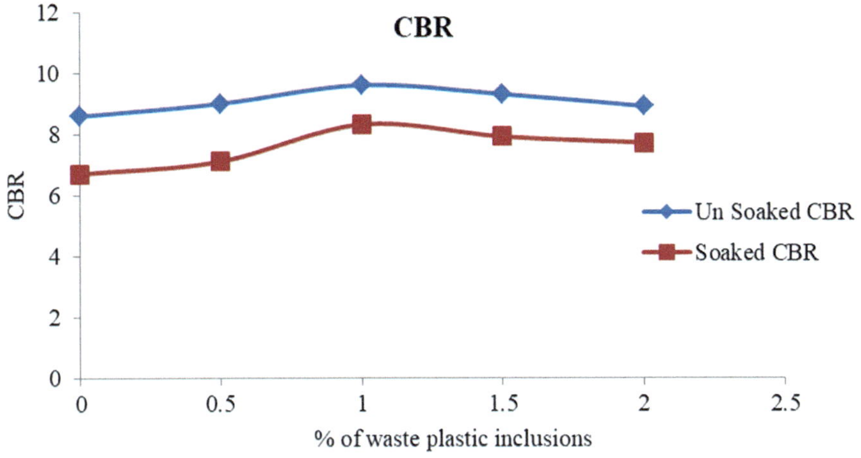

Fig. 9 Variation of CBR with % of different waste plastic inclusions at 20% VPW as replacement and 6% addition of lime

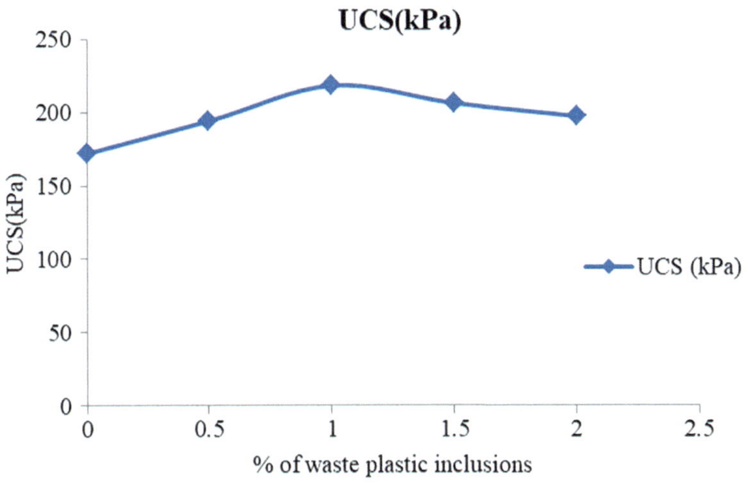

Fig. 10 Variation of UCS with % of different waste plastic inclusions at 20% VPW as replacement and 6% addition of lime

4 Conclusions

This study was carried out to investigate the performance of materials mixed with expansive clay by performing laboratory tests. The following conclusions have been made based on the results obtained from the experimental work:

- It is noticed that the expansive clay taken was problematic clay and it has high swelling and plasticity properties. It was also observed that VPW is effective to increase the strength properties up to 20% replacement.
- There is a moderate improvement in maximum dry density with percentage replacement of up to 40%. It is found that for 20% replacement of VPW, the maximum dry density is about 2.75%.
- Further when lime was added to optimum content of VPW from 0 to 6%, there is a significant improvement in maximum dry density and also corresponding strength characteristics. The increment is about 191.5% for CBR and 73.7% for UCS values.
- Further mixed with different waste plastics as reinforcement with 0–2% with an increment of 0.5%. It is observed that CBR and UCS values of expansive clay are increased about 23% and 26.7%, respectively.
- Finally it can be concluded that the VPW showed a significant improvement in compaction, strength and penetration characteristics to some extent and on further mixed it with lime and waste plastic inclusions the required strength characteristics are achieved. From this research, it is noticed that there is twofold advantage of minimizing the disposal of waste problem and also increment in the strength properties of expansive clay.

References

1. Umar SY et al (2015) Hydraulic conductivity of compacted lateritic soil partially replaced with Metakaolin. J Environ Earth Sci 5(4)
2. Konstantinos et al (2013) Mechanical properties of soilcrete mixtures modified with metakaolin. Constr Build Mater 47:1026–1036
3. Binici H (2007) Effect of crushed ceramic and basaltic pumice as fine aggregates on concrete mortars properties. Constr Build Mater 21:1191–1197
4. Koyuncu et al (2004) Utilization of ceramic wastes in the construction sector. Key Eng Mater 264–268:2509–2512
5. Rani G et al (2014) Strength behavior of expansive soil treated with tile waste. Int J Eng Res Dev 10(12):52–562
6. IS 2720-Part VII: Methods of test for soils—Determination of Water Content—Dry Density relation using Light Compaction, Bureau of Indian Standards, New Delhi (1980)
7. IS 2720-Part 16: Methods of test for soils—Laboratory Determination of CBR, Bureau of Indian Standards, New Delhi (1987)
8. IS 2720-Part 10: Methods of test for soils—Determination of Unconfined Compressive Strength, Bureau of Indian Standards, New Delhi (1991)

Performance of Stone Column Subjected to Dynamic Loading in Sand Bed

D. V. Khedkar, A. I. Dhatrak, and S. W. Thakare

1 Introduction

Stone columns are densified granular material columns inserted in fine-grained material to transfer load down to the bearing strata or to strengthen the soil to improve liquefaction resistance. Stone columns are frequently used to improve the engineering characteristics of soft soils in order to sustain weakly and moderately loaded structures such as highway embankments and large diameter storage tanks.

A modern improvement in the stone column technique is reinforcing the column with horizontal layers of reinforcement or encasing the individual stone column with geosynthetics over the full or partial height of the column. The geosynthetics' encasement multiplies the load-bearing capability of stone columns. To improve load-carrying capacity and reduce stone column settlement, the granular bed can be combined with more reinforced geogrid. Ordinary stone column (OSC), geosynthetic encased stone column (GESC), and horizontal encased stone column + vertical encased stone column (HESC + VESC) are the three types of stone columns shown in Fig. 1.

Stone columns were installed using vibratory floats mounted on hydraulic excavators. The float is inserted into the ground to the desired depth using water and air jets. The void formed at the design depth is filled with granular material. This can be done with a bottom feed method or top feed method. Areas of applications of stone columns are infrastructure, mining, construction, and liquefaction mitigation.

D. V. Khedkar · A. I. Dhatrak (✉) · S. W. Thakare
Civil Engineering Department, Government College of Engineering, Amravati, India
e-mail: dclaus37@gmail.com

B. T. Jose et al. (eds.), *Proceedings of the Indian Geotechnical Conference 2022 Volume 4*, Lecture Notes in Civil Engineering 479,
https://doi.org/10.1007/978-981-97-1753-8_10

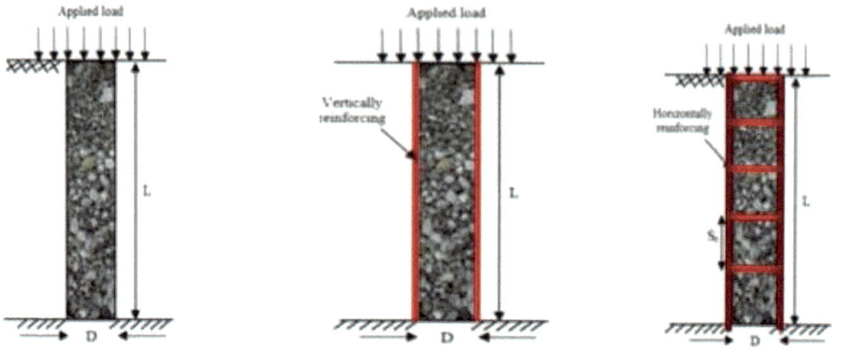

(OSC) (GESC) (HESC + VESC)

Fig. 1 Different configurations of stone column

2 Literature Review

A brief review of literature on experimental studies on stone column is presented as follows:

Salem et al. [1] presented the findings of a study on the efficiency of stone columns as liquefaction remediation. The research was based on twenty-four case studies in which SPT and CPT tests were done before and after stone column strengthening. The mechanisms of densification and stiffening were studied, as well as their individual and combined impacts.

Kolekar et al. [2] conducted laboratory experiments to better understand how stone columns erected in marine clay behave when subjected to cyclic pressures. For single stone column testing, the unit cell method was used. To make a slurry, water equal to 1.5 times the clay's liquid limit was mixed with marine clay. The clay bed was consolidated from slurry under a pre-consolidation pressure of 18 kPa, which was determined through laboratory odometer measurements, and the clay bed thickness after consolidation was 500 mm. A replacement approach was used to install a stone column with a diameter of 100 mm in the clay bed's center. Stone columns were made with aggregates ranging in size from 2 to 8 mm. The behavior of a stone column was investigated using both static and cyclic load tests. A total of four tests were performed: two static and two cyclic. One static load test was performed on an unreinforced clay bed, while the other was performed on a stone column reinforced clay bed. Only the reinforced clay bed was subjected to cyclic loading. In the case of static load testing, strain-controlled loading was used at a rate of 1.2 mm/min. One test was conducted in two stages for the cycle test on reinforced clay beds. Stage 1 consisted of a sinusoidal cyclic load equal to 35% of the static failure load of a reinforced stone column, followed by a sinusoidal cyclic load with a frequency of 0.1 Hz and 500 cycles. After 500 cycles, the same specimen was subjected to a static failure load of a reinforced stone column, followed by cyclic loading at the same frequency of 0.1 Hz, dubbed stage 2.

During and after seismic excitations, Cengiz et al. [3] conducted experimental tests on the behavior of geosynthetic encased stone columns (GECs) and ordinary (conventional) stone columns (OSCs). Well-instrumented GECs and OSCs were installed in kaolinite clay beds that were consolidated in a big steel tank for this purpose. Surcharge loads were used to model the seismic behavior of columns supporting an embankment, and the experimental equipment was put through large-scale shaking table testing. The clay bed was made from kaolinite clay, which has a specific gravity (Gs) of 2.62 and plastic and liquid limits of 26% and 49%, respectively. The clay slurry was made with a water content of 75%, which was nearly 1.5 times the material's liquid limit. The GECs were encased in three different geotextiles. The first geotextile was Tencate Polyfelt TS 10, a spun-bonded non-woven geotextile that was commercially accessible (designated as GT1). Sefitec PP 50 and Stabilenka 100 were the second and third geotextiles, respectively (designated as GT2 and GT3, respectively). These samples were tensile strength tested on 200 mm broad samples in line with DIN EN ISO 10319.

In their experimental tests, Adalier et al. [4] conducted centrifuge testing to evaluate the performance as a liquefaction countermeasure. Rather than the drainage impacts, the total site stiffening effects due to the stone column placement were evaluated. The behavior of a saturated silt layer was studied under baseline conditions. Circumstances of dynamic excitation. That stratum was evaluated in a series of four independent model experiments, first without stone columns, then with stone columns in a free-field setting, and finally with a surface foundation surcharge. Based on the recorded dynamic responses, the underlying mechanism and effectiveness of the stone columns were discussed. We examined and reanalyzed the effect of the installed columns on excess pore pressures and deformations.

Nima Mehrannia et al. [5] conducted experimental studies on soil improvement with stone columns and granular blankets. The bearing capacity of stone columns, granular blanket, and a combination of both methods in reinforced and unreinforced modes were studied using scaled physical models.

3 Experimental Investigations

About 1-g shaking table tests were carried out on performance of stone column in sand bed subjected to dynamic loading using acrylic tank of 0.6 m × 0.4 m × 0.6 m and the frequency of 2 Hz for the amplitude of ± 2 mm and ± 3 mm, respectively. Total four different series of shaking table tests were conducted: Series A: test on sand bed without stone columns, Series B: test on sand bed with OSC, Series C: test on sand bed with ESC, and Series D: test on sand bed with RSC.

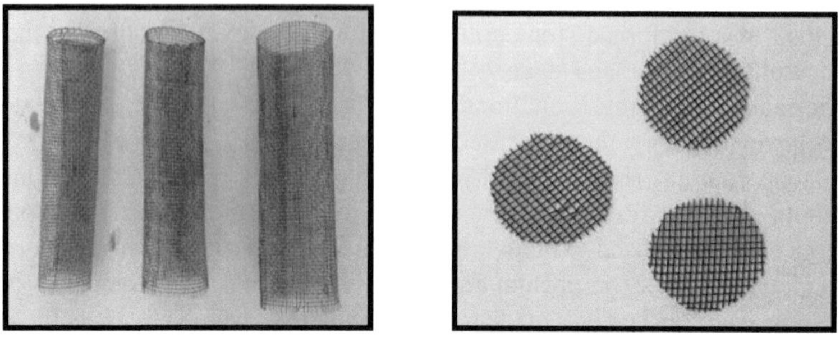

(a) (b)

Fig. 2 Steel mesh used for **a** encasing **b** reinforcement of stone column

3.1 Materials Used for Experimental Investigations

Soil

For the experimental investigation, Kanhan sand (cohesionless, dry, and clean) available in Nagpur region of Vidarbha (MH) was used as the foundation soil. The test sand was angular and of uniform yellow color, with small proportion of black flint stones. The particle size of sand decided for the test was passing through 2 mm IS sieve.

Crushed Stone Aggregate

Crushed stone chips passing 6.3 mm I.S sieve and retained on 2 mm I.S sieve were used to form the stone column. The particle sizes for the column were as per the guidelines of Nayak (1983), which suggest that the particle size should be in the range of 1/6–1/7 diameter of the column (i.e., 30 mm, 36 mm, and 45 mm, respectively). The specific gravity of crushed stone aggregates was (S.G) 2.75.

Steel Wire Mesh

The reinforcements for horizontal and vertical encasings of stone column formed using a steel wire mesh had an aperture size 1 mm × 1 mm (Fig. 2).

The properties of sand, crushed Stone aggregates, and steel wire mesh are shown in Table 1.

4 Experimental Set-Up

The experimental set-up consists of shaking table, acrylic tank, data logger, control panel, accelerometer, LVDT, and motor. The 3D view of experimental set-up is shown in Fig. 3.

Table 1 Properties of sand, crushed stone aggregates, and steel wire mesh

S. No.	Materials	Properties	Values
1	Sand	IS classification	SP
		Specific gravity	2.638
		Maximum void ratio (e(max))	0.79
		Minimum void ratio (e(min))	0.59
		Relative density	36%
		Angle of internal friction (ϕ)	34°
		Coefficient of uniformity (Cu)	2.22
		Coefficient of curvature (Cc)	1.4
2	Crushed stone aggregates	Specific gravity	2.75
3	Steel wire mesh	Aperture size (mm)	1 × 1

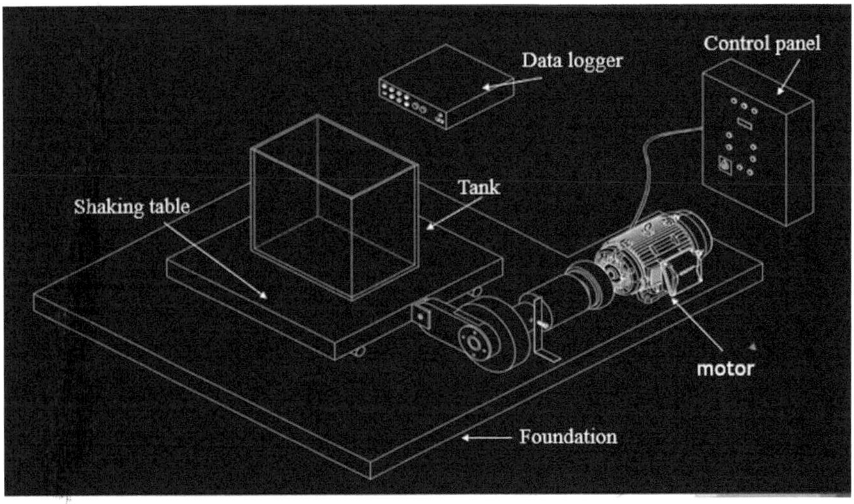

Fig. 3 Three-dimensional view of shaking table

5 Experimental Test Procedure

Using the gravity raining technique, the sand was filled with a relative density (Dr) of 36%, with a height of fall of 10 cm up to the height of H. (total height of sand bed). The sand bed was prepared to a depth of 1/3rd the height from the top of the respective stone column, and Accelerometer 1 (i.e., A1) was placed at that level. After that, the remaining sand bed was prepared up to height H. After filling the tank, a plane wooden plate and a spirit level were used to level the bed's surface. The model footing, which was attached to the superstructure, was then placed in the center of the sand bed. The LVDT was mounted on the top surface of the superstructure to measure upward settlement as shown in Fig. 4a. The Accelerometer 2 (i.e., A2) was fixed on the base of the shake table to measure acceleration coming to the shake table.

The displacement method was used to build the stone columns. For various experiments, the stone column diameters (dsc) were 30, 36, and 45 mm. The stone column's length-to-diameter ratio (lsc/dsc) was kept at 5.0. The sand was filled up to the lowest level of the stone column, and a PVC pipe with a diameter equal to the stone column's diameter was installed in the center of the bed at that level. The sand bed preparation was completed up to the 2/3rd height of the stone column, when Accelerometer 1 (i.e., A1) was installed. The sand bed was then raised to the desired height, and PVC pipe of the required height for each stone column was installed. The quantity of stone aggregate required to build the stone column was premeasured and put into the casing pipe in three layers, each of (lsc)/3 mm thickness, when the tank was filled. After adding stone aggregates in each layer, it was compressed by using a tamping rod with a 20 mm diameter and 1600 gm weight that was dropped freely

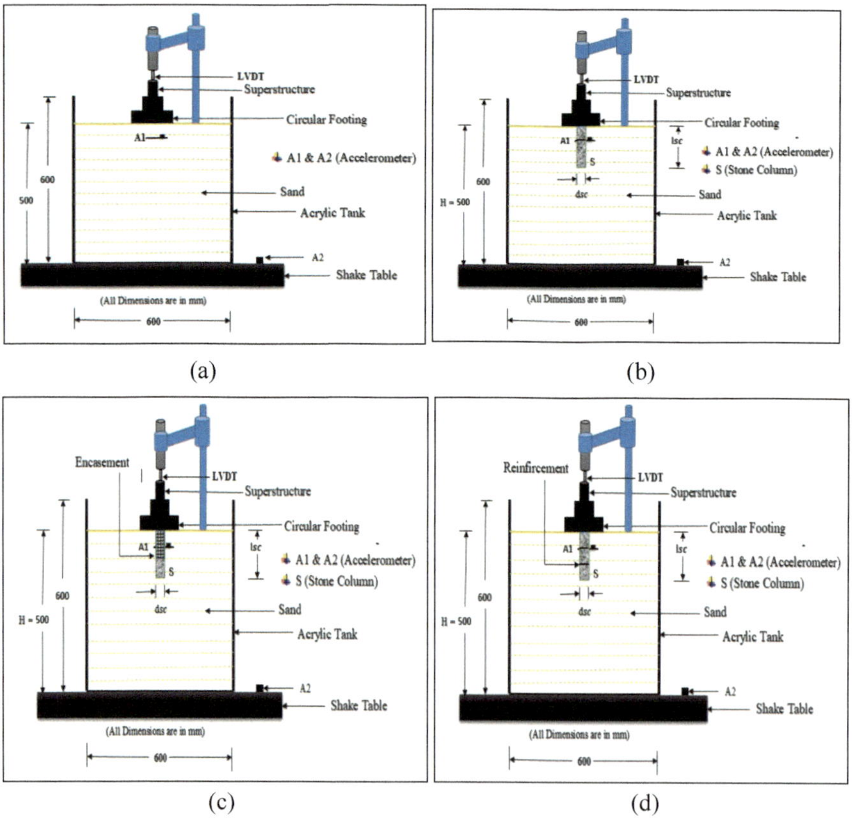

Fig. 4 Schematic view of experimental test procedure

from a height of 50 mm for ten strikes. After each layer of aggregate was placed, the casing pipe was slowly pushed up for a distance equivalent to the thickness of that layer. Using the same method, the next layer of aggregates was compacted until the complete length of the stone column was created. The stone column's upper surface was leveled. Then, in the center of the sand bed, a footing attached to the superstructure was built. The LVDT was fixed on the base of the shake table to measure downward settlement, and the Accelerometer 2 (i.e., A2) was mounted on the top surface of the superstructure to measure upward settlement. A stone column with a footing and a superstructure is represented schematically which is shown in Fig. 4b.

The placement of the encased stone column in the sand was identical to that of a regular stone column. While constructing the sand bed, steel wire mesh with a height of 100 mm from the top was joined to PVC pipe casing in a circular shape to provide encasement to the column. The schematic view of an incased stone column is shown in Fig. 4c (ESC). Crushed stone of a specific gradation was then poured in the same manner as a regular stone column. In all situations, the height of the encased chosen for experimental research was 100 mm, which was the "optimal height of encasement.

In the case of a reinforced stone column (RSC), the technique was similar to that of an ordinary stone column, with the exception that following compaction of one layer of stone column, a circular steel mesh was placed above that layer and on top of the prepared stone column. The schematic view of RSC is shown in Fig. 4d. The schematic perspective of the experimental test technique is shown in Figs. 4, and 5 shows the complete sequence of making ordinary stone column in sand bed.

Results

Experimental investigations were carried out on the loose sand bed with ordinary stone column, sand bed with encased stone column, and sand bed with reinforced stone column subjected to dynamic loading with the constant natural frequency of machine foundation 2 Hz and amplitude of $\pm$ 2 and $\pm$ 3 mm. The results of all tests on the sand bed were compared in terms of the settlement and acceleration. Figure 6 shows typical settlement verses time graphs obtained from data logger.

The settlement values of foundation on soil bed, soil bed provided with OSC, soil bed provided with RSC, and soil bed provided with encasement for $\pm$ 2 mm amplitude and $\pm$ 3 mm amplitude are shown in following Tables 2 and 3, respectively.

The performances of sand bed provided with OSC, ESC, and RSC were analyzed in terms of percentage reduction in settlement of footing on sand bed without stone column. The effect of various parameters such as length and diameter of stone column and amplitude of vibrations on the settlement of footing were examined.

Table 4 and Figs. 7 and 8 show % reduction in settlement of footing on sand bed with OSC, RSC, and ESC of different diameters and different amplitudes of vibrations, as compared with that of footing on sand bed without stone column.

The percentage reduction value of settlement for sand bed with OSC, sand bed with RSC, and sand bed with ESC was more when it was compared with sand bed without stone column.

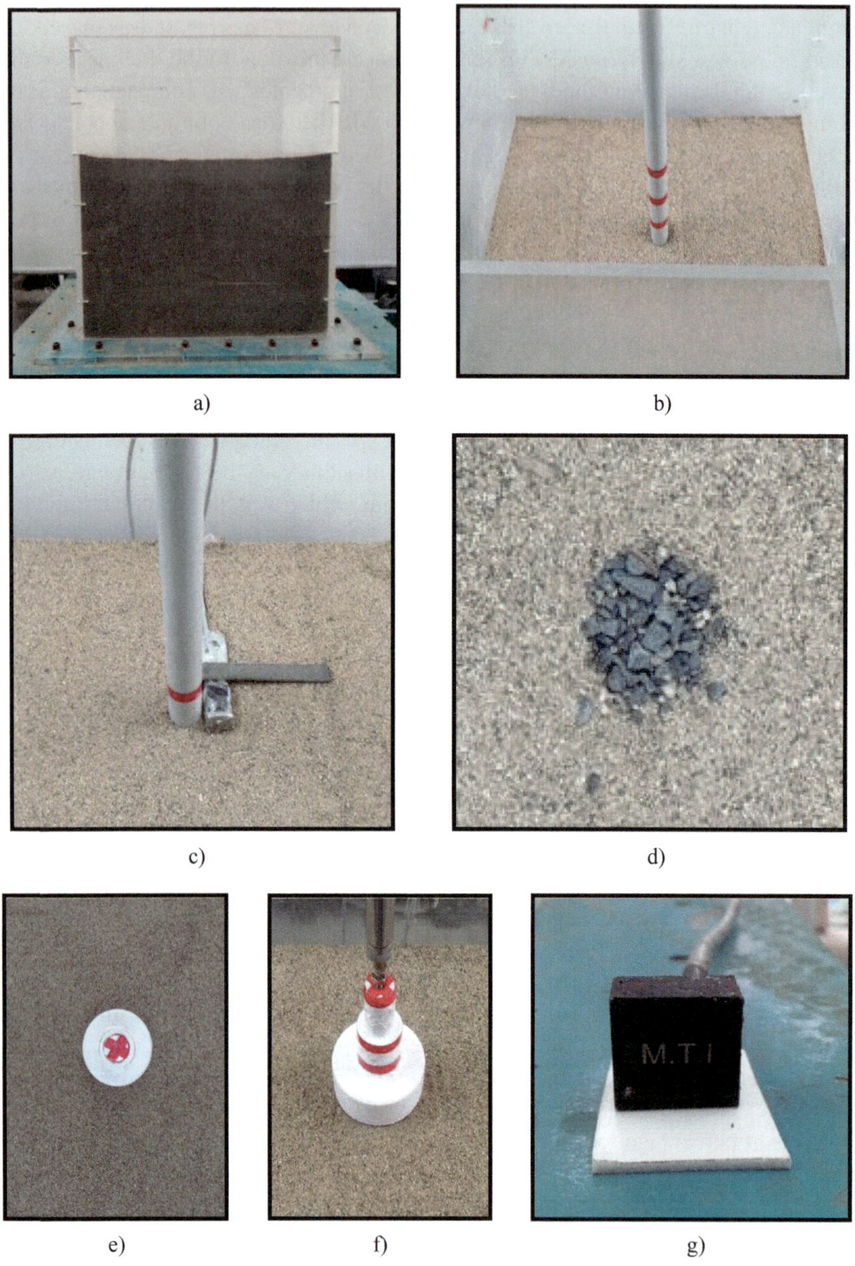

Fig. 5 Complete sequence of making ordinary stone column in sand bed represented by: **a** sand filled up to H—height of respective stone column; **b** placement of marked PVC pipe casing for stone column; **c** placement of accelerometer A1 on sand bed at 1/3 height of stone column from top of sand; **d** top view of formed stone column; **e** top view of placement of footing and superstructure on stone column; **f** placement LVDT on superstructure; **g** accelerometer A2 mounted on base of shake table

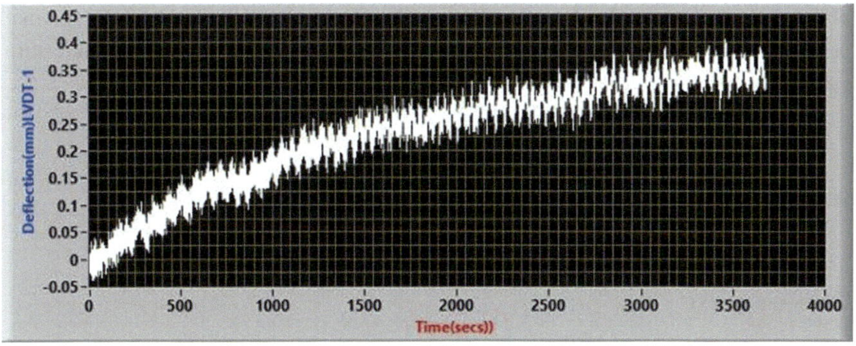

Fig. 6 Graph of settlement verses time (LVDT placed above superstructure)

Table 2 Settlement values for dynamic loading with ± 2 mm amplitude

Description	Diameter of stone column (cm)	Settlement (mm)
Sand bed	–	0.36
	–	0.36
	–	0.36
	3	0.24
Sand bed with OSC	3.6	0.19
	4.5	0.14
	3	0.127
Sand bed with RSC	3.6	0.12
	4.5	0.112
	3	0.088
Sand bed with ESC	3.6	0.067
	4.5	0.024

From the result, it is evident that encased stone columns are more effective as compared to ordinary stone column and reinforced stone column, toward reducing the settlement of footing subjected to dynamic loading. The settlement was observed to be reduced up to 1400% by providing encased stone column.

Reinforced stone columns are, however, more effective than ordinary stone columns, toward reducing the settlement of footing subjected to dynamic loading. The settlement was observed to be reduced up to 250% by providing encased stone column.

Table 3 Settlement values for dynamic loading with ± 3 mm amplitude

Description	Diameter of stone column (cm)	Settlement (mm)
Sand bed	–	0.43
	–	0.43
	–	0.43
Sand bed with OSC	3	0.32
	3.6	0.22
	4.5	0.175
Sand bed with RSC	3	0.142
	3.6	0.128
	4.5	0.117
Sand bed with ESC	3	0.122
	3.6	0.113
	4.5	0.086

Table 4 % reduction in settlement of foundation due to provision of OSC, ESC, and RSC

Amplitude of vibration (mm)	Diameter of stone column (cm)	% Reduction in settlement		
		Sand bed with OSC	Sand bed with RSC	Sand bed with ESC
+ 2 mm	3	50	186.46	309.09
	3.6	89.47	200	437.313
	4.5	157.14	221.42	1400
+ 3 mm	3	34.37	202.81	252.45
	3.6	95.45	235.93	280.53
	4.5	145.71	267.52	400

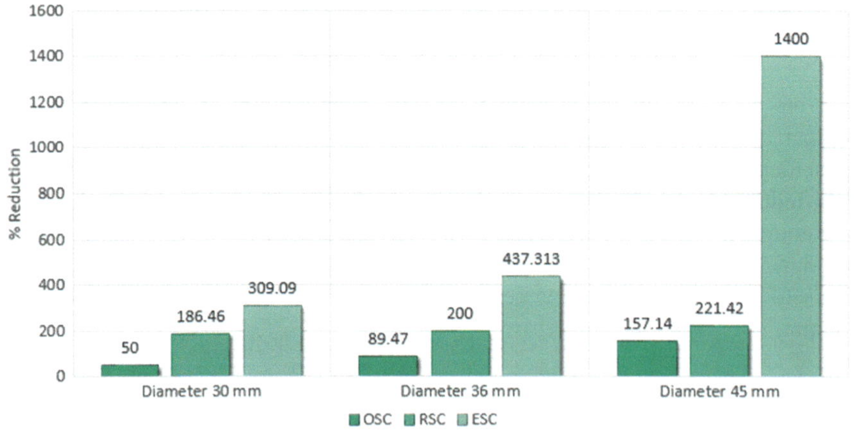

Fig. 7 % reduction in settlement of foundation on soil provided with OSC, RSC, and ESC for + 2 mm amplitude

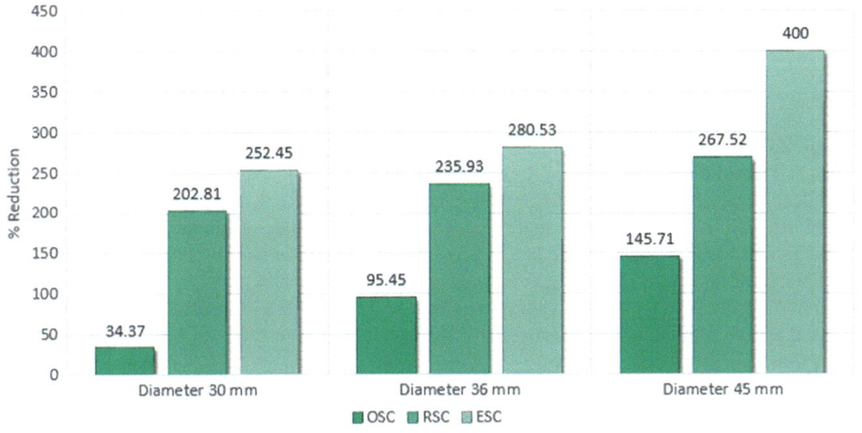

Fig. 8 % reduction in settlement of foundation on soil provided with OSC, RSC, and ESC for + 3 mm amplitude

6 Conclusions

Experimental investigations on footing supported on sand bed and provided with ordinary, encased, and reinforced stone column and subjected to dynamic loading using shake table were performed and settlement of footing was determined in each case. Percentage reduction in settlement was determined to judge the performance of these systems.

Based on the experimental investigations on the behavior of model footing sand bed provided with ordinary stone column (OSC)/encased stone column (ESC)/ reinforced stone column, following broad conclusions are drawn:

1. Encased stone column is more effective as compared to OSC and RSC.
2. Settlements of footing on sand bed provided with encased stone columns may be reduced up to 1400%.
3. Settlements of footing on sand bed provided with reinforced stone columns may be reduced up to 250%.
4. Settlements of footing on sand bed provided with ordinary stone columns may be reduced up to 150%.
5. The effect of providing stone column in soil bed and subjected to dynamic loading is more improved with increase in the diameter of stone column.
6. As the amplitude of vibration increases, the performance of stone column in soil bed and subjected to dynamic loading is more enhanced.

References

1. Akimusuru JO, Akinbolade JA (1981) Stability of Loaded footing on reinforced soil. J Geotech Geo Environ Eng ASCE
2. Baez JI (1995) A design model for the reduction of soil liquefaction by vibro-stone columns. Ph.D. Dissertation, University of Southern California
3. Delgadillo R, Zonno G, Boroschek R, Lourenco PB, Aguilar R, Study of the environmental influence on the dynamic behaviour of adobe walls: preliminary test in labor atory specimens. In: Proceedings of the international conference on experimental vibration analysis for civil engineering structures, San Diego
4. FHWA, Design and Construction of Stone Column, FHWA/Rd-83/026 (1983) Federal highway administration office of engineering and highway operation research and development Washington, D.C. 20590
5. Goughnour RR, Pestana JM (1998) Mechanical behavior of stone columns under seis- mic loading. In: 2nd International conference on ground improvement techniques

Utilization of Brine Sludge to Improve the Strength and Compressibility Characteristics of Soft Clays

Merin Kuriakose, K. N. Athira, Benny M. Abraham, and Sobha Cyrus

1 Introduction

Increase in population coupled with rapid industrialization has led to extensive generation of industrial wastes. Globalization has also triggered considerable development in infrastructure in urban areas recently. Due to the scarcity of land, the soils having low bearing capacity and high compressibility characteristics are also being considered for construction purposes. Hence, much priority has been placed to modify and stabilize these soils. Scientists, all over the world, are trying to convert the industrial waste to value-added materials. Over the past decade, several attempts have been made to utilize the industrial waste in the construction industry. Brine sludge is an industrial waste product, precipitated during chlorine and caustic soda production through the electrolysis of brine. India's Caustic Soda plant capacity is predicted to grow from 3.751 million tons per annum (mtpa) in 2017 to 5.032 mtpa in 2022, as per the findings of GlobalData, a leading data and analytics company [1]. Substantial levels of dissolved metals and impurities may be present in brine sludge. These toxic substances that get precipitated from the brine affect the environment adversely, and hence, timely and proper disposal of brine should be advocated by this

M. Kuriakose (✉) · K. N. Athira · B. M. Abraham · S. Cyrus
CUSAT, Kochi, India
e-mail: merinthaliyachira93@gmail.com

K. N. Athira
e-mail: athirakn29@gmail.com

B. M. Abraham
e-mail: bennymabraham@gmail.com

S. Cyrus
e-mail: sobharoymthomas@gmail.com

B. M. Abraham
AISAT, Kochi, India

B. T. Jose et al. (eds.), *Proceedings of the Indian Geotechnical Conference 2022 Volume 4*, Lecture Notes in Civil Engineering 479,
https://doi.org/10.1007/978-981-97-1753-8_11

industrial sector. Several researchers have investigated the reuse of brine sludge, in addition to cement and fly ash, in non-structural building components such as paver block and bricks [2]. The percentage of brine sludge was optimized to be 35% for the construction of non-traffic (M30) paver blocks [3]. A sustainable way has been conceived to address environmental hazards of brine sludge by developing non-toxic, geo-polymeric (cement-free) materials. The fabrication of advanced composite was successfully done by incorporating geo-polymerized brine sludge matrix [4].

No research has been reported regarding the utilization of brine sludge in soil stabilization. In light of this, the aim of the present study is to examine the potential of brine sludge as an additive to stabilize a soft clayey soil such as Cochin marine clay, known for its low shear strength and high compressibility [5]. The research also paves way to understand whether the inclusion of lime can further advance the properties of soil.

2 Experimental Investigations

Materials Marine Clay

The soil used in the present study was marine clay, which was collected from a site near Kundanoor, located in Ernakulam district, Kerala. It was extracted from a depth of 13 m below the ground level. The collected soil was stored in polythene bags, without losing the moisture content. The soil was grayish black in color and had a natural moisture content of 94% (Table 1).

Brine Sludge (BS)

The brine sludge was collected from Travancore Cochin Chemicals, Ernakulam. The brine sludge was analyzed for its chemical constituents by quantitative analysis (Table 2).

Lime

Table 1 Physical properties of the soil in moist condition

Property	Test values
Natural moisture content (%)	94
Liquid limit (%)	110
Plastic limit (%)	39
Plasticity index (%)	71
Shrinkage limit (%)	27
Free Swell index (cc/g)	4.8
Clay size (<0.002 mm) %	49
Silt size (<0.002–0.075 mm) %	25
Sand size (0.075–4.75 mm) %	26
Unconfined compressive strength at NMC (kPa)	4.1

Table 2 Physical and chemical properties of brine sludge

Property	Test values
Color	Light gray
Physical state at room temperature	Semi-solid
pH	10
Water content	65%
$CaCO_3$	32.5%
$BaSO_3$	67.5%

For the preparation of lime, precisely chosen uniform shells were used. It was preserved in airtight multilayer polythene bag. On the day of preparation of lime-treated samples, sufficient water was sprinkled over the shells which were stored in these polythene bags. It was sieved through IS 425 micron sieve when all the shells crumble to fine powder.

Specimen Preparation and Testing

In an attempt to study the effectiveness of brine sludge and its performance in the presence of additive, soil specimens were prepared by mixing various ingredients in concentrations as given in Table 3. Some amount of moist clay was taken and its moisture content was determined. The moist soil was mixed with predetermined quantity of brine sludge as percentage by dry weight of soil. Varying percentages of brine sludge, 10, 20, 30, and 40%, of the dry weight of soil were added to marine clay. Lime was also used as an additive to study its effect on brine sludge-treated clayey sample. Samples were prepared with 30% of brine sludge + 3% of lime by dry weight of soil. Five different combinations were used for the preparation of various soil specimens and they are listed as follows:

1. Clay + 10% brine sludge.

Table 3 Experimental program for the study

Soil	Brine sludge (%)	Additive lime (%)	Curing period (days)	Tests conducted
	0	0	0	UCC test, liquid limit, plastic limit, free swell index, and consolidation tests
Marine clay				
	10	0	0, 7	UCC test
	20	0		
	30	0		
	40	0		
	30	0	0, 7, 30, 60,	UCC test,
	30	3	90, and 120	Liquid limit, plastic
				Limit, free swell index, and consolidation

2. Clay + 20% brine sludge.
3. Clay + 30% brine sludge.
4. Clay + 40% brine sludge.
5. Clay + 30% brine sludge + 3% lime.

Soil samples required for unconfined compressive tests and consolidation tests were prepared in respective molds and stored in sealed polythene bags under controlled temperature and humidity. These samples were tested at predetermined curing periods of 7, 30, 60, 90, and 120 days. The samples were covered with moist cotton cloth throughout its curing period. Unconfined compressive tests and consolidation tests were conducted on the soil specimens in accordance with the Indian Standards [6]. The prepared soil specimens were also tested for obtaining Atterberg limits and Free Swell Index values at the same curing periods. The Free Swell Index was determined by using the method suggested by Sridharan and Rao [7]. The experimental program for the study is given in Table 3.

3 Results and Discussions

Effect of Brine Sludge on Unconfined Compressive Strength

The effect of different percentages of brine sludge on unconfined compressive strength of soil at curing periods, 0th day and 7 days, is presented in Table 4. The results from the experiments performed demonstrate that the soil sample treated with 30% of brine sludge has yielded highest strength at the end of 7 days of curing as compared to other soil specimens. Therefore, 30% of brine sludge was taken as the optimum dosage for further investigations.

Effect of Additive on Unconfined Compressive Strength of Brine Sludge-Treated Samples

Although the UCC strength is slightly increasing, it does not show considerable strength gain. Hence, attempts were made to see that whether small percentages of lime, if mixed with the brine sludge-treated soil, could bring in substantial improvement in strength. The effect of lime was determined by blending the marine clay with 30% of brine sludge and 3% of lime (by dry weight of soil). Thirty % brine sludge was chosen as the optimum dosage from the preliminary study by considering

Table 4 Variation of unconfined compressive strength with curing period

Type of sample	Unconfined compressive strength (kPa) for a curing period of	
	0 day	7 days
10% brine sludge	9	9.5
20% brine sludge	9.5	10.4
30% brine sludge	8.5	11.3
40% brine sludge	8.4	10.2

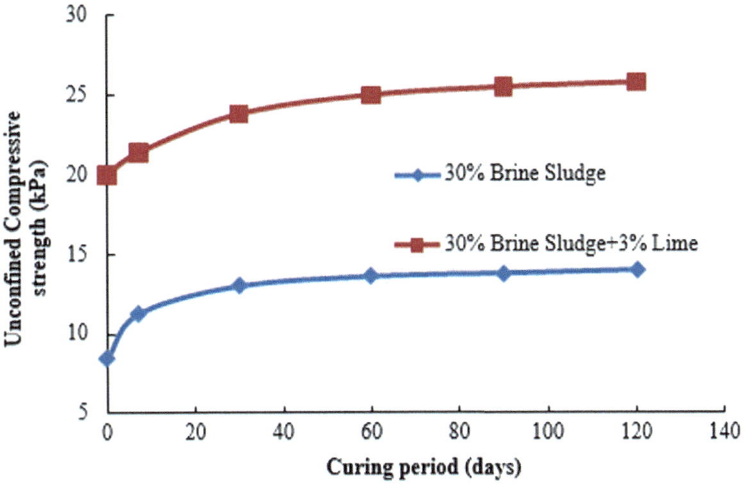

Fig. 1 Effect of brine sludge and lime on unconfined compressive strength with curing period

the increase in the percentage of strength gain at 7 days of curing. The effect of the addition of brine sludge and lime on unconfined compressive strength is presented in Fig. 1.

Hydration reactions and development of strength in lime-treated soil take place immediately after mixing. With the use of lime, it was observed that the strength of soil samples increased significantly. The unconfined compressive strength of lime-treated samples increased to 20.33 kPa at 0th day and continuously increased up to 25.8 kPa at the end of 120 days of curing. The increase in strength with the addition of lime can be due to the development of gelatinous compounds that improves the binding strength. These gelatinous compounds are formed when the pozzolanic reaction takes place between the soil and the stabilizer [8]. The strength gain of soil also depends on the curing period in all the cases.

The test result implies that with the increase in curing period, the strength of soil also increases. Therefore, it can be inferred that brine sludge along with small percentage of lime can be effectively used to stabilize soft clayey soils.

Effect of Brine Sludge and Lime on Liquid Limit of Soil

Liquid limit of the soil at its natural state was 110%. The liquid limit of treated clay sample seems to increase initially, when compared to that of untreated sample, and thereafter, it decreases with curing period. The liquid limit of soil when lime was added to the clay treated with 30% of brine sludge increased to more than 120% immediately on mixing, and then, it decreased with respect to the increase in curing period. At the end of 120 days of curing, the liquid limit continuously decreases to a value of 108%, as shown in Fig. 2. It was observed that as the curing period increases, the liquid limit shows a decreasing trend. The decrease in double-layer thickness and the formation of aggregates in treated soil may be concluded as the reason to the decrease in liquid limit [9, 10].

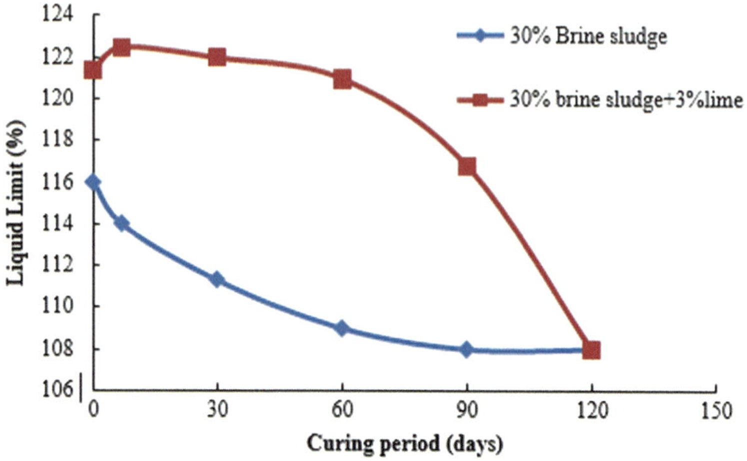

Fig. 2 Effect of brine sludge and lime on liquid limit of soil

Effect of Brine Sludge and Lime on Plasticity Index of Soil

The Plasticity Index of soil in its natural state was 71%. The effect of brine sludge and lime on Plasticity Index of the soil is depicted in Fig. 3.

It was observed that the incorporation of brine sludge into clay sample decreases the Plasticity Index with respect to the curing time. The Plasticity Index of soil, with the addition of lime, increases slightly in the initial stage and then decreases, with respect to the increase in curing period.

Effect of Brine Sludge and Lime on Free Swell Index

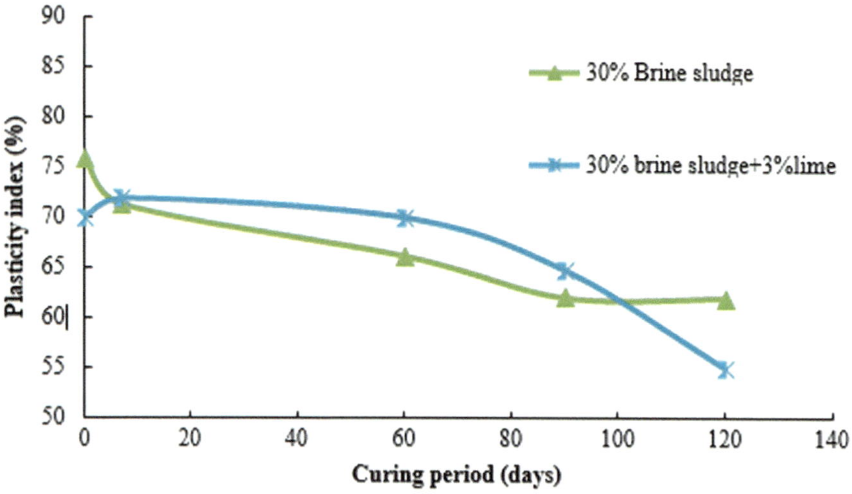

Fig. 3 Effect of brine sludge and lime on plasticity index of brine sludge-treated soil

Free Swell Index of soil in its natural state was 4.28 cm^3/g. Table 5 displays the results of Free Swell Index on treatment with different percentages of brine sludge with curing period. The Free Swell Index values of all treated soil samples decrease marginally as the curing period increases. The reduction in swelling potential of all the treated clayey samples can be attributed to the formation of pozzolanic compounds, which helps in cementing the soil particles together [11].

Effect of Brine Sludge and Lime on Compressibility Characteristics of Soil

In order to determine the effectiveness of calcium-based stabilizers on brine sludge-treated samples, the compressibility characteristics of brine sludge-treated samples with lime is studied by performing one-dimensional consolidation tests. The specimens were cured for 0 day, 1 month, and 3 months and were subjected to consolidation test by giving different pressure increments.

Figure 4 compares the e-log p curves for samples treated with 30% of brine sludge at 0, 30, and 90 days of curing with e-log p curve of untreated marine clay. It can be seen that the clay treated with 30% of brine sludge shows lesser compressibility compared to that untreated clay, indicating the effectiveness of brine sludge in reducing the compressibility characteristics of clay. The result also throws light on the significance of curing as brine sludge-treated clay at 1 month of curing period yields higher compressibility with respect to samples at the end of 3 months of curing. Further, the effectiveness of lime is investigated by conducting compressibility studies on lime with brine sludge-treated soil.

Figure 5 displays the e-log p curves for samples treated with 30% of brine sludge + 3% lime at 0, 30, and 90 days of curing with e-log p curve of the untreated marine clay. It depicts the effect of curing on void ratio–pressure relationships for clay stabilized with lime and brine sludge. It is observed that addition of lime can decrease the compressibility quite effectively. The interactions in lime–soil matrix have been divided as long-term and short-term reactions [12]. The index properties of the soil are mainly affected due to short-term reactions such as cation exchange, flocculation, pH, carbonation, and lime migration, whereas the changes in the strength and compressibility characteristics are primarily accomplished from the long-term reactions, which involves the development of various reaction products.

With the inclusion of lime and brine sludge in the soil matrix, the compression loading can be suppressed in a much better way, and as a result, it shows lesser compressibility. The results also clearly indicate that the presence of lime in cured samples helps to resist the axial load very effectively resulting in flatter load–compression curves.

Table 5 Effect of brine sludge and lime on Free Swell Index with curing period

Samples	Free Swell Index (cm^3/g) for a curing period of				
	0 day	7 days	1 month	3 months	4 months
Clay + 30% brine sludge	4.19	4.18	4.16	4.14	3.26
Clay + 30% brine sludge + 3%lime	4.59	5.52	4.28	4.07	3.88

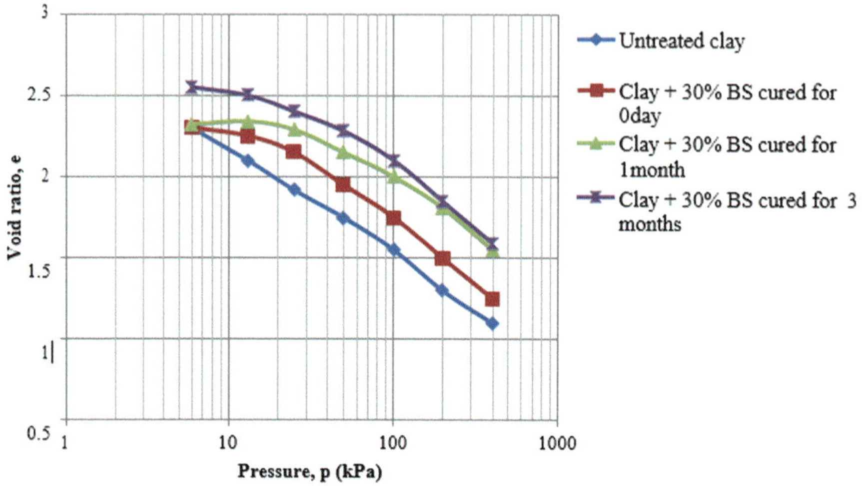

Fig. 4 Comparison of e-log *p* curves of untreated clay and clay treated with 30% brine sludge at different curing periods

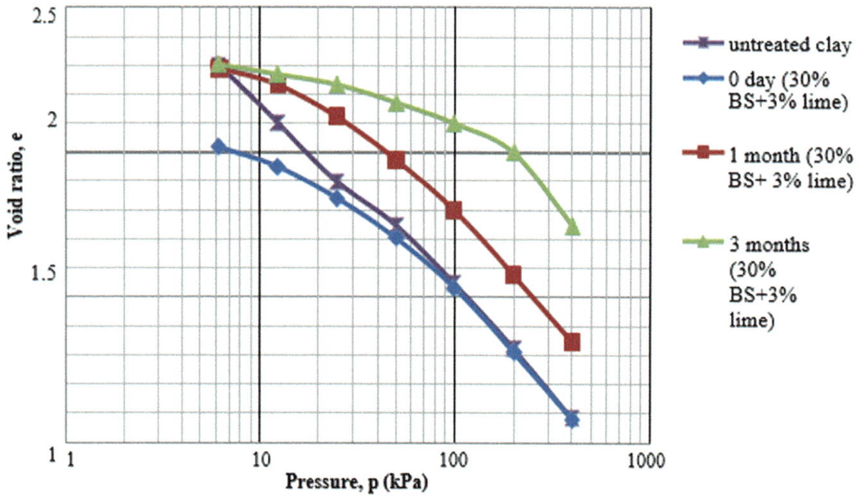

Fig. 5 Comparison of e-log *p* curves for 30% brine sludge + 3% lime at different curing periods

Figure 6 compares the compressibility curves of brine sludge-treated samples with and without lime. It clearly shows the effectiveness of lime in reducing the compressibility behavior of brine sludge-treated clay.

The variation of de/d(log p) with curing period for treated clay is presented in Figure 7. By addition of lime with brine sludge, de/d(log p) seems to decrease drastically, clearly showing the effectiveness of lime in reducing the settlement of the brine sludge-treated marine clay.

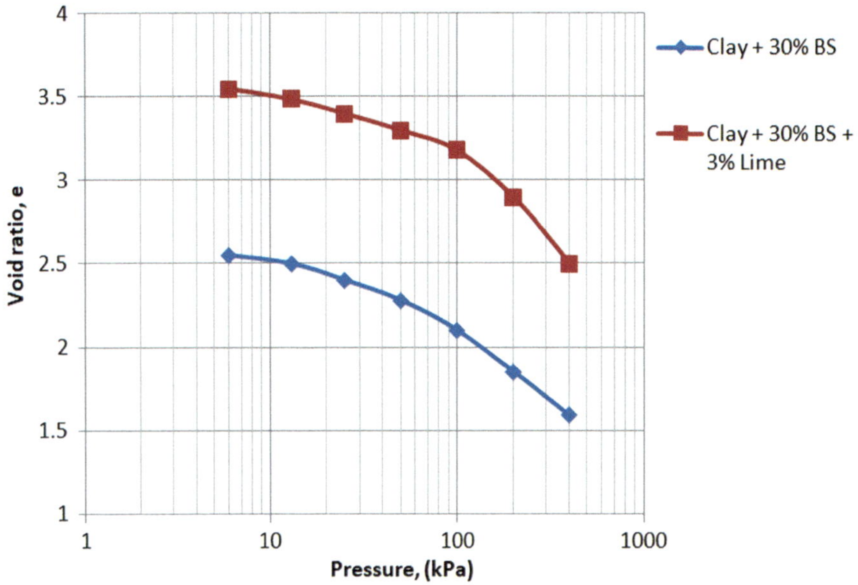

Fig. 6 Comparison of e-log p curves of 30% brine sludge and 30% brine sludge + 3% lime (curing period: 3 months)

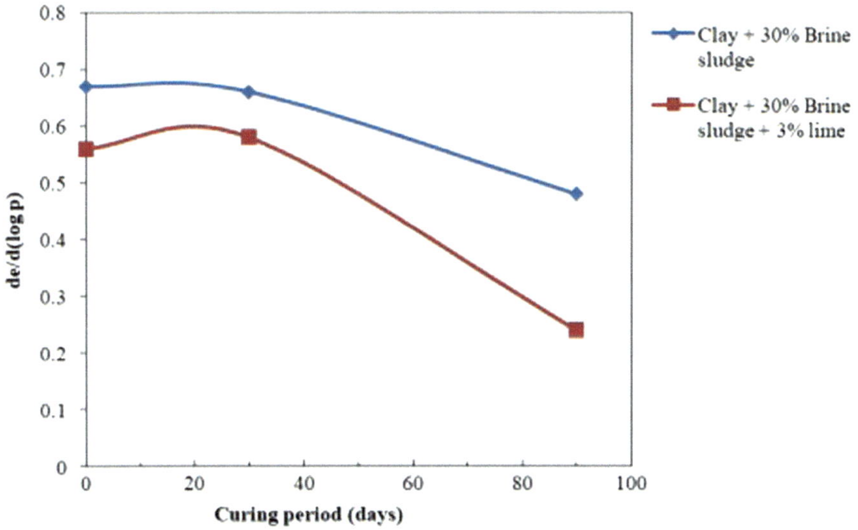

Fig. 7 Variation of de/d(log p) with curing period for treated clay (pressure range = 50–100 kPa)

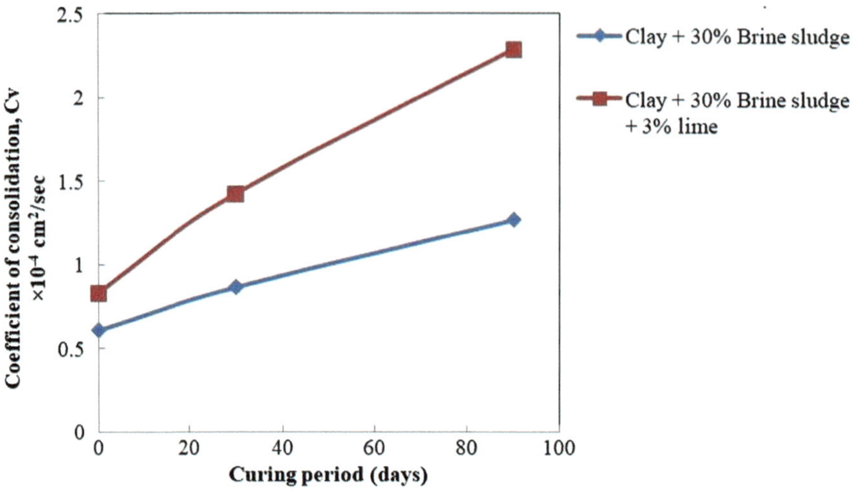

Fig. 8 Variation of coefficient of consolidation with curing period for treated clay (pressure range = 50–100 kPa)

Figure 8 shows the variation of coefficient of consolidation with curing period for 30% brine sludge and 30% brine sludge + 3% lime at a pressure range of 50–100 kPa.

From the figure, it can be inferred that the incorporation of lime into brine sludge-treated sample resulted in higher coefficient of consolidation at 3 months of curing period, which indicates that the treated soil undergoes consolidation at a faster rate than untreated clay. Therefore, this will help in reducing the settlement of the soil during its design life.

4 Conclusions

In the framework of experiments conducted on the marine clay sample, the following conclusions are drawn:

1. The studies carried out have addressed the environmental concern of brine sludge from chloral alkali industry by utilizing it for soil stabilization.
2. The incorporation of brine sludge has improved the index properties of clayey soil. The liquid limit has showed a decreasing trend with the increase in curing period. When lime is used as an additive, the liquid limit decreases with curing period. Plasticity Index displays a decreasing trend in the case of both brine sludge-treated marine clay with or without additive.
3. The Free Swell Index tends to decrease with curing period for all the treated clayey soils.

4. Even though the inclusion of brine sludge provides increase in unconfined compressive strength, a considerable increase can be brought about only with the addition of small percentages of lime. With the increase in curing period, the strength seems to increase significantly.
5. The compressibility characteristics get improved on addition of lime to the brine sludge-treated marine clay. With the addition of lime with brine sludge, de/d(log p) seems to decrease drastically, clearly showing the effectiveness of lime in reducing the settlement of the brine sludge-treated marine clay. The increase in the coefficient of consolidation values indicates that the treated soil undergoes consolidation at a faster rate than untreated clay.

Hence, it can be concluded that, brine sludge, an industrial waste material, can be effectively utilized for improving the strength characteristics of soft marine clays.

References

1. https://www.globaldata.com/media/press-release/caustic-soda-capacity-in-india-expected-to-witness-moderate-growth-by-2022-says-globaldata/
2. Garg M, Pundir A (2014) Utilization of brine sludge in nonstructural building components: a sustainable approach. J Waste Manage 2014:1–7. https://doi.org/10.1155/2014/389316
3. Pathak G, Saxena AK (2016) Utilization of brine sludge in non-traffic paver blocks. Int J Sci Res Develop 4(10):431–434
4. Verma S, Amritphale SS, Khan MA, Anshul A, Das S (2017) Development ofadvanced geopolymerized brine sludge based composites. J Polym Environ 25:999–1010
5. Jose BT, Sridharan A, Abraham BM (1988) A study of geotechnical properties of cochin marine clays. Mar Geotechnol 7(3):189–209
6. IS 2720: Part 10 (1991) Methods of test for soils: Part 10 Determination of uncon- fined compressive strength
7. Sridharan A, Rao SM (1985) Free swell index of soils—a need for redefinition. Indian Geotechn J 15(2):94–99
8. Sharma LK, Sirdesai NN, Sharma KM, Singh TN (2018) Experimental study to examine the independent roles of lime and cement on the stabilization of a mountain soil: a comparative study. Appl Clay Sci 152:183–195
9. Wang J, Mateos M, Davidson D (1963) Comparative effects of hydraulic, calcitic and dolomitic limes and cement in soil stabilization. In: Stabilization of soilwith lime and fly ash. bull. highway res. board, Washington, 29, pp 42–54
10. Jan MA, Walker RD (1963) Effect of lime, moisture and compaction on a clay soil. Highway Res Rec No. 29:1–12
11. Al-Rawas AA, Hago AW, Al-Sarmi H (2005) Effect of lime, cement and Sarooj(artificial Pozzolan) on the swelling potential of an expansive soil from Oman. Building Environ 40(50):681–687
12. Rajasekaran G (2005) Sulphate attack and ettringite formation in the lime and cement stabilized marine clays. Ocean Eng 32:1133–1159

Application of Biopolymers in Geotechnical Engineering Practices: An Eco-Friendly Approach

Rakesh Pydi, Laxmikant Yadu, and Sandeep Kumar Chouksey

1 Introduction

To achieve the desired and ideal performance of the week soil, ground improvement is the targeted enhancement of the geotechnical characteristics of the week soil. A geotechnical engineer must develop the most economical and effective approach to soil treatment. Based on the type of soil, the desired level of improvement and environmental friendliness of the method are among the most important parameters to be considered. The selection of the appropriate treatment method is the most effective part of a geotechnical engineer to stabilize the week soil [1]. At the beginning of human civilization, a wide range of materials and methods have been used to improve the properties of soils [2]. Rome and Roman concrete are made from volcanic ash, aggregates, and binders (i.e., gypsum or lime) [3]. Portland cement became the most widely used substance for the construction and treatment of soils, especially after the Industrial Revolution [2].

In the building and construction industry, Portland cement is a substance that offers a number of well-known advantages, including affordability, durability, and workability. The manufacture and use of cement, however, are linked to some environmental concerns. Global warming and climate change are the most concerning issues in recent decades. According to reports, one of the main contributors to worldwide CO_2 emissions, cement use accounts for approximately 8% of all CO_2 emissions [1, 4]. The scientific community has been creating new, better applications

R. Pydi · L. Yadu (✉) · S. K. Chouksey
Department of Civil Engineering, NIT-Raipur, G.E Road, Chhattisgarh, Raipur, India
e-mail: lkyadu.ce@nitrr.ac.in

R. Pydi
e-mail: rpydi.phd2021.ce@nitrr.ac.in

S. K. Chouksey
e-mail: schoksey.ce@nitrr.ac.in

B. T. Jose et al. (eds.), *Proceedings of the Indian Geotechnical Conference 2022 Volume 4*, Lecture Notes in Civil Engineering 479,
https://doi.org/10.1007/978-981-97-1753-8_12

with lower pollution output in an attempt to reduce these environmental effects [5]. Sustainable techniques have been used to create environmentally friendly materials, notably microbial biopolymers, for application (i.e., high molecular polysaccharides produced by microbes). As an alternative to traditional soil treatment methods, these biopolymers have been introduced [6, 7]. Due to direct ionic bonding with small particles or the development of a continuous biopolymer matrix between coarse particles, several investigations have demonstrated that the soil is significantly strengthened after the application of biopolymers. To resolve soil problems, Geotechnical Engineering has identified numerous potential applications for these biopolymers [2].

Most studies on the use of biopolymers have recently been published with experimental data, characterization assessments, and, in a few cases, case studies of actual application. The most frequently mentioned biopolymers in the literature are guar gum, polyacrylamide, xanthan gum, gellan gum, agar, and polyacrylamide. In this study, the xanthan gum-treated soil's index properties, unconfined compressive strength, and resilient modulus are assessed.

2 Materials in Present Study

2.1 Selection of Biopolymers

In the present study, biopolymer, i.e., xanthan gum (XG) is procured from PC industries, Jodhpur, India. With a global industrial output of 30,000 tons per year, XG is an EPS produced by the bacterium Xanthomonas campestris [8]. The strong demand for XG in the food industry led to the development of a number of alternative techniques to create it, including the fermentation of bacteria and yeast as well as the use of other low-cost substrates, nutrients, and raw materials [8, 9]. By adding a little amount of XG, an aqueous system's viscosity can be increased, creating a pseudoplastic solution [10]. Therefore, with a significant increase in strength, these natural biopolymers make it easier to bind the non-cohesive and dispersive geomaterials [11]. The characteristics and environmental impacts of xanthan gum are listed in Fig. 1, Table 1.

2.2 Collection of Soil Sample

The soil sample is collected from Antagarh town, from Kanker district located in the state of Chhattisgarh, India. The basic geotechnical and strength properties of the collected soil sample are evaluated as per Indian Standards which have been tabulated in Tables 2 and 3.

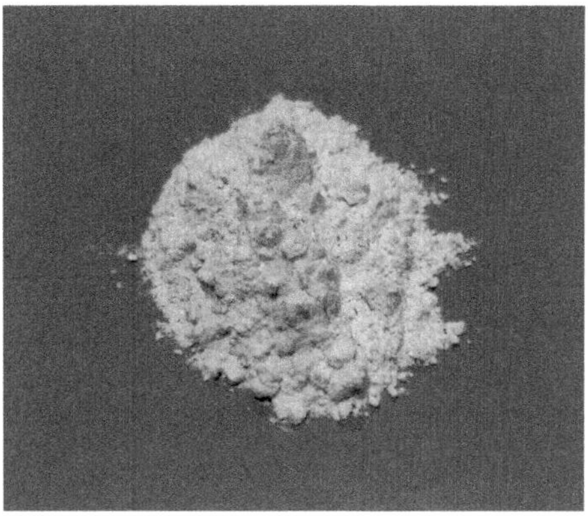

Fig. 1 Xanthan gum powder

Table 1 Characteristics of XG utilized for soil treatment and their environmental impacts [12]

Source	Glucose or sucrose fermentation by the Xanthomonas campestris bacterium
Charge type	Anionic
Water solubility	Soluble
Potential environmental impact	0.1 kg CO_2 for production of 1 kg biopolymer, no land use anionic

3 Experimental Methodology

In this study, 0.5, 1, 1.5, 2, and 2.5% of biopolymer, i.e., xanthan gum (XG), were used to investigate the index and strength properties of the soil bound with biopolymer.

The commercially available dry powdered biopolymer is mixed dry before being introduced. Prior to the addition of water, dry mixing involves combining the biopolymer and soil in the form of a proportion based on the weight of the soil. The biopolymers are confined during dry mixing. The mentioned procedure is the process adopted for the preparation of soil samples. The demolded soil samples were individually kept for 7 and 28 days of curing, which were tightly wrapped in polyethylene bags to minimize moisture loss before being stored in a desiccator. With the intention of keeping the relative humidity inside the desiccator at a steady state, a tiny amount of water was retained at the bottom of the device. The desiccator was covered and stored in a room set at about 21°C [13, 14].

Table 2 Basic geotechnical properties of the soil

Name of the test		Test property	Test method	Test result
Grain size analysis (%)	Coarse sand fraction	Particle size analysis	IS 2720 (Part-IV) 1985 Reaffirmed 1995	22.8%
	Medium sand fraction			32.2%
	Fine sand fraction			17.5%
	Silt + clay			27.5%
Atterberg's limits		Liquid limit (W_L), %	IS 2720 (Part-V) 1985 Reaffirmed 1995	32.0%
		Plastic limit (W_P), %		20.0%
		Plasticity Index (I_P), %		12.0%
		Shrinkage limit, %	IS 2720 (Part-VI)1972 Reaffirmed1995	8.6%
Free Swell Index		Swell potential, %	IS 2720 (Part-XL) 1977 Reaffirmed 1997	31%
Specific gravity (G)			IS 2720 (Part III/Sec1 & Sec2) 1980 (Reaffirmed 1997)	2.47
Soil classification as per IS 1498–1970				SC

Table 3 Strength properties of soil

Name of the test	Strength property	Test method	Test result
Modified proctor test	Maximum dry density, kN/m^3	IS 2720 (Part-VIII)1980 (Reaffirmed 1997)	16.25
	Optimum moisture content, %		21
Unconfined compressive strength (UCS) test	Compressive strength, Mpa	IS: 2720 (Part-X) 1991(Reaffirmed 1995)	0.60
Resilient modulus	Modulus of elasticity, Mpa	AASHTO T307	115.32

4 Results and Discussions

4.1 Index Properties of Soil Treated with Biopolymer

It can be observed from Fig. 2 that the liquid limit of the biopolymer-treated soil varies in a range of 35–38% and the plastic limit in a range of 23–25%. There is an increase in both the liquid and plastic limit values as the percentage of biopolymer increases. However, the type, content, and material characteristics of the biopolymer used were related to the rate of fluctuation. The addition of a biopolymer modifies the liquid limit through a variety of important mechanisms. Inclusion of biopolymers decreases flowability by making the pore solution more viscous. Additionally, the added biopolymer interacts with the soil matrix and free water to create a cross-linked hydrogel network or hydrogel bonding [15]. The creation of hydrogels helps to boost the liquid limit and plastic limit value by increasing the shear resistance of biopolymer-modified soil. In the case of virgin soil, the liquid limit was found to be 32% and the plastic limit was 20%.

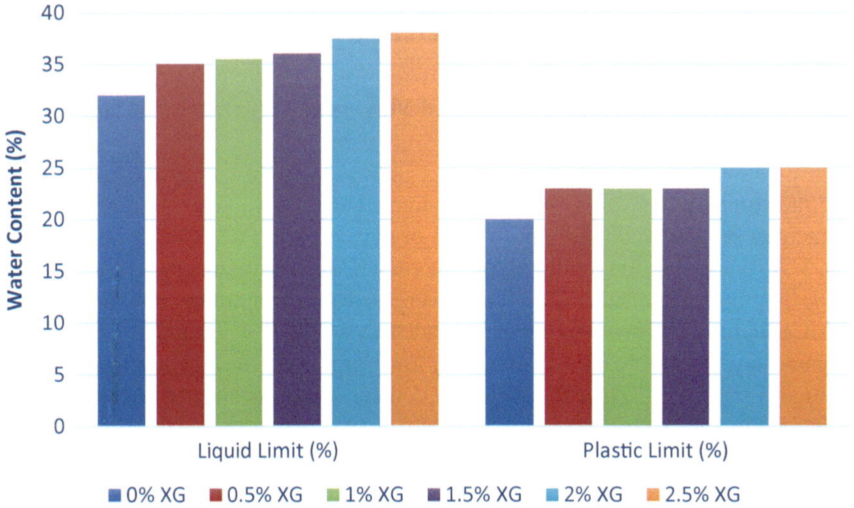

Fig. 2 Liquid and plastic limits of soil treated with biopolymer

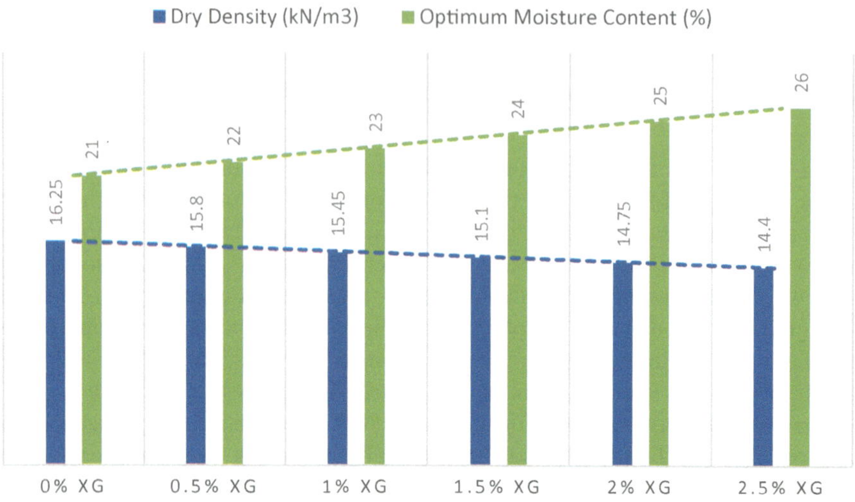

Fig. 3 Maximum dry density and optimum moisture content of biopolymer-treated soil

4.2 *Strength Properties of Soil Treated with Biopolymer*

4.2.1 Optimum Moisture Content and Maximum Dry Density of Soil Treated with Biopolymer

The moisture content of the biopolymer-treated soil varies in a range of 22–26%, an increase is observed, and a decrement of dry density is observed as stated from Fig. 3. As the proportion of biopolymer increased, the maximum dry density (MDD) of the biopolymer-treated soil gradually decreased and the optimum moisture content (OMC) increased. The biopolymer is added to soil and water to create a highly viscous solution that resists compaction, lowering MDD and increasing OMC for biomodified soil [14]. The decrease in dry density of biopolymer-treated soils ranges from 15.8 to 14.4 kN/m^3.

4.2.2 Unconfined Compressive Strength of Biopolymer-Treated Soil

The biopolymer-treated soil's unconfined compressive strength ranges from 0.4 to 1.7 MPa for 7 days; the peak strength for 7 days is 1.7 Mpa at a 1.5% concentration of xanthan gum. The unconfined compressive strength ranges from 1.08 to 4.6 Mpa after 28 days of curing; the peak strength for the entire 28-day period is 4.6 Mpa at a 1.5% concentration of xanthan gum. More biopolymer concentration leads the solution to become more viscous and less workable [15]. It might be a significant factor in the peak strength at 2.5% of concentration being lower than at 1.5–2%.

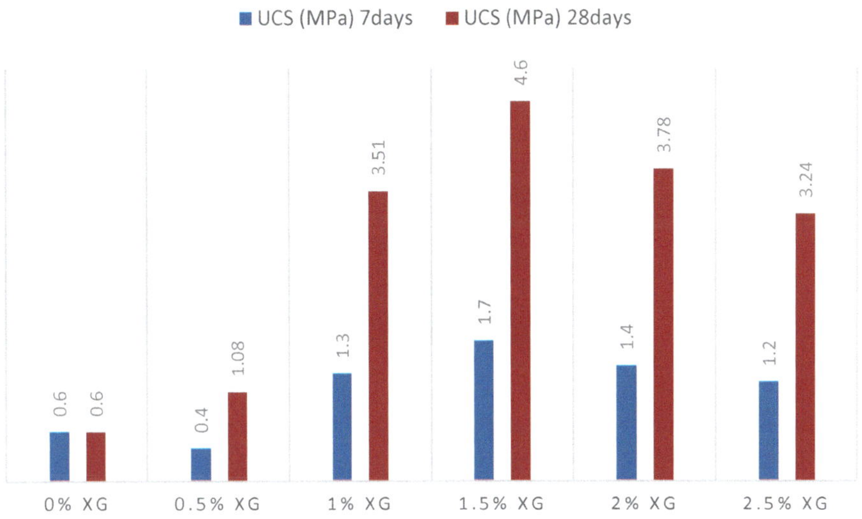

Fig. 4 Unconfined compressive strength (UCS) of biopolymer-treated soil for 7 and 28 days

The unconfined compressive strength of a material is a crucial geotechnical parameter to assess its suitability for construction purposes and to understand how the addition of a biopolymer affects the material's shear strength. Figure 4 displays the results of the unconfined compressive strength test conducted on soil treated with a biopolymer. This improves the soil strength after the biopolymer and soil are mixed. During curing, the hydrophilic biopolymers absorb the water and form hydrogel networks in the soil pores. Biopolymer that covers the soil grains connects the loose particles by building bridges between them, which enhances the strength properties of the weak soil. Lower biopolymer concentration causes hydration process more quickly, and a high saturation rate causes the peak strength to decrease [16]. As the biopolymer concentration rises, the peak stress rises along with the corresponding strain, which indicates a strain-hardening characteristic and ductile failure. It is concluded that treated soil behaves better than untreated soil and can withstand loads to a greater extent.

4.2.3 Resilient Modules for Biopolymer-Treated Soil

Resilient modulus is regarded as the appropriate input for the linear elastic theory selected in these guidelines for the analysis of flexible pavements because it is measured in a repeated load test while only accounting for the elastic (or resilient) component of the specimen's deformation (or strain). The repeated triaxial test described in AASHTO T307-99 can be used to measure the resilient modulus of soils in a laboratory setting [17] (Fig. 5).

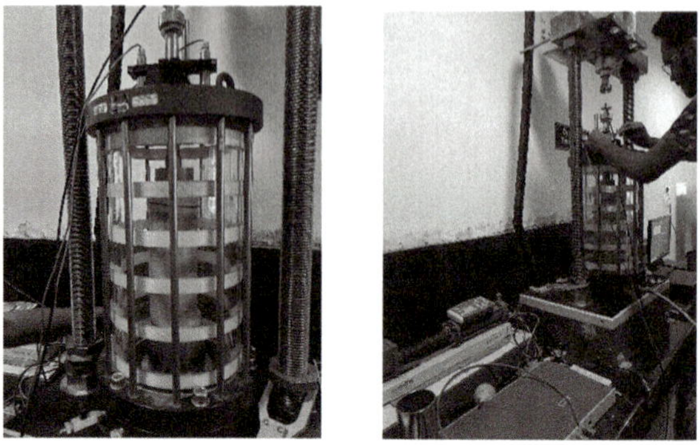

Fig. 5 Resilient modules testing for biopolymer-treated soil

Resilient modulus is the most important parameter for the evaluation of the stabilized pavement layer. The better performance of the biopolymer-treated soil is due to the formation of a gelling structure that fills the pores of the soil and holds the soil particles together. From Fig. 6, it is observed that the resilient modulus of biopolymer-treated soil varies from 86.73 to 258.32 Mpa for 7 days, with a 1.5% concentration of xanthan gum producing the highest modulus value 258.32 Mpa. The resilient modulus of soil treated with biopolymers for 28 days ranges from 183.96 to 687.32 Mpa, with 1.5% xanthan gum concentration producing the highest modulus value of 687.32 Mpa.

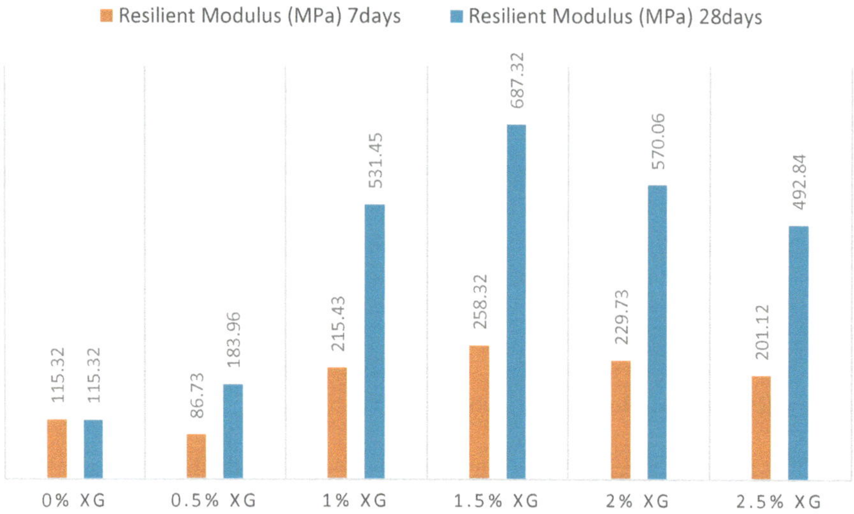

Fig. 6 Resilient modulus of biopolymer-treated soil for 7 and 28 days

5 Conclusion

The addition of biopolymer as a binder is one of the most effective applications for soil treatment. In this study, xanthan gum is used for week soil improvement and a change in the geotechnical properties of the soil is observed.

- The addition of biopolymer has an impact on the physical characteristics of soil, resulting in higher liquid and plastic limit. This may be due to adsorption of polymer and cross linkage with soil particles which plays an important role in biomodification of week soil.
- Due to the biopolymer concentration and the soil type, both the ideal moisture levels and the maximum dry densities are varying. Moisture content is a crucial parameter for improved soils to achieve the desired compaction conditions.
- Upon curing for 7 and 28 days, it is observed that the strength parameters are increasing by 2.5 time.
- The results of unconfined compressive strength values of biopolymer-treated soil depict that the ideal biopolymer percentage is observed as 1.5% by weight of the soil.
- The results of resilient modulus values observed from the biopolymer-treated soils depict that the treated soil satisfied the criteria of low and high-volume roads in accordance with IRC 37–2018 codal provision.
- Thus, this study assists that the biopolymers are eco-friendly material that can replace cement in ground improvement, which decreases the carbon footprint.
- The comparatively low percentage needed to obtain a compressive strength that is equivalent to more conventional materials like cement and lime is one benefit of utilizing a biopolymer as a binder.
- Finally, this study observed that biopolymer additive gives required strength characteristics at minimal percentages as compared with conventional additives like cement or lime, etc. However, this study may further be extended to other soils at different moisture levels.

References

1. Nicholson PG (2014) Soil improvement and ground modification methods. Butterworth-Heinemann
2. Ivanov V, Chu J (2008) Applications of microorganisms to geotechnical engineering for bioclogging and biocementation of soil in situ. Rev Environ Sci Biotechnol 7:139–153
3. Brune P, Perucchio R, Ingraffea AR, Jackson MD (2010) The toughness ofimperial roman concrete. In: Proceedings of the 7th international conference on fracture mechanics of concrete and concrete structures, Jeju Island, Korea
4. Murphy EM, Ginn TR (2000) Modeling microbial processes in porous media. Hydrogeol J 8:142–158
5. Basu D, Misra A (2013) Sustainability in geotechnical engineering. In: Proceedings of the 18th international conference on soil mechanics and geotechnical engineering, Paris, France

6. Narjary B, Aggarwal P, Singh A, Chakraborty D, Singh R (2012) Water availability in different soils in relation to hydrogel application. Geoderma 187–188:94–101
7. Murthy VNS (2016) Geotechnical engineering: principles and practices of soil mechanics and foundation engineering; Marcel Dekker, Inc.: New York, NY, USA
8. Palaniraj A, Jayaraman V (2011) Production, recovery and applications of xanthan gum by Xanthomonascampestris. J Food Eng 106(1):1–12
9. Silva MF, Fornari RCG, Mazutti MA, Oliveira DD, Padilha FF, Cichoski AJ, Cansian RL, Luccio MD, Treichel H (2009) Production and characterization of xantham gum by Xanthomonascampestris using cheese whey as sole carbon source. J Food Eng 90(1):119–123
10. Chen R, Zhang L, Budhu M (2013) Biopolymer stabilization of mine tailings. J Geotech Geoenviron 139(10)
11. Dehghan H, Tabarsa A, Latifi N, Bagheri Y (2018) Use of xanthan and guar gums in soil strengthening. Clean Technol Environ Policy 21:155–165. https://doi.org/10.1007/s10098-018-1625-0
12. (2007) Determining the resilient modulus of soil and aggregate materials AASHTO Designation: T 307–99
13. Kaniraj R, Shenbaga R, Gayathri V (2003) Factors influencing the strength of cement fly ash base courses. J Transport Eng 129.5:538–548
14. Turkane Sagar D, Sandeep KC (2022) Design of flexible pavement thickness using stabilized high plastic soil by means of fly ash-based geopolymer. Int J Pavement Eng 1–15
15. Fatehi H et al (2021) Biopolymers as green binders for soil improvement in geotechnical applications: a review. Geosciences 11.7: 291
16. Guidelines for the design of flexible pavements IRC: 37–2018
17. Mahamaya M et al (2021) Interaction of biopolymer with dispersive geomaterial and its characterization: an eco-friendly approach for erosion control. J Clean Product 312:127778

Utilization of Gondegaon Coal Mine Overburden Dump as a Construction Material

Atul D. Gautam and L. K. Yadu

1 Introduction

Mining plays a prominent role in the economy worldwide since human efforts have been more concentrated in extraction than recycling of natural resources. Waste generated due to mining activity poses a serious predicament due to the substantial amounts generated and is often associated with the risk posed by its storage and environmental management. The mining industry triggers intense environmental impacts.

The propitious management of the overburden dump generated during mining activities is one of the formidable issues faced by the industry [1]. It has been observed that the residual friction angle of this mining OB material is one of the major aspects during mining activities [2]. After detailed investigation, this OB material generated during mining can be utilized for filling the voids generated after mining operation [3]. In India, humungous quantity of mining waste is generated; hence, this material should be utilized in construction activities or in manufacturing of construction materials [4].

The effective utilization of this large amount of waste generated during mining activities is the main objective of this study. This aspect will help in minimizing the geoenvironmental issues in the vicinity.

A. D. Gautam (✉) · L. K. Yadu
Department of Civil Engineering, NIT Raipur, Raipur 492010, India
e-mail: atuldevendragautam@gmail.com

B. T. Jose et al. (eds.), *Proceedings of the Indian Geotechnical Conference 2022 Volume 4*, Lecture Notes in Civil Engineering 479,
https://doi.org/10.1007/978-981-97-1753-8_13

Fig. 1 Sample location Gondegaon mine

2 Site Details

2.1 Location

The Gondegaon Extension Opencast (OC) Coal Mine Expansion project is also known as Amalgamated Gondegaon-Ghatrohan OC mine of 3.5 Million Tons Per Annum (MTPA). Fig. 1 represents the OB location in village Gondegaon, Tehsil Parseoni, District Nagpur, in the State of Maharashtra, India. Gondegaon opencast mine located approximately 25 km away from Nagpur is selected for the study. It is situated in Parseoni district. The mine was started in 2009 under the sponsorship of Western Coalfields Limited. The coal obtained from this mine is used in Khaparkheda thermal power plant, Koradi thermal power plant and other nearby small-scale industries. Figure 1 shows the sample location of Gondegaon mines.

2.2 Geology of the Area

At this location, cohesive soil is perceived in the upper layer, which is followed by thick layer of sandstone and clay. Occurrence of coal is observed in different seams up to a large depth.

2.3 Description of Material

Based on field observations, cohesive as well as cohesion-less OB dump samples are collected from different overburden dumps for research work. Cohesive material is obtained from the upper layer deposits excavation, whereas the cohesion-less material is formed by the disintegration of coarse-grained sandstone during mining operations. This formation is highly weathering in nature resulting in disintegration while performing mining activities.

3 Methodology

The experimental program for the OB dump and fly ash is carried out in the following progression:

a. Index and engineering properties of the cohesive and cohesion-less OB sample have been determined during laboratory investigation.
b. Fly ash has been mixed with cohesive and cohesion-less OB samples in varying doses of 10, 20, 30 and 40%, and following tests have been performed on the mix with abovementioned doses.

    (i) Atterberg's limits
    (ii) Compaction test
    (iii) Direct Shear test
    (iv) UCS test
    (v) CBR test

c. Compaction test has been performed on cohesive and cohesion-less OB samples in the laboratory. The determined OMC and MDD value is used for UCS and CBR tests.

3.1 Testing on OB Sample

The collected cohesive and cohesion-less overburden dump samples are tested in laboratory for various parameters as per available Indian standards. Figs. 2 and 3 show samples in laboratory after performing the specified tests with addition of fly ash.

Fig. 2 UCS test on OB sample

Fig. 3 Shrinkage limit test on OB sample

3.1.1 Sieve Analysis

Particle size distribution of the accumulated OB samples has been carried out as per IS: 2720 part 4-1985 [5]. The graphical demonstration Figs. 4 and 5 show the gradation results that the gravel fraction is nil in both types of samples. Silt and clay percentage is more in cohesive sample, whereas in cohesion-less sample, the sand content is high.

3.1.2 Specific Gravity

Specific gravity test is performed according to IS 2720 part 3, 1980 [6]. In recent analysis, the average specific gravity of the cohesive and cohesion-less OB material is obtained as 2.635 and 2.648, respectively.

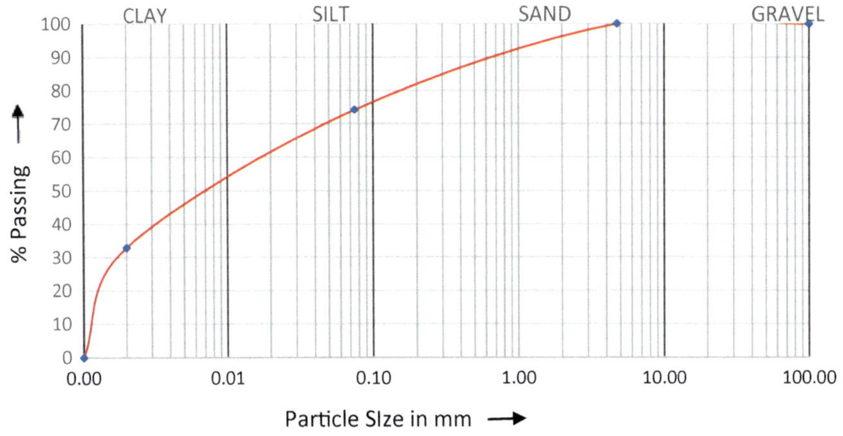

Fig. 4 Particle size distribution of cohesive OB sample

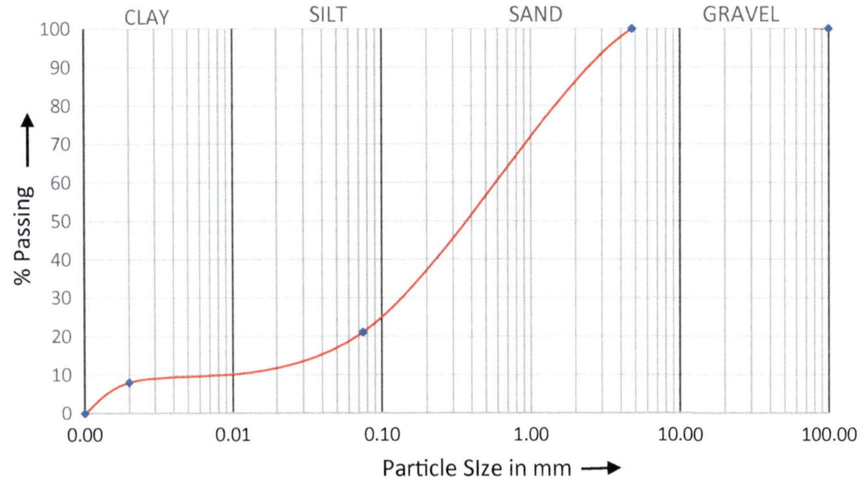

Fig. 5 Particle size distribution of cohesion-less OB sample

3.1.3 Atterberg's Limit

This test has been done as per the IS 2720 part 5, 1985 [7]. As per the test result indicated in Fig. 6, the liquid limit of the cohesive and cohesion-less OB material is obtained as 35.8 and 16.51%, respectively. It has been observed that the plasticity properties are decreasing with increase in dose of fly ash.

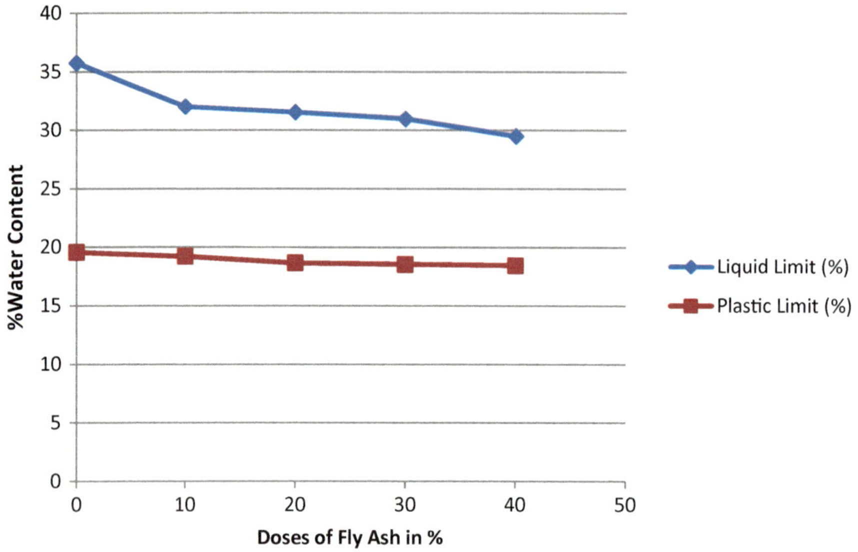

Fig. 6 Liquid limit and plastic limit of cohesive overburden dump sample

3.1.4 Compaction Test

Compaction test is performed as per the IS 2720 part 8, 1983 [8]. The compaction test results displayed in Figs. 7, 8, 9, and 10 point out that the optimum moisture content (OMC) and maximum dry density (MDD) of cohesive as well as cohesion-less OB samples are decreasing with increase in dose of fly ash.

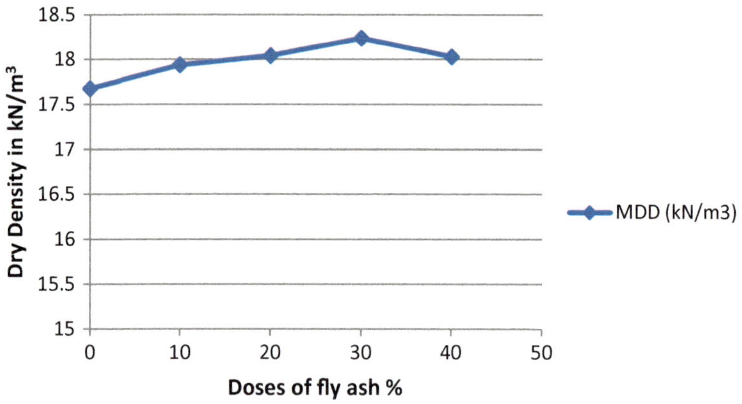

Fig. 7 MDD of cohesive OB sample

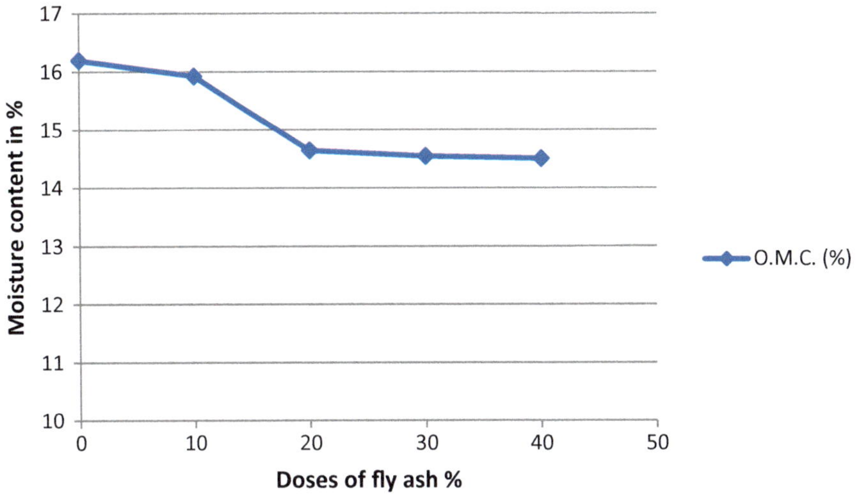

Fig. 8 OMC of cohesive OB sample

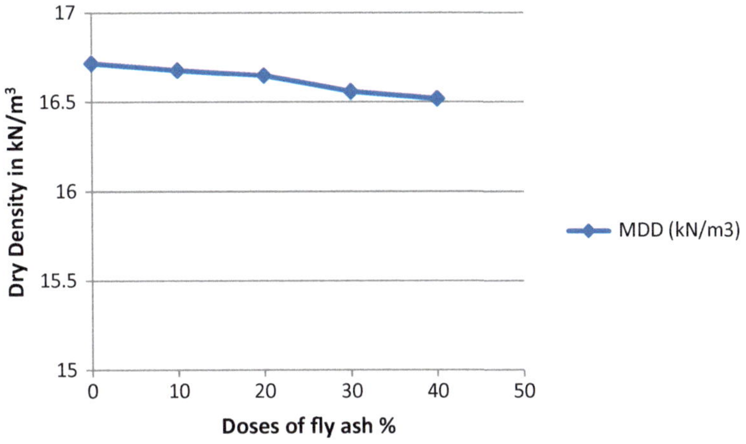

Fig. 9 MDD of cohesion-less OB sample

3.1.5 UCS Test

Unconfined compressive strength (UCS) test has been performed on the cohesive OB material as per IS 2720 part 10, 1991 [9]. The soil samples were placed in desiccator at 100% humidity and were removed from the desiccator after specific curing time period. The soil samples mixed with fly ash were cured to analyze the pozzolanic reaction and improvement in UCS of the soil samples. Figure 11 indicates that the value of UCS after seven days is increasing with increase in fly ash doses.

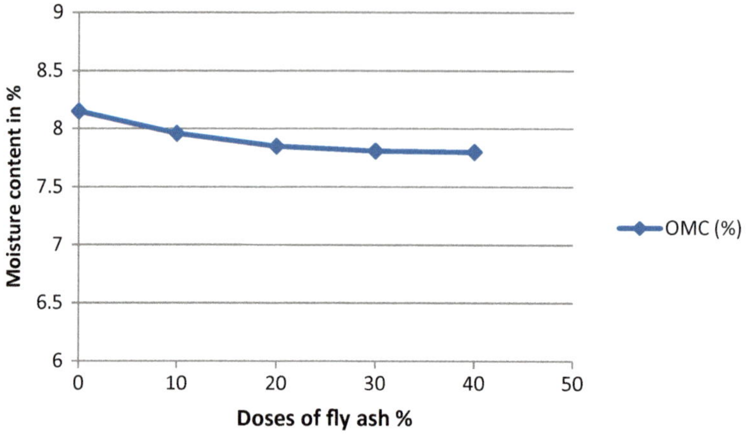

Fig. 10 OMC of cohesion-less OB sample

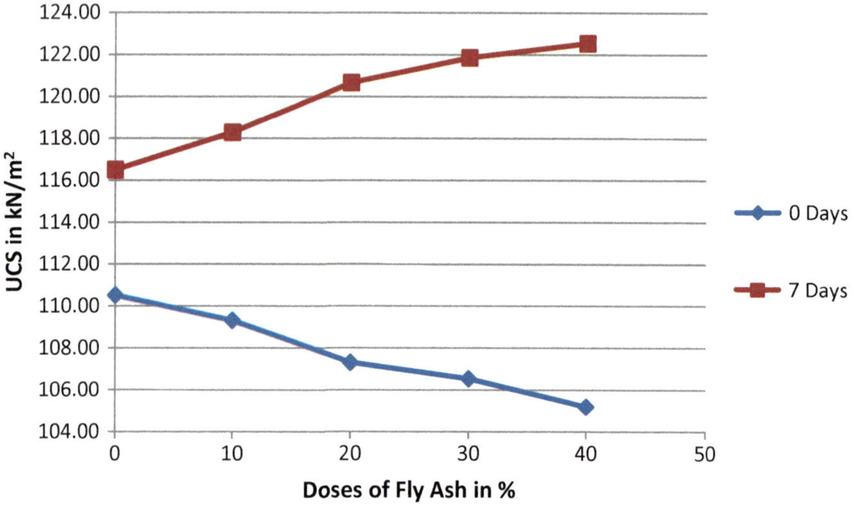

Fig. 11 UCS of cohesive OB sample

3.1.6 Direct Shear Test

Direct shear test has been done on the collected cohesion-less OB dump sample as per IS 2720 part 13 1986 [10]. The test results obtained from Figs. 12 and 13 indicate that the value of cohesion is increasing with increase in doses of fly ash, whereas the angle of shearing resistance is decreasing.

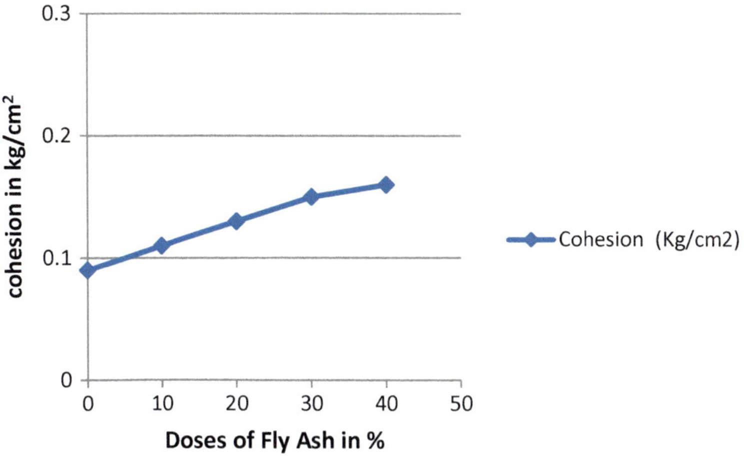

Fig. 12 Cohesion ‘c’ value of cohesion-less OB sample

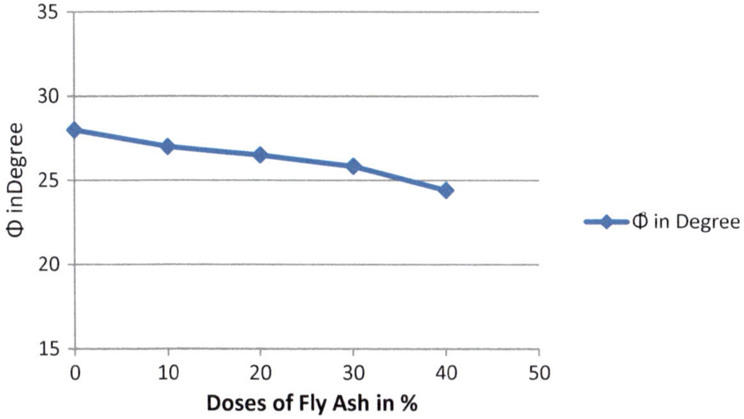

Fig. 13 Angle of internal friction Φ value of cohesion-less OB sample

3.1.7 California Bearing Ratio Test

California bearing ratio (CBR) test has been performed on the collected OB sample as per IS 2720 part 16, 1987 [11]. Figures 14 and 15 represent the compaction test and CBR test performance on the OB sample. The test result indicates that the CBR value of cohesive as well as cohesion-less sample is increasing significantly with increase in doses of fly ash successively (Figs. 16 and 17).

Fig. 14 Compaction test on OB sample

Fig. 15 CBR test on OB sample

3.1.8 Free Swell Index

Free swell test has been performed on the collected OB sample as per IS 2720 Part-XL, 1997 [12]. In recent analysis, the free swell index value of the cohesive OB material is obtained as 21.74%. Table 1 shows the test results conducted on cohesive and cohesion-less OB samples.

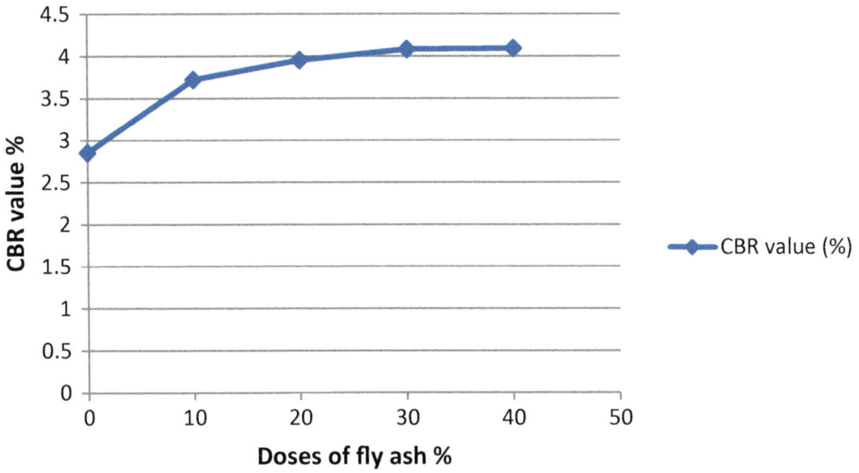

Fig. 16 CBR value of cohesive OB sample

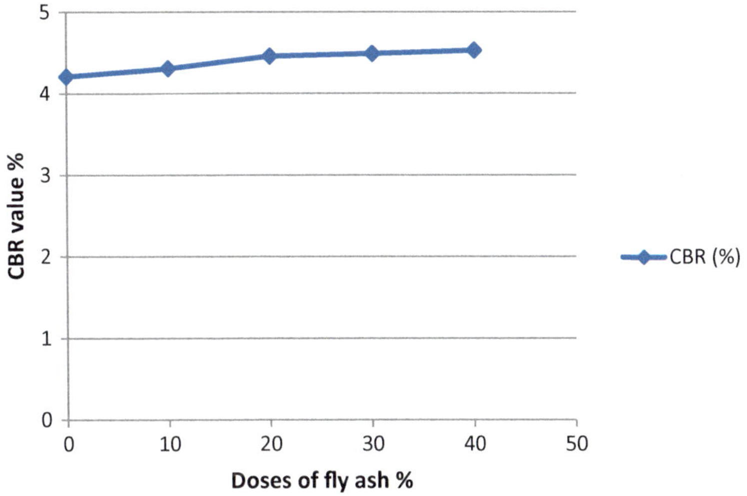

Fig. 17 CBR value of cohesion-less OB sample

4 Discussion

Based on the overburden dump material characteristics collected from Gondegaon mines, it has been accentuated that cohesion-less OB material supplemented with different doses of fly ash is showing considerable enhancement in CBR values and compaction parameters. In case of cohesive overburden sample, the notable amelioration in unconfined compressive strength value has been observed. Hence, it has been

Table 1 Properties of OB sample

Parameter	Test method	Cohesive OB sample	Cohesion-less OB sample
Grain size analysis	2720 (Part-IV)		
Gravel		NIL	NIL
Sand		25.77%	78.91
Silt		41.21	12.97
Clay		33.02	8.12
Specific gravity	2720(P-III)	2.635	2.648
Free swelling index (FSI)	IS 2720 (Part-XL)	21.74%	–
Liquid limit (%)	IS 2720—(Part V)	35.8	16.51
Plastic limit (%)		19.57	NP
Plasticity index		16.23	Nil
Shrinkage limit (%)	IS—2720 Part VI	13.95	–
OMC (%)	IS—2720 (Part VIII)	16.2	8.15
MDD (kN/m^3)		17.68	16.72
C (kg/cm^2)	IS—2720 (Part XIII)	0.27	0.09
Φ In degree		9	28
UCS (soil) kN/m^2 0 Days	IS—2720 (Part X)	110.51	–
UCS (soil) kN/m^2 7 days		116.52	–
CBR soak (%)	IS—2720 (Part XVI)	2.85	4.21

recommended that these overburden dump materials, which are impediment for the mining industry can be efficiently brought into play by the construction industry, in addition of nearby industrial by-product, i.e., fly ash collected from Khaparkheda thermal power plant. This effort will lead to the sustainable development of the region.

5 Conclusions

The following conclusion can be drawn, from the test results:

- After detailed investigation, with hindsight, this overburden material can be used in combination with nearby industrial by-products like fly ash, in variable percentages in myriad construction activities, i.e., pavement construction, embankments, etc.

- Strength parameters of the OB material are improving considerably after adding different doses of fly ash. This will fructify the effective land utilization as well as industrial by-product, with appropriate safety analysis, which is indispensable desideratum. Furthermore, environmental hindrances with their feasible solutions can be proposed to the authorities in charge keeping in mind further research work of overburden dump.

References

1. Karzyńska S, Krystyna M (1995) Reuse of coal mining wastes in civil engineering-part 2: utilization of minestone. Waste Management 15(2):83–126
2. Poulsen B, Manoj K, Manohar Rao A, Deepak A, Balusu R (2014) Mine overburden dump failure: a case study. Geotech Geolog Eng 32(2):297–309
3. Gupta AK, Paul B (2015) A review on utilisation of coal mine overburden dump waste as underground mine filling material: a sustainable approach of mining. Int J Mining Mineral Eng 6(2):172–186
4. Pappu A, Saxena M, Asolekar SR (2007) Solid wastes generation in India and their recycling potential in building materials. Build Environ 42(6):2311–2320
5. IS:2720 (Part 4) (2006) 1985 Indian standard, methods of test for soils, part 4: grain size analysis. Bureau of Indian standards, New Delhi, India., Reaffirmed, 1–38
6. Siebel EA, Kott AE (1937) Determination of specific gravity. J AOAC Int 20(3):535–542
7. IS: 2720 (part 5): (2006) 1985: Determination of liquid limit and plastic limit of soil. Bureauof Indian Standards, New Delhi, India., Reaffirmed(2006), 1–16
8. IS 2720 (Part 8) (1983) Determination of water content-dry density relation using heavy compaction. Bureau of Indian Standards, New Delhi, 3562–3577
9. IS: 2720 (Part 10):1991. (1991) Indian Standard: Methods of test for soils, Part 10: Determination of unconfined compressive strength. Bureau of Indian Standards, New Delhi, Reaffirmed 2006, 1–6
10. IS, I. S (1986) Methods of test for soils, Part 13: Direct shear test. Bureau of Indian Standards, New Delhi, India., Reaffirmed, 1–12
11. IS:2720(Part XVI)-1987 (1987) Method of test for soils, Part 16: Laboratory determination of CBR. Bureau of Indian Standards, Reaffirmed(2002), 1–10
12. IS2720(Part-40) (1977) Indian standard, methods of test for soils, part 40: determination of free swell index of soils [CED 43: soil and foundation engineering. Bureau of Indian Standards 2720, no. Reaffirmed 1997

Comparison of Different Ground Improvement Techniques for the Road Construction Over Kuttanadu Clay Strata

Abhijeet Kumar and Parvathi Geetha Sreekantan

1 Introduction

State highway 11, which is popularly known as AC road (Alappuzha-Changanassery) road that starts from Kalarcode, Alappuzha and ends in Perunna, Changanassery, Kerala, India. This is a double-laned road of length 24.3 km. The road alignment is parallel to AC (Alappuzha-Changanassery) canal. Since the road was made through the paddy fields of the Kuttanadu area, where paddy cultivation is done below sea level, the whole road is a reclaimed land made on the soil collected from the paddy fields during the construction of the canal. Thus, the road is mostly bounded by paddy fields on one side and the canal on the other side. The construction of the canal, which was aimed to divert the excess water from the Kuttanadu region to Vembanadu Lake, remained incomplete even after several decades. Frequent rainfall and incomplete canal construction resulted in the submergence of the AC road even during moderate flood seasons. To overcome this difficulty, it was decided to elevate the level of AC road, with due consideration to the present and anticipated future vehicular traffic. However, this is a difficult task owing to the very soft clayey subsoil strata beneath the road embankment.

Several studies regarding road embankment construction over soft subsoils were reported across the world. The most common technique adopted was prefabricated vertical drains with or without applying vacuum pressure [1, 2], followed by granular or piled columns [3–6]. The popular method of analysis and design of such

A. Kumar (✉)
Department of Civil Engineering, National Institute of Technology Karnataka, Mangaluru, Surathkal 575025, India
e-mail: abhijeetkumar.202gt001@nitk.edu.in

P. G. Sreekantan
Geotechnical Engineering Division, CSIR—Central Road Research Institute, New Delhi 110025, India

B. T. Jose et al. (eds.), *Proceedings of the Indian Geotechnical Conference 2022 Volume 4*, Lecture Notes in Civil Engineering 479,
https://doi.org/10.1007/978-981-97-1753-8_14

ground improvement techniques followed calculations based on consolidation, area replacement, and stress transfer between subsoil and columns. The use of finite element analysis tools facilitated the detailed study of these consolidation and stress transfer mechanism, thereby optimizing the solutions [7, 8]. The present study reports the possible solutions for the same considering the stability, total settlement, and construction time for the embankment raising. To this end, different solutions such as prefabricated vertical drains and lightweight geofoam embankments were designed and analyzed using 2D finite element modeling. The results were also compared with that of the conventional embankment raising technique.

2 Site Characteristics

The AC road is located in the Kuttanadu region which is surrounded by lakes and waterways. The aerial image of the Kuttanadu region along with the elevation profile of the road is shown in Fig. 1. The profile indicates that the initial 2 km stretch of the road has a higher elevation at 6 m after which it remains at an elevation of −1 to 2 m from mean sea level (MSL). The high flood level recorded during the 2018 flood level (HFL) varied from 0.521 to 0.764 m [9]. This indicates that the road has many stretches which are below the high flood level.

A typical bore log data of the road stretch is shown in Fig. 2. The top layer consists of filled soil up to a maximum depth of 4.30 m. This is followed by very loose clay with an average N value of zero and medium/stiff clay up to a depth of 50 m with an

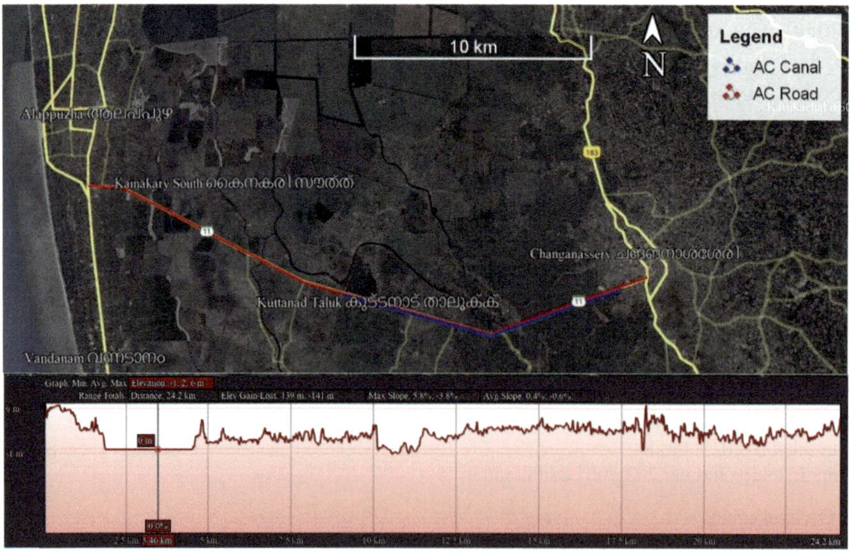

Fig. 1 Aerial image and elevation profile of AC road (Ref. Google earth, 2022)

average N value of 40. The harder layers can be observed after 50 m, having clayey sand up to a depth of 60 m with an average N value greater than 50 and sand layer with some silt with an average N value greater than 50. The groundwater table was found at a depth of 1.50 to 2.50 m from the ground surface, during the non-monsoon period. The liquid limit of clay layers varies from 35 to 150, indicating high compressibility. The unit weights of the clay layers are also low indicating high void ratios.

The height increment proposed for the road ranges from 0.3 to 1.2 m. However, the most critical section of this road is considered at a chainage of 14/710, where the proposed embankment height is 0.805 m. This section comprises a 20 m deep soft clay layer below the existing embankment. Thus, this cross-section was used to analyze various ground improvement techniques. Figure 3 shows the detailed cross-section profile.

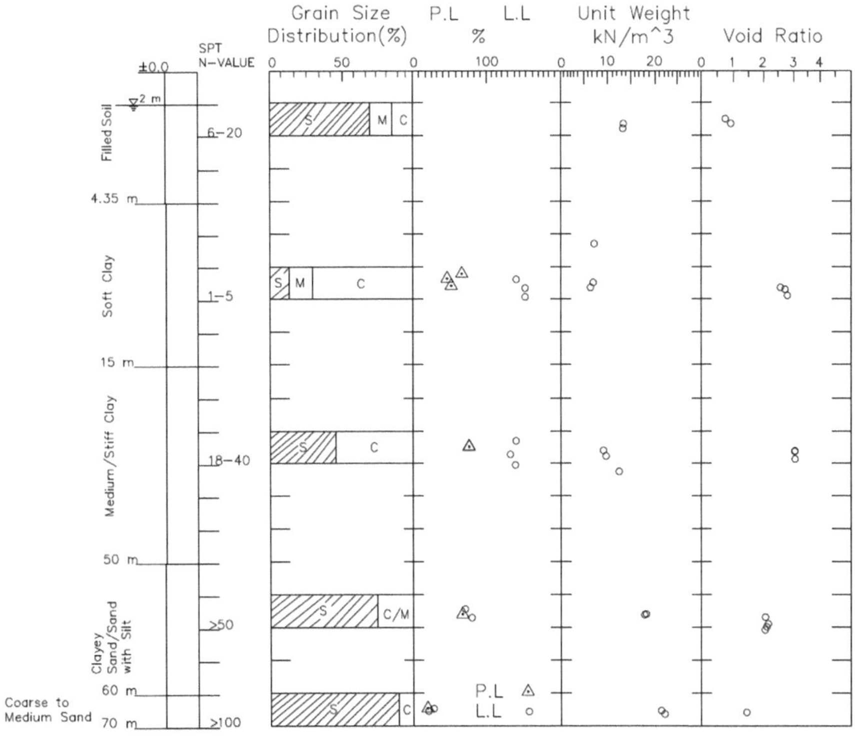

Fig. 2 Bore log data

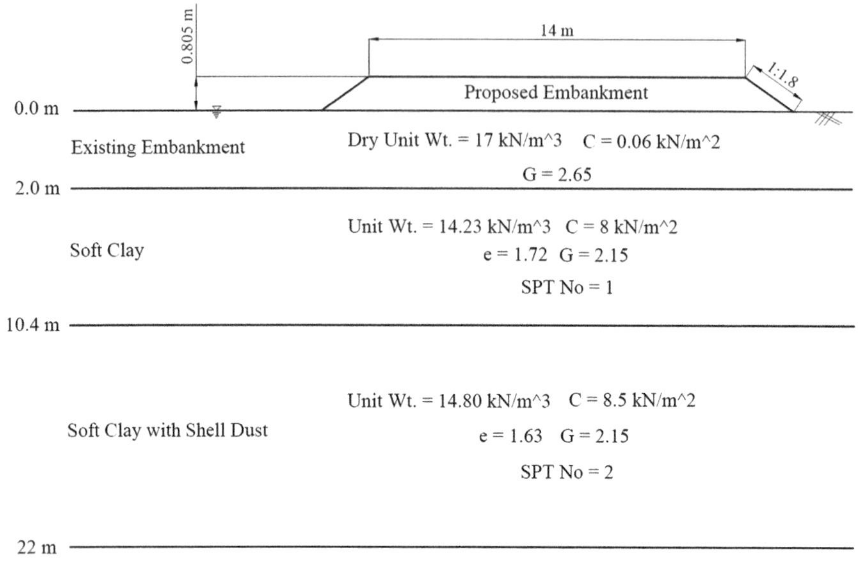

Fig. 3 Cross-section profile

2.1 *Modeling*

To analyze various probable solutions for embankment construction, various options were considered as described below:

- Case a: Soil embankment without improvement.
- Case b: Soil embankment with improvement (PVD).
- Case c: Lightweight fill embankment (geofoam).
- Case d: Lightweight fill embankment (geofoam) with soil replacement for 0.5 m depth.

The height of new embankment construction remains 0.805 m for cases a, b, and c. However, after soil replacement, the height of the new embankment for case d increases to 1.305 m. One-dimensional consolidation and two-dimensional finite element method were considered for the study. The computer program used for modeling the stability of road embankment is MIDAS soilworks, which is widely used for geotechnical modeling program. The constitutive models for soft clay are Mohr–Coulomb and modified Cam-Clay, respectively, for 1D consolidation and 2D-FEM analysis. For all other soil layers, Mohr–Coulomb constitutive model was adopted. However, for geofoam, a linear elastic model was chosen [10]. The input parameters for the soil embankment fill and subsoil layers are shown in Tables 1 and 2, respectively.

Different properties of proposed embankment soil, existing embankment soil, soft clay, and prefabricated vertical drain and lightweight fill used in 1-D consolidation

Table 1 Soil, existing filled soil and geofoam properties

Property	New embankment fill	Existing embankment	Geofoam
Bulk unit weight (kN/m^3)	20	19.03	0.294
Saturated unit weight (kN/m^3)	21	20.57	0.294
Cohesion, C (kPa)	0.5	0.06	59.75
Internal friction angle, Φ (°)	40	39	6
Modulus of elasticity (kPa)	30,000	29,000	7400
Poisson ratio	0.3	0.3	0.1
Earth pressure coeff., K_o	0.36	0.39	0.33
Permeability coeff. (m/min)	3×10^{-5}	3×10^{-5}	1×10^{-10}

Table 2 Soft clay properties

Property	Layer 1	Layer 2
Modulus of elasticity (kPa)	3000	3300
Saturated unit weight (kN/m^3)	14.23	14.80
Poisson ratio	0.38	0.38
Earth Pressure Coeff., K_o	1	1
Cohesion, C (kPa)	8	10
Void ratio	1.72	1.63
Horizontal permeability (m/min)	1.3×10^{-7}	1.3×10^{-7}
Vertical permeability (m/min)	1.3×10^{-7}	1.3×10^{-7}
Pre-consolidation pressure, P_c (kPa)	38.06	64.06
Compression index (C_c)	0.784	0.800
Over consolidated ratio (O.C.R)	1	1
Sec. consolidated coefficient	0.003	0.003
Strength increase ratio	0.26	0.26
Coeff. of consolidation (C_v)	1.4×10^{-6} m^2/min	1.5×10^{-6} m^2/min

limit equilibrium method and finite element method analysis for the calculation of construction time, settlement, and stability are shown below.

For case b, prefabricated vertical drains (PVD) of width (D_w) 0.066 m having a Permeability of PVD is 0.06 m/min was considered for the analysis. For the smear zone, a permeability ratio (K_h/K_s) of 2 and a Coefficient of consolidation ratio of 0.33 were considered during modeling. To derive the optimal spacing of PVD, varying center to center distance of the PVD from 0.8 m to 1.5 m was considered during consolidation analysis. To include the possible field variations in the horizontal consolidation coefficient, a varying ratio of consolidation ratio (Cv/Ch) from 0.2 to 1 was considered for the consolidation analysis.

For the 2D FEM finite element analysis, AutoCAD interface of the software was used to model the geometry. The subsoil strata are defined as 2D plain strain elements.

Six noded triangular elements of 0.5 m size were used for the surface discretization of the model. The mesh generation was followed by defining the loading conditions. Self-weight was defined initially for all cases. A pavement and traffic load of 24 kPa was applied over the top of the embankment as a line load in 2D.

Two-dimensional non-linear stress analysis using finite element method was carried out for different construction stages mentioned below. The different stages of analysis for all the cases considered for analysis are given below:

Stage 1: Existing embankment with traffic load.

Stage 2: Embankment height increment (with or without PVD).

Stage 3: Opening of traffic.

The initial stage involves the calculation of initial stress conditions. It also includes ground support, drainage conditions, self-weight, and live load (traffic). Further, in the next stage, at the end of construction of the embankment after improvement, includes the embankment, pavement and traffic load. In case d, the removal of 0.5 m of the existing embankment and filling of the same and further embankment height with lightweight fill (geofoam) was considered. The last stage which is the opening of traffic includes the live load (traffic). The optimal spacing of PVD derived from consolidation analysis was used for FEM analysis.

3 Results and Discussions

3.1 One-Dimensional Consolidation

One-dimensional consolidation analysis revealed that the total consolidation settlement for (case a) is 0.641 m. The improved shear strength of soft soil layers 1 and 2 are 8.4 kPa and 18.3 kPa, respectively. The time required to achieve 90% degree of Consolidation is dependent on the consolidation ratio, and spacing of PVD. The variation of the same is indicated in Figs. 4 and 5. It can be observed that as the horizontal consolidation coefficient increases, the time required for achieving 90% degree of consolidation also decreases. It can be observed that as the PVD spacing increases, the time required for achieving 90% degree of consolidation increases. The results indicated that the average time for 90% consolidation is 340, 130 and 80 days for PVD spacing of 1.5, 1, and 0.8 m, respectively, with the constant coefficient of consolidation ratio of 0.33. However, optimal spacing of 1 m requires 130 days for 90% consolidation. The results are compiled in Table 3.

Meanwhile, with improvements using geofoam, the total consolidation significantly reduces to 0.487 and 0.171 m for case c and case d, respectively (Table 4). However, for case, d, there is a very significant acceleration of the consolidation time as to the other three cases. The time for 90% consolidation is derived as 225 days for case d.

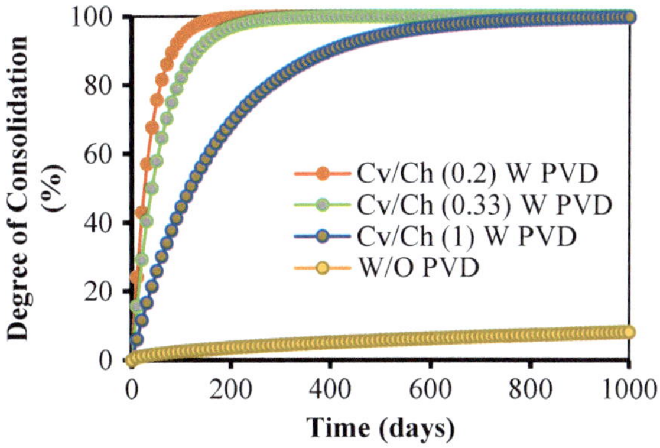

Fig. 4 Degree of consolidation for varying consolidation ratio (PVD spacing = 1 m)

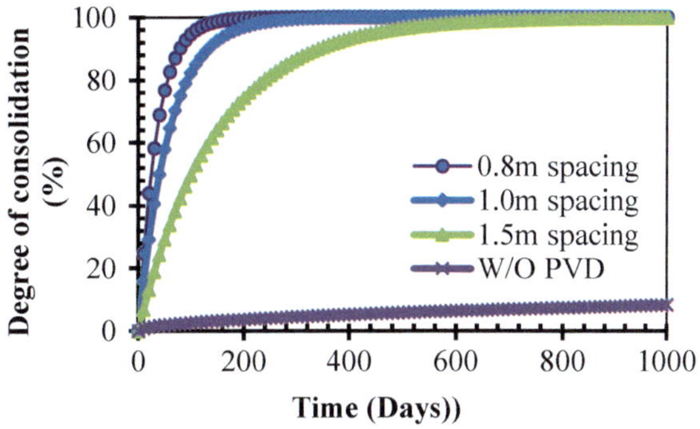

Fig. 5 Degree of consolidation for varying consolidation ratio (C_v/C_h = 0.33)

Table 3 Consolidation analysis results

Case	Cohesion (kPa)	Cv/Ch	Total settlement (m)	t_{90} (days)	Improved cohesion (kPa)	
					Layer 1	Layer 2
Without Improvement	8	-	0.641	> 10 yrs	8.4	18.3
0.8 m c/c PVD		0.2		50		
		0.33		80		
		1		230		
1 m c/c PVD		0.2		80		
		0.33		130		
		1		400		
1.5 m c/c PVD		0.2		210		
		0.33		340		
		1		1000		

Table 4 Comparison of results

Case	Total 1D consolidation settlement (m)	Maximum settlement attained (m)	Consolidation time (years)	Degree of consolidation (%)
Case a	0.641	0.4	50.1	62
Case b	0.641	0.4	2.6	62
Case c	0.487	0.4	10.5	89
Case d	0.171	0.16	0.61	90

3.2 *Finite Element Analysis*

Finite element analysis was carried out for a time period of 250 days. It was revealed that for case a and case b, the maximum settlement attained after stage 3 is only 0.12 m and 0.35, respectively, as shown in Fig. 6. This is different from consolidation analysis, where settlement was estimated at 0.03 m and 0.48 m, respectively, for case a and case b. This difference can be attributed to the immediate settlement and PVD smear effect considered in the FEM analysis. For case c and case d, the vertical settlement after stage 3 is observed as 0.18 and 0.16 m, respectively. It needs to be noted that the degree of consolidation attained is maximum for case d, followed by case b, as shown in Fig. 7. This can be attributed to the fact that the increase in vertical stress is very minimal for case d, whereas accelerated consolidation using PVD is considered for case b. Thus, long-term settlement is negligible for case d and minimal for case b.

As per [11], the tolerable residual settlement for embankment is limited to 0.3 m. However, this considers only consolidation settlement. During FEM analysis, the

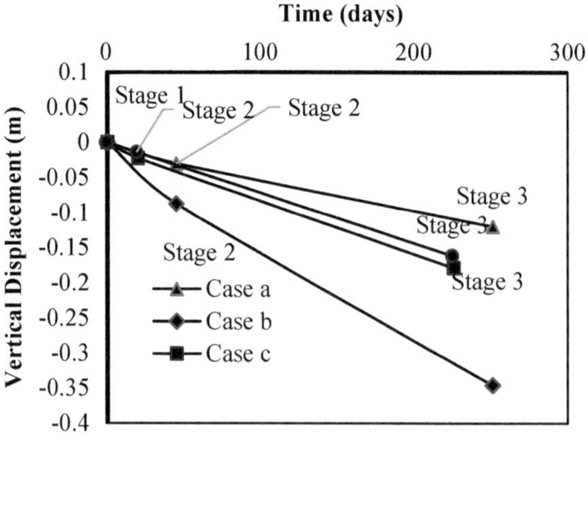

Fig. 6 Variation of vertical settlement for different stages

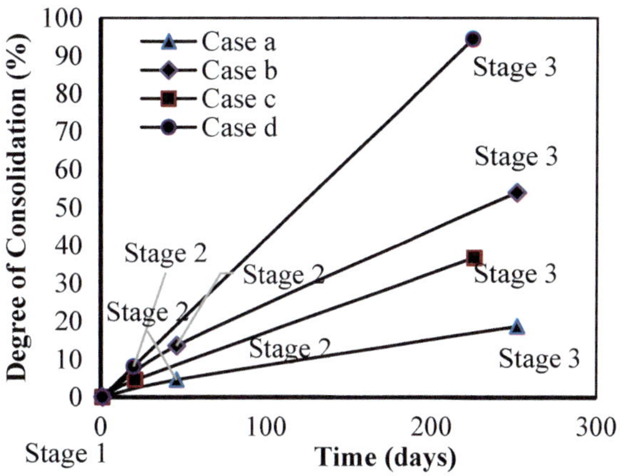

Fig. 7 Variation of degree of consolidation for different stages

elastic settlement of the subsoil and the embankment fill is also compounded in the total vertical settlement. Thus, for the finite element analysis, a tolerable total settlement of 0.4 m can be considered reasonable. To this end, FEM analysis was carried out till a limiting displacement of 0.4 m, to analyze the time required for attaining this settlement. The results are indicated in Table 4. It can be inferred that case a and c undergoes long-term settlement for more than 10 years. However, the rate of settlement is very slow for case a (8 mm per year). This is an advantageous situation. However, the rate of settlement is the maximum for case b. For case d, the long term is negligible.

The vertical settlement profile after stage 3 is indicated in Fig. 8. It can be inferred that the vertical settlement is maximum at the center of the road for all the cases. However, for case b, significant vertical settlement is observed in the neighboring areas as well. This may affect the nearby structures. Except for case b, the effect of settlement extends to a depth of 5 m. For case b, a deeper extent till 10 m can be observed due to the lateral drainage of PVDs. The lateral displacement profile is shown in Fig. 9. The maximum lateral expansion of the left side below the existing embankment changes from 0.022 to 0.035 m, respectively, for case a and case b. This indicates that after the installation of PVD due to an increase in the settlement there is a minor increase in lateral expansion as well. However, lateral expansion is observed at a depth of 2 m. However, lateral displacements in the range of 0.03 m are also observed at the embankment side slope. For case c and case d, the maximum lateral displacement of the slide slope of the embankment slope is in the range of 0.03 m. This lateral displacement at the embankment side slope in all the cases considered can be attributed to the rotational movement of the embankment along the slip circle. Even though this lateral displacement is very minimum and does not affect the serviceability of the structure, it is recommended to provide basal reinforcement using high-strength geogrid to arrest further movement due to the increase of traffic load in future.

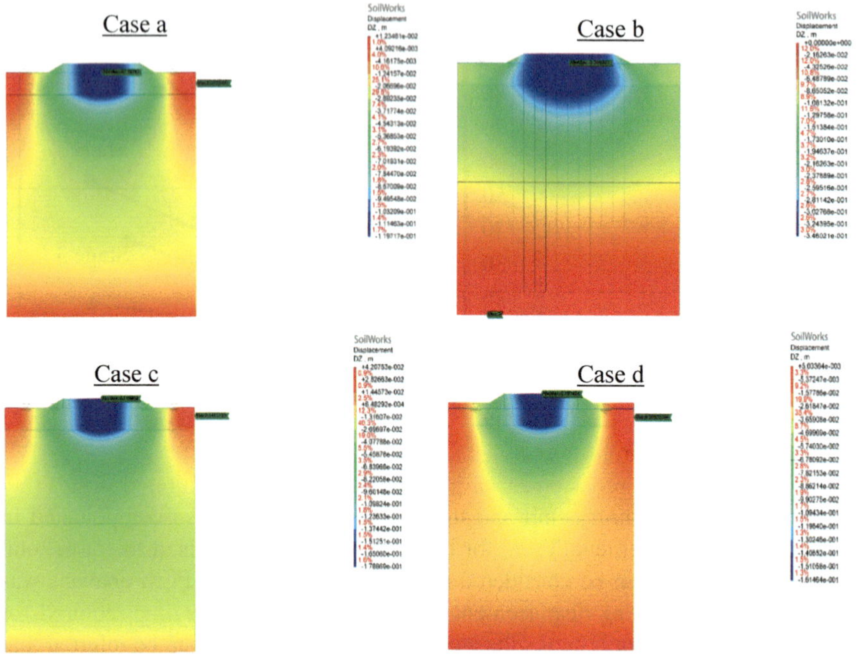

Fig. 8 Vertical displacement profile

4 Conclusions

In the present study, a comparison of different ground improvement techniques was made in terms of stability, total settlement, and construction time using 1-D consolidation and finite element analysis. The results were compared with the embankment construction without any ground improvement technique. The conclusions arrived at from the study are summarized in the following points.

- The time required to achieve 90% degree of consolidation is dependent on the consolidation ratio and spacing of PVD. The optimal spacing of PVD was derived as 1 m, which requires 130 days of waiting period before laying the pavement. However, the improvement in shear strength after accelerated consolidation using PVD is very minimal for the first 10 m of soft clay.
- The degree of consolidation is maximum for the geofoam embankment with 0.5 m soil replacement, indicating faster pavement construction and negligible long-term settlement. In this case, the overall settlement is also within the tolerable settlement recommended by the IRC code. The long-term settlement is maximum when no ground improvement technique is used. However, the rate of this settlement is extremely low.

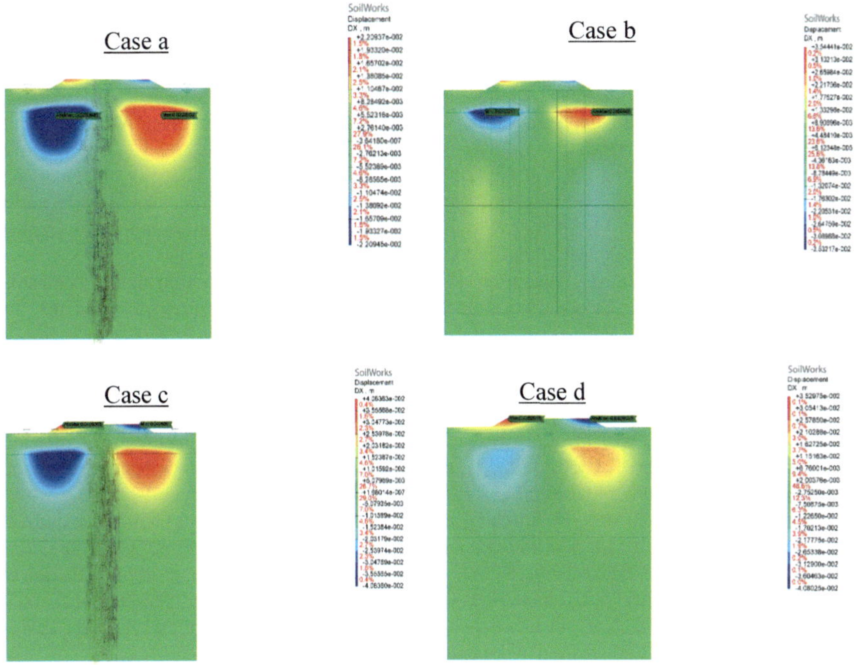

Fig. 9 Lateral displacement profile

- FEM analysis indicated that the vertical settlement for the nearby areas is maximum for the embankment construction with PVD. For all other cases, settlement is limited to the area below the embankment.
- Lateral extrusion is observed for soft clay after a depth of 2 m for embankment construction with conventional soil (with or without PVD). However, lateral displacement of 0.03 m on the road shoulders, due to rotational slip circle movement can be anticipated for all cases.
- Considering these observations, it is recommended to adopt the embankment construction using geofoam along with the removal of the top 0.5 m of the existing embankment layer. However, since the rate of settlement is extremely low (8 mm per year) conventional soil embankment without ground improvement may also be adopted provided, that lateral displacement is arrested using appropriate techniques.

References

1. Bergado DT, Balasubramaniam AS, Fannin RJ, Holtz RD (2002) Prefabricated vertical drains (PVDs) in soft Bangkok clay: a case study of the new Bangkok International Airport project. Can Geotech J 39(2):304–315
2. Shen SL, Chai JC, Hong ZS, Cai FX (2005) Analysis of field performance of embankments on soft clay deposit with and without PVD-improvement. Geotext Geomembr 23(6):463–485
3. Almeida MSS, Ehrlich M, Spotti AP, Marques MES (2007) Embankment supported on piles with biaxial geogrids. Proceed Inst Civil Eng Geotech Eng 160(4):185–192
4. Liu H, Gangqiang K, Jian C, Xuanming D Grouted gravel column-supported highway embankment over soft clay: case study. Canad Geotechn J 52(11):1725–1733. https://doi.org/10.1139/cgj-2014-0284
5. Liu HL, Ng CW, Fei K (2007) Performance of a geogrid-reinforced and pile-supported highway embankment over soft clay: case study. J Geotech Geoenviron Eng 133(12):1483–1493
6. Nunez MA, Briançon L, Dias DCFS (2013) Analyses of a pile-supported embankment over soft clay: full-scale experiment, analytical and numerical approaches. Eng Geol 153:53–67
7. Chai J, Igaya Y, Hino T, Carter J (2013) Finite element simulation of an embankment on soft clay–Case study. Comput Geotech 48:117–126
8. Huang W, Fityus S, Bishop D, Smith D, Sheng D (2006) Finite-element parametric study of the consolidation behavior of a trial embankment on soft clay. Int J Geomech 6(5):328–341
9. Kerala Public Works Department, KPWD (2019) Detained Project Report on 'upgradation of Alappuzha-Changanassery Road (CH:0/000—CH:24/164). Unpublished internal document
10. Stark TD, Arellano D, Horvath JS, Leshchinsky D (2004) Geofoam applications in the design and construction of highway embankments. NCHRP Web Doc 65:792
11. IRC-75 (2015) Guidelines for the design of high embankments. Indian Roads Congress. New Delhi

Efficacy of Cross-Linking of Biopolymers in Soil—Stabilization

M. Ashok Kumar, Arif Ali Baig Moghal, Jagadeesh Bommisetty, and Nuruddin Mohammad

1 Introduction

Mechanical, biological, and chemical soil improvement technologies are the three main categories of soil improvement technology. Cement and lime are the two most popular chemical stabilizing substances used to stabilize the soil chemically [1]. Although the traditional techniques are well known for their capability of improving the soil properties, they also result in the emission of greenhouse gases and recycling costs for cement and concrete. Recently, eco-friendly alternatives to conventional soil improvement techniques in geotechnical engineering have been investigated using biological methodologies including reactive enzymes, microbial polymers like biopolymers, and microbially induced calcite precipitation [2, 3]. Due to their natural origins, biopolymers manufacturing procedures are frequently less damaging to the environment and its biota. They can provide the same level of strengthening at far lower concentrations and are more environmentally benign than many other chemical soil additives. More significant usage of biopolymers could reduce the 1.25 tonnes of carbon dioxide that can be released into the atmosphere for every tonne of cement

M. Ashok Kumar (✉) · A. A. B. Moghal · N. Mohammad
Department of Civil Engineering, National Institute of Technology Warangal, Warangal, Telangana 506004, India
e-mail: m.ashokkumar@cvr.ac.in; ma21cerep04@student.nitw.ac.in

A. A. B. Moghal
e-mail: baig@nitw.ac.in

N. Mohammad
e-mail: mnce20224@student.nitw.ac.in

M. Ashok Kumar · J. Bommisetty
Department of Civil Engineering, CVR College of Engineering, Hyderabad, Telangana 501510, India
e-mail: b.jagadeesh@cvr.ac.in

B. T. Jose et al. (eds.), *Proceedings of the Indian Geotechnical Conference 2022 Volume 4*, Lecture Notes in Civil Engineering 479,
https://doi.org/10.1007/978-981-97-1753-8_15

produced. Liner contaminant sorption and anti-seepage potentials are prime essential in the design of landfill. The clay minerals have high sorption capacity, long-term structural stability, and low permeability, which aids their usage as an efficient landfill liner. Bentonite clay has gained wide attention in the design of landfill liner systems due to its low hydraulic conductivity, high swelling capacity, and adsorption properties. According to the researchers, the soil stabilized with biopolymers has gotten industrial attention as a competent landfill liner due to its low permeability [4, 5].

Muhammad and Marri [6] observed from their investigations that adding stabilizing agents like lime and wheat straw to the clayey soil helps in enhancing the shrinkage and swelling properties. The nanomaterials were also used as stabilizing agents that showed an improvement in the shrinkage properties of the soil [7]. Imane et al. [8] investigated the effect of wheat straw addition on the swelling nature of clay soil. Obtained results of the investigation have shown that there is a considerable decrease in compressibility index and swelling index with an increase in the wheat straw content.

Researchers mentioned that the locally available soil could be used as a low-cost landfill liner material if it is stabilized by adding biopolymers [4, 5, 9]. By evaluating the parameters like unconfined compressive strength and hydraulic conductivity, they have assessed the suitability of a soil as a liner in landfill construction. Two biopolymers namely xanthan and guar gum, were added to the expansive soil in different dosages and their effect on hydraulic conductivity and unconfined compressive strength was evaluated. The results showed that the values of hydraulic conductivity decreased with an increase in consolidation pressure and biopolymer content. Reliability-based design optimization methodology was used to assess the performance of the soils. Their investigations focused mainly on the efficacy of soil modified by biopolymer as a substitute liner material for MSW. Appreciable improvement in anti-seepage has been observed in the soil after blending it with guar and xanthan gums. Dosages of the biopolymers needed to modify the cohesive soils at different curing periods were done by using target reliability-based design optimization and deterministic design. The results of these methods showed that when the guar gum concentration rises from 0.5 to 4%, the reliability index and factor of safety against failure increase drastically. On the other side, the dependability index and factor of safety continued to rise notably when the cohesive soils were blended with guar gum at a dose of 0.5 to 2.9%. Although the reliability index and element of safety are significantly decreased when guar gum is added to more than 2.9%, results show that it is the optimum dosage [5]. Studies like Fourier transform infrared spectroscopy and electron microscopy on unamended and biopolymer-treated soils revealed that cross-linking of soil particles with gum strands has improved the strength properties of soil [9].

Chang and Cho [10] studied to check the suitability of using environment-friendly biopolymers for soil stabilization. A commercial β-1,3/1,6-glucan polymer product was used to stabilize the hwangtoh, a Korean residual soil. The residual soil was mixed with liquid type β -1,3/1,6-glucan polymer solutions and cured at different temperatures. Time-dependent compressive strength was determined to assess the strengthening process. A gain in the compressive strength of soil was observed, and it is due to the particle surface sorption of biopolymer. However, the gain in strength was very high at a curing temperature of 60^0C. From a quick environmental/economic study, it was proved that the biopolymer treatment is more beneficial compared to conventional methods like cement stabilization and reinforcement. Chang et al. [11] investigated the usage of microbial polysaccharides called gellan gum to strengthen sandy soil. The effect of gellan gum on sandy soil was evaluated by performing different tests under various moisture conditions. They observed an enhancement of bonding between the sand particles with hydrogel condensation, which is due to the development of artificial cohesion by the microbial agent. Three different pore fluids, namely deionized water, kerosene, and 2 M-NaCl brine were used to obtain the liquid limit values of sand and biopolymer-treated clay mixture. Adding xanthan gum to the soil can decrease the liquid limit attributed to particle aggregation or increase the liquid limit due to hydrogel formation [12].

Variation of adhesion and friction angle in the sand treated by xanthan gum due to the construction of subsequent structural members was explained by Lee et al. [13]. From the results, they concluded that a solid intergranular bonding was observed due to condensed biopolymer biofilm. They also stated that physical condition through the surface affects the interfacial shear. However, the rise of moisture content results in the reduction of intergranular bonding. Kwon et al. [14] investigated the significance of exocultered biopolymers in controlling the erodibility coefficient and critical shear stress of soil with different particle size distribution using an erosion functioning apparatus and ultrasonic P-wave reflection monitoring device. Chen et al. [15] performed UC tests to assess the possibility of replacing cement stabilization with the combined action of biopolymer and fiber inclusion in the soil. Results from the UC tests revealed that the combination of biopolymer and fiber inclusion was ineffective in managing the compressive strength and ductility of soil. Biopolymers were found effective in increasing the unconfined compressive strength, and fiber inclusion helped increase the soil ductility [15].

Vydehi et al. [16] mentioned that biopolymers could be used as an effective soil stabilizer in place of cement and fly ash to reduce the negative impact on the environment. By adding xanthan and guar gum biopolymers at different dosages to the soil, they evaluated shrinkage characteristics along with curling and cracking. Possible mechanisms behind the cracking phenomenon, particle arrangement, and curling state were explained.

Muguda et al. [17] mentioned that the hydrogels are responsible for the strength improvement of the soil. The mechanical properties and physical integrity of hydrogel depend on the properties of the biopolymer used and the time taken for the dilution of a hydrogel. They have concluded that cross-linking of biopolymer can be done to delay or avoid the dilution of hydrogels. He also mentioned that biopolymers xanthan and guar gum can instantly react and form cross-links without any initiation.

With the benefits of biopolymers and their cross-linking, it becomes a prospective area for the study of biopolymers and their effects on various soil characteristics. The main goal of the present work is to study the efficacy of cross-linking of select biopolymers in enhancing the geotechnical properties. The targeted geotechnical properties in clude Atterberg limit, standard Proctor compaction test, unconfined

Table 1 Properties of unamended soil

S. No.	Property	Value
1	Liquid limit	63.5%
2	Plastic limit	30.5%
3	Compressive strength	1.48 kg/cm^2
4	Optimum moisture content	26%
5	Max. dry density	1.47 g/cc
6	Hydraulic conductivity	5.24×10^{-3} cm/s

compression, and hydraulic conductivity. Results are then compared and interpreted to better understand the effect of cross-linking of biopolymers on the geotechnical properties.

2 Materials and Methods

To determine the effectiveness of cross-linking of biopolymers for stabilizing the soil, black cotton soil passing through a 4.75 mm sieve was used. Two biopolymers namely, xanthan gum and guar gum were added in different mix proportions to the soil for cross-linking.

2.1 Black Cotton Soil

Black cotton soil collected from the Ibrahimpatnam lake located beside Hyderabad—Nagarjuna Sagar Road (17.19°N, 78.63°E) is used for investigation. The physical properties of the soil such as Atterberg limits and compaction characteristics were determined according to IS 2720(Part V) and IS 2720(Part VII), respectively. Using the plasticity chart as a reference and based on the values of Atterberg limits, the soil was classified as clay with high compressibility (CH). Table 1 presents the physical properties of the soil used for investigation.

2.2 Biopolymers

The biopolymers used in this investigation namely xanthan gum and guar gum were purchased online from Nature's Velvet Lifecare, Hyderabad. Basically, xanthan gum comes under the category of anionic-based polysaccharide, which is a microorganism based and can be produced by fermentation of Xanthomonas campestris bacterium. Xanthan gum is being used widely in the oil, cosmetics, and paper industries as a thickening agent. Galactose and mannose are the two sugars we find in guar gum, and it is extracted from the guar plant. It is available in both the cationic and anionic states and helps in hydrogel formation among all the particles of soil through bonds of hydrogen.

2.3 *Cross-Linking of Biopolymers*

The hydrogels formed in the soil after adding biopolymers are responsible for improving soil strength. Hydrogel formations connected with the soil particles change their nature from a glassy to a rubbery state. The mechanical properties and integrity of the hydrogel depend on the properties of biopolymers added to the soil [5]. The outer chains of biopolymer absorb and holds water which results in the dilution of the hydrogel. The hydrogels formed start to dissolve in the water after desiccation and the time required for dissolving depends on the chemical properties of the biopolymer. To improve the mechanical properties and physical integrity of hydrogels, cross-linking of biopolymers is done where the outer chains of biopolymers react with the polymeric chains of another biopolymer [17]. The literature recommends that polysaccharide galactomannans like guar gum can generate cross-links instantaneously with helix-forming polysaccharides like xanthan with no need for any initiation.

Cross-linking of biopolymers can be done in various methods such as physical, chemical, and enzymatic methods. Even though the properties of biopolymers can be improved by cross-linking, most of the cross-linkers may end up with unwanted changes in functionality or may lead to cytotoxicity. Physical cross-linking is used in this investigation as potential toxicity can be avoided and biomedically safe [4].

Biopolymers were mixed into the soil in different combinations as shown in Table 2. M0 refers to the unamended soil, M1 refers to 1% XG (Xanthan Gum) + 2% GG (Guar Gum), and M2 refers to 2% XG (Xanthan Gum) + 1% GG (Guar Gum) to the dry weight of soil.

Table 2 Combinations of biopolymers in soil

S. No.	Soil description	Dosage of biopolymers	
		Guar gum	Xanthan gum
1	M0	0	0
2	M1	2.0	1.0
3	M2	1.0	2.0

2.4 Tests Conducted

Atterberg Limits. The unamended soil mixture was sieved using a 425-micron sieve and kept in an oven for 24 h at a temperature of 100 ± 55 °C [18]. Oven-dried soil was used for conducting the Atterberg limit tests using Casagrande liquid limit apparatus (Fig. 1).

Compaction Tests. Compaction characteristics of unamended and biopolymer-treated soil were determined using the test procedure in IS 2720–7 [19]. Biopolymers were added to the soil in the amounts mentioned in Table 3 before adding the water and thoroughly mixed to obtain a uniform mixture. The soil was compacted into a lightweight compaction apparatus, and the values of optimum moisture content (OMC) and maximum dry density (MDD) were determined (Fig. 2).

Compression Strength Tests. Unconfined compression strength (UCS) tests were conducted for all the soil mixes M0, M1, and M2 as per IS 2720-part10 [20]. Three cylindrical specimens of size 38 mm in diameter and 76 mm in length (l/d ratio = 2)

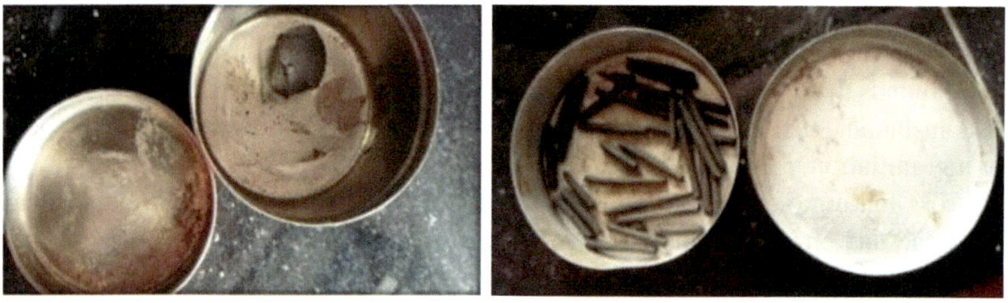

Fig. 1 Atterberg limit tests

Table 3 Variation of atterberg limits

S. No.	Soil description	Liquid limit (%)	Plastic limit (%)
1	M0	63.5	30
2	M1	111.7	41.9
3	M2	144	54.6

Fig. 2 Standard Proctor compaction test setup

Fig. 3 Unconfined compression test setup

were prepared for each trial mix by adding water content equal to optimum moisture content (OMC). Specimens were tested in the compression testing machine and sheared at a rate of 0.6 mm/min, and the values of peak stresses were noted (Fig. 3).

Hydraulic Conductivity Test. The soil was compacted into the permeameter mold by adding water equal to OMC and compacted it to the required density. Water was allowed to pass through the soil compacted in the permeameter to make it saturated [21]. The variable head permeability method was used in this investigation, and the hydraulic conductivity was calculated using the appropriate formula (Fig. 4).

Fig. 4 Permeability test setup

3 Results and Discussions

3.1 Variation of Atterberg Limits

Table 3 shows the values of Atterberg limits of unamended and biopolymer stabilized soil. Values of liquid and plastic limits for the soil mix M0 were 63.5% and 30.5%, respectively. After the stabilization of soil by cross-linking the biopolymers, the liquid limit, and plastic limit values were increased. For the soil mix M1, the liquid limit value increased to 111.7% and the plastic limit increased to 41.93%. For the soil mix M2, the liquid and plastic limit values were increased to 144% and 54.6%, respectively. From the values of plastic and liquid limits of soil after cross-linking with biopolymers, it can be noted that the effect of cross-linking on the plastic limit and liquid limit is considerable.

By observing the values of Atterberg limits for unamended soil and modified soil, it can be noted that the plasticity index is drastically increased. The increase in plasticity index of stabilized soil for the soil mix M1 is around 110%, and for the soil mix M2 is 170%. Increment of GG content in the soil resulted in the rise of the plasticity index. From Fig. 5, it can be observed that there is an effective change in the values of plastic and liquid limit after adding biopolymers to the soil and cross-linking them.

The rise in values of liquid limit is a result of the chemical properties of biopolymers, and the chemical reaction they undergo with the clay particles. Added to the soil matrix and mixed with it, hydroxyl ions in guar gum interacts in two essential

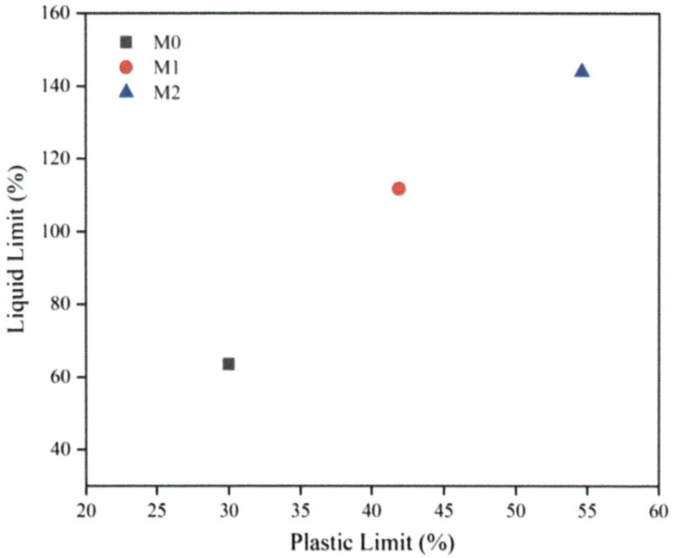

Fig. 5 Variation of Atterberg limits at different biopolymer combinations

Table 4 Variation of compaction characteristics

S. No.	Soil combination	Max. dry density (g/cc)	Optimum moisture content (%)
1	M0	1.47	26
2	M1	1.25	28
3	M2	1.23	29

ways. Initially, the pore water in the soil interacts with hydroxyl ions of guar gum and its viscosity gets increases. Due to this, the liquid limit of the soil also increases. Secondly, soil agglomerations occur when biopolymer chains interact with clay particles and is due to the reduction of soil particle surface area. Soil's liquid limit is determined by the combination of these two interactions. Being a neutral polysaccharide, guar gum interacts with clay particles in soil to form hydrogen bonds. Thus, the soil with higher guar gum has a higher liquid limit than soils with other stabilizing agents [5].

3.2 Variation of Compaction Characteristics

Table 4 shows the variation of Maximum dry density (MDD) and optimum moisture content (OMC) for unamended and modified soil. There is an increase in optimum moisture content and a decrease in dry density values for the soil mix M1.

A slight increment in the value of optimum moisture content and fall in the dry density of soil was observed after increasing the amount of Xanthan gum in the soil to 2% in the soil mix M2. The reason for the reduction of MDD after adding biopolymers to soil is attributed to the increase of biopolymers filling the void spaces and adhering to soil particles. This improves the resistance of the resultant soil–biopolymer matrix against applied compaction energy [9]. An increase in optimum moisture content is due to the adsorption of water which helps to dissolve the biopolymer content [9]. It may also be due to the higher affinity toward the water and the viscous nature of xanthan gum. As shown in Fig. 6, considerable change was observed in both optimum moisture content and maximum dry density after adding biopolymers to stabilize the soil.

3.3 Variation of Unconfined Compressive Strength

Table 5 shows the variation of compressive strength of soil for the unamended and soil modified by adding biopolymers.

The average compressive strength of unamended soil is 1.48 kg/cm^2. By the inclusion of biopolymers in the soil as per the mix M1, the compressive strength of the soil was increased to 1.862 kg/cm^2, and it may be due to the formation of hydrogel

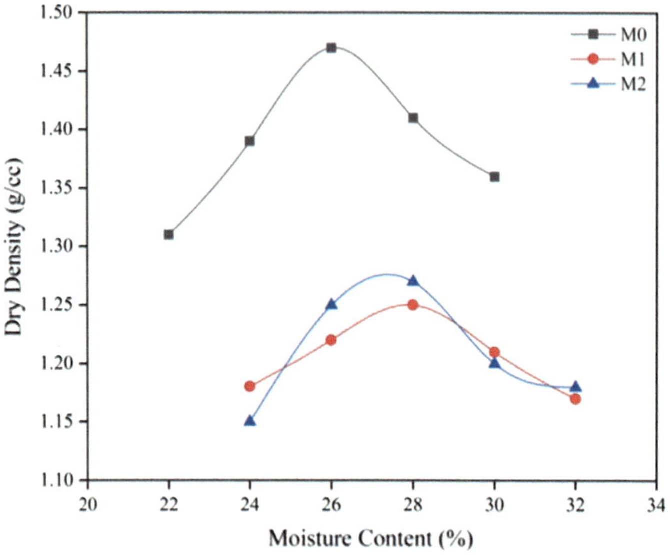

Fig. 6 Variation of MDD and OMC for different combinations of biopolymers

Table 5 Compressive strength characteristics of soil

S. No.	Soil combination	Compressive strength (Kg/cm^2)
1	M0	1.48
2	M1	1.862
3	M2	1.815

and soil suction. Guar gum added to the soil helps transform the hydrogels to a glassy state on drying and is also a reason for the improvement of compressive strength. No consider able change in the compressive strength of soil is observed after increasing the xanthan gum and decreasing the guar gum to 2% and 1%, respectively (Fig. 7).

3.4 Variation of Hydraulic Conductivity

Table 6 shows the values of the coefficient of permeability for unamended and amended soils. A considerable decrease in the value of the hydraulic conductivity is seen after adding the biopolymers to the soil.

The hydrogels formed in the soil after adding biopolymers to the soil fill the pore spaces available in the soil and thus decrease the tendency of water to flow between the soil particles. The biopolymer added to the soil reacts with water and turns it into leachate and decreases the permeability of the soil. The clay particles surrounded by

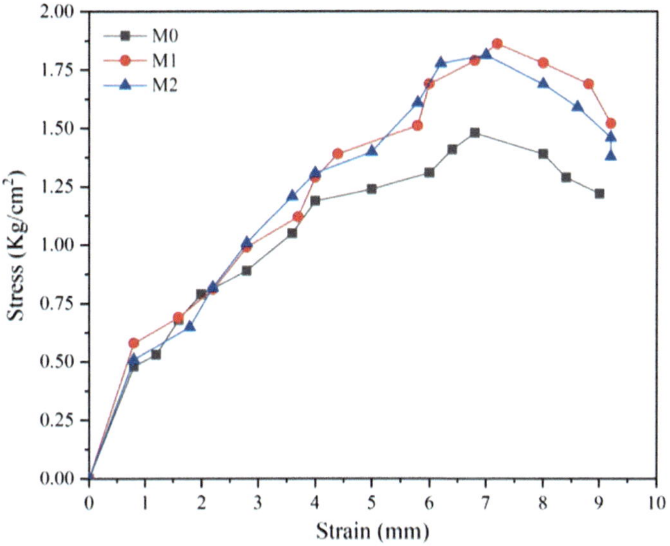

Fig. 7 Variation of stress and strain in UCS test for different combinations of Biopolymers

Table 6 Permeability characteristics of soil

S. No.	Soil combination	Coefficient of permeability (cm/s)
1	M0	5.24×10^{-3}
2	M1	1.25×10^{-3}
3	M2	1.23×10^{-3}

water molecules will create a narrow flow path to the water and result in a decrement in the soil's hydraulic conductivity [22, 23].

4 Conclusions

In this study, variation of strength and permeability characteristics of high compressible clay after adding biopolymers has been investigated. Both guar gum and xanthan gum instantly interact with soil particles and form cross-links without adding any initiating agents. Both guar and xanthan gum were added to the soil at different dosages, and tests like Atterberg limits, standard Proctor compaction, unconfined compression, and permeability were conducted. Following were the conclusions drawn from the investigation.

- The addition of biopolymers to soil was found to increase in both liquid and plastic limits. For the soil mix M1, the liquid limit was 111.7% and it is 75% more than the soil mix M0. The plastic limit was 41.9%, and it is 40% more than the soil mix M0.
- Both liquid and plastic limits of soil were increased when the biopolymers were mixed as per the soil mix M2. The liquid limit was 144%, and the plastic limit was 54.6%. Rise in the liquid and plastic limits was 126% and 82%, respectively.
- The rise in plasticity index of soil mix M1 is around 110%, and for the soil mix M2 is around 170%.
- MDD of unamended soil M0 is 1.47 g/cc. Inclusion of the biopolymers in the soil has shown a fall in the values of soil MDD. For the soil mix M1 and M2, values of MDD were 1.25 g/cc and 1.23 g/cc, respectively. This could be due to the resistance offered by the soil–biopolymer mix to the applied compaction energy.
- The inclusion of biopolymers in the soil mix has shown an increase in compressive strength. For unamended soil mix M0, compressive strength was 1.48 kg/cm^2. For the soil mix M1 and M2, compressive strength was 1.862 kg/cm^2 and 1.815 kg/cm^2, respectively. The rise in compressive strength of soil mix M1 after adding biopolymers was around 25%.
- No appreciable change in compressive strength was observed for the soil mix M2 when compared to the soil mix M1.
- Hydraulic conductivity of unamended soil (M0) is 5.24×10^{-3} cm/s. Cross-linking of biopolymers in the soil to the dosage of M1 has reduced the hydraulic conduc tivity to 1.25×10^{-3} cm/s and no change was observed when the dosage of both guar gum and xanthan gum were interchanged in the soil mix M2. For this reason, higher dosages of xanthan gum are not required to improve the compressive strength and permeability of the soil.

By analyzing the results from this investigation and available literature on methods of stabilization, it can be concluded that cross-linking of biopolymers is more effective in improving the geotechnical properties and is environmentally sustainable. More research needs to be done on cross-linking of biopolymers to understand the mechanism involved in it and to check the durability of biopolymers in the long run to replace the existing biopolymers.

References

1. Moghal AAB, Vydehi V, Moghal MB, Almatrudi R, AlMajed A, Al-Shamrani MA (2020) Effect of calcium-based derivatives on consolidation, strength, and lime-leach- ability behavior of expansive soil. J Mater Civ Eng 32(4):04020048. https://doi.org/10.1061/(ASCE)MT.1943-5533.0003088
2. Amulya G, Moghal AAB, Almajed A (2021) A state-of-the-art review on suitability of granite dust as a sustainable additive for geotechnical applications. Crystals 11(12):1526. https://doi.org/10.1061/(ASCE)MT.1943-5533.0003088
3. Ashok Kumar M, Baig Moghal AA, Bommisetty J, Nuruddin M, Mohammad N, Moghal AAB, Rasheed RM, Almajed A (2022) Critical review on the ef- ficacy of electrokinetic techniques

in geotechnical and geoenvironmental applications. Arab J Geosci 15(8):781. https://doi.org/10.1007/s12517-022-10037-1

4. Vydehi KV, Moghal AAB, Basha BM (2022) Reliability-based design optimization of biopolymer-amended soil as an alternative landfill liner material. J Hazard Toxic Radioact Waste 26(3):04022011. https://doi.org/10.1061/(ASCE)HZ.2153-5515.0000697
5. Vydehi KV, Moghal AAB, Basha BM (2022) Target reliability-based design of embankments using biopolymer-modified cohesive soil. Int J Geomech 22(8):04022115. https://doi.org/10.1061/(ASCE)GM.1943-5622.0002429
6. Muhammad G, Marri A (2018) Immediate and long-term effects of lime and wheat straw on consistency characteristics of clayey soil. Geomech Eng 16(3):217–231. https://doi.org/10.12989/GAE.2018.16.3.217
7. Harsh H, Moghal AAB, Rasheed RM, Almajed A (2022) State-of-the-Art review on the role and applicability of select nano-compounds in geotechnical and geoenvironmental applications. Arab J Sci Eng. https://doi.org/10.1007/s13369-022-07036-5
8. Imane I, Souhila RB, Riad B, Inas B (2021) Use of biopolymers in the stabilization of clay soils. Select Sci Pap J Civil Eng 16(2):37–53. https://doi.org/10.2478/sspjce-2021-0015
9. Vydehi KV, Moghal AAB (2022) Effect of biopolymeric stabilization on the strength and compressibility characteristics of cohesive soil. J Mater Civ Eng 34(2):04021428. https://doi.org/10.1061/(ASCE)MT.1943-5533.0004068
10. Chang I, Cho G-C (2012) Strengthening of korean residual soil with β-1,3/1,6-glucan biopolymer. Constr Build Mater 30:30–35. https://doi.org/10.1016/j.conbuildmat.2011.11.030
11. Chang I, Im J, Cho G-C (2016) Geotechnical engineering behaviors of gellan gum biopolymer treated sand. Can Geotech J 53(10):1658–1670. https://doi.org/10.1139/cgj-2015-0475
12. Chang I, Kwon Y-M, Im J, Cho G-C (2019) Soil consistency and interparticle characteristics of xanthan gum biopolymer–containing soils with pore-fluid variation. Can Geotech J 56(8):1206–1213. https://doi.org/10.1139/cgj-2018-0254
13. Lee M, Im J, Cho G-C, Ryu HH, Chang I (2020) Interfacial shearing behavior along xan- than gum biopolymer-treated sand and solid interfaces and its meaning in geotechnical engineering aspects. Appl Sci 11(1):139. https://doi.org/10.3390/app11010139
14. Kwon Y-M, Ham S-M, Kwon T-H, Cho G-C, Chang I (2020) Surface-erosion behaviour of biopolymer-treated soils assessed by EFA. Géotechnique Lett 10(2):106–112. https://doi.org/10.1680/jgele.19.00106
15. Chen C, Wei K, Gu J, Huang X, Dai X, Liu Q (2022) Combined effect of biopolymer and fiber inclusions on unconfined compressive strength of soft soil. Polymers 14(4):787. https://doi.org/10.3390/polym14040787
16. Vydehi KV, Moghal AAB, Mariyam Rasheed R (2022) Shrinkage characteristics of bi-opolymer treated expansive soil. In: Geo-Congress 2022. American society of civil engineers, charlotte, North Carolina, pp 92–99. https://doi.org/10.1061/9780784484012.009
17. Muguda S, Hughes PN, Augarde CE, Perlot C, Walter Bruno A, Gallipoli D (2022) Cross-linking of biopolymers for stabilizing earthen construction materials. Build Res Inform 50(5):502–514. https://doi.org/10.1080/09613218.2021.2001304
18. IS 2720-Part 5 (1985) Methods of test for soils—determination of liquid and plastic limit, Bureau of Indian Standards, New Delhi
19. IS 2720-Part 8 (1983) Methods of test for soils—determination of water content—dry density relation, Bureau of Indian Standards, New Delhi
20. IS 2720- Part 10 (1991) Methods of test for soils—Determination of unconfined compressive strength, Bureau of Indian Standards, New Delhi
21. IS 2720-Part 17 (1986) Methods of test for soils—Laboratory determination of permeability, Bureau of Indian Standards, New Delhi
22. Anandha Kumar S, Sujatha ER (2021) Compaction and permeability characteristics of biopolymer-treated soil. In: Ramanagopal S, Gali ML, Venkataraman K (eds) Sustainable practices and innovations in civil engineering. Springer Singapore, Singapore, pp 107–117
23. Moghal AA, Vydehi V, Moghal MB, Almatrudi R, Almajed A, Al-Shamrani MA (2020) Effect of Calcium-based derivatives on consolidation, strength, and lime-leach- ability behavior of expansive soil. J Mater Civil Eng 32. https://doi.org/10.1061/(ASCE)MT.1943-5533.0003088

Shear Strength Behavior of the Flyash-Treated Lime Columns on Expansive Soils

G. Gnana Prasanna, M. Manoj Kumar, A. Dhanaraj, and K. Raja Sekhar

1 Introduction

Due to rapid increase in population and pollution, it is becoming very difficult to find soils whose strength is sufficient to bear a structure. Developing countries like India, second-highest populated country over the world facing number of problems related to soils. Expansive soil is one of the problematic soils that changes in volume with relation to changes in water content. Expansive soils are a worldwide problem, posing challenges for civil engineers. They are considered a potential natural hazard which can cause extensive damage to civil engineering structures. In India 2% of total area is covered with expansive soils. In those areas, civil engineers have to face many problems due to the behavior of soil. Bell [1, 2], Broms [3], Rajasekaran [4], and etc., are done extensive research on lime treated and lime column treated soils. Reddy [5], Basumajumdar [6], and many other researchers worked on lime–flyash columns.

2 Materials Used

The following materials are used to study shear strength parameters of the flyash-treated lime columns:

1. Expansive soil.
2. Quick lime.
3. Flyash.

G. Gnana Prasanna (✉) · M. Manoj Kumar · A. Dhanaraj · K. Raja Sekhar
Department of Civil Engineering, Sri Venkateswara College of Engineering, Karakambadi, Tirupati 517507, India
e-mail: gnanaprasanna.24@gmail.com

B. T. Jose et al. (eds.), *Proceedings of the Indian Geotechnical Conference 2022 Volume 4*, Lecture Notes in Civil Engineering 479,
https://doi.org/10.1007/978-981-97-1753-8_16

Table 1 Physical and index properties of soil

S. No.	Property	Value
1	In situ water content	28%
2	In situ bulk density	1.7 g/cc
3	Fine fraction	65%
4	Coarse fraction	35%
5	Specific gravity	2.5
6	Liquid limit	63%
7	Plastic limit	20.75%
8	Shrinkage limit	18%
9	Plasticity index	42.25%
10	Indian standard soil classification system	CH
11	Free swell index	61%
12	Maximum dry density	1.486 g/cc
13	Optimum moisture content	21%

4. Steel containers.

2.1 Expansive Soil

Expansive soils are found in various parts of our country. It occupies about 40% of land in our country. In this experiment, we had used black cotton soil which is collected from Chandragiri locality. The soil sample collected is 1.5 m below the ground level. The physical and index properties of the natural clay sample collected are given in Table 1.

2.2 Quick Lime

Hydrated lime is dry form of lime. The lime shown in the Fig. 1 is used for our work and the lime collected from a shop in Tirupati. The chemical and physical properties of quick lime used in this study is given in Table 2.

2.3 Flyash

Flyash is a byproduct obtained from chemical industries. It is collected from brick manufacturing industry in Tirupati. Figure 2 shows the flyash used for study. The flyash used in the study is classified as class F based on the manufacturer input.

Fig. 1 Quick lime used for the study

Table 2 Physical and chemical properties of quick lime

Property	Value
CaO	82.2%
MgO	3.1%
Other compounds	12.5%
Loss on ignition	1.25
Particle size	< 90 μm
Particle density	1.15 g/cc

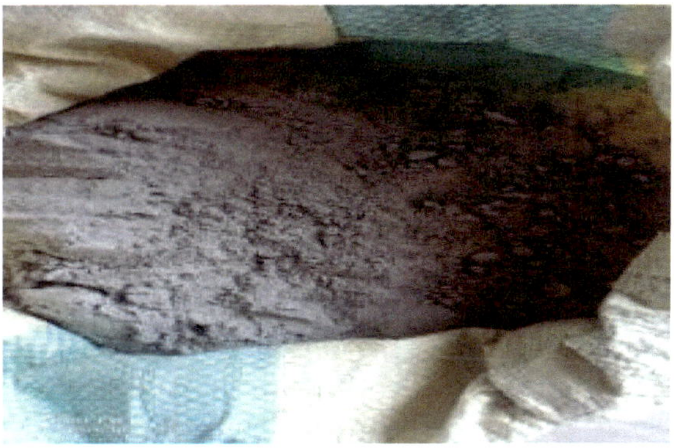

Fig. 2 Flyash used for the study

2.4 *Test Tanks*

Iron cylindrical containers of diameter 40 cm and depth 40 cm are used for filling the soil and determining the shear strength of the soil after installing the lime columns and flyash–lime columns.

3 Experimental Setup

The natural clay was oven dried and the mechanically ground to avoid the lumps. The soil was then mixed with water to have the in situ moisture content. The desired amount of soil to be filled in test tanks is determined and compacted to have the in situ bulk density to simulate the natural field conditions.

The soil was compacted in a circular tank of 40 cm in both diameter and height. Soil was filled in four equal layers of 9 cm each, and compaction is done manually with tampers.

3.1 Installation of Lime Piles

In field applications, a hollow tube was forced into the ground to desired depth and binding agent is applied with pressure. The same technique is used in this study, the installation of five columns in compacted soil block with equal distance from center pile. Columns are made by hollow PVC pipe of 3 cm diameter and 30 cm height, open at both ends. PVC pipe is driven by manually into the soil block.

Then, each column was filled with powdered quick lime of uniform mass in one test Tank 1, and in test Tank 2, powdered lime and flyash are mixed with equal proportion of weights and were filled in the holes made. This application was in three successive uniform layers with lightly compaction on each layer. In the test tanks, 4 cm in height was left above the soil block. The following Figs. 3, 4, and 5 show the procedure we followed during the study.

3.2 Curing

After installation of lime piles, setup was covered with a thin cloth, and it is covered with a layer of sandy soil of particle size 2 mm up to 2 cm thick as shown in Fig. 6 and then supplied water to maintain the natural water content of soil block. The soil blocks were left to react chemically with piles for a curing period of 28 days. The following Fig. 7 shows the test tanks covering with polythene covers and kept for curing.

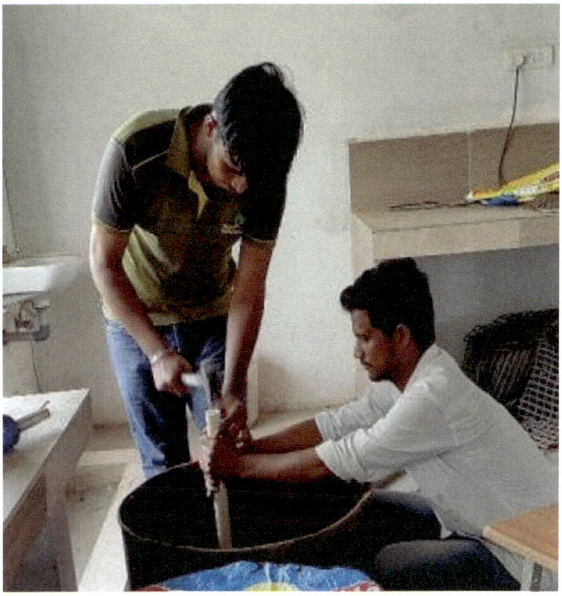

Fig. 3 Making piles in soil block

Fig. 4 Soil block ready for lime treatment

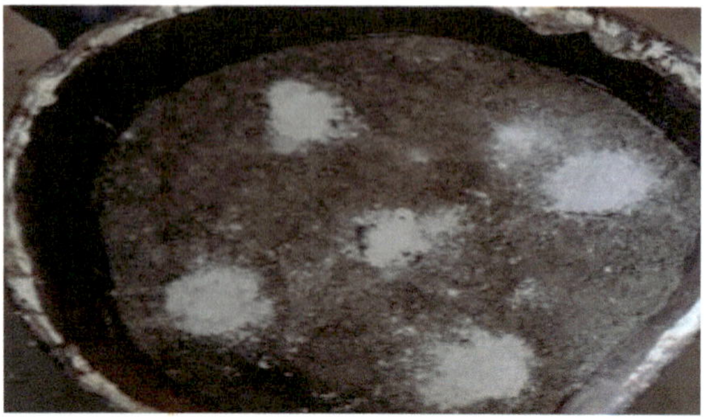

Fig. 5 Surface of soil block after lime piles installation

a b

Fig. 6 **a** Cloth covering soil block. **b** Sand layer on the cloth

Fig. 7 Soil with lime columns in tank 1 and soil with flyash–lime columns in tank 2 are placed for curing

Table 3 Comparison of UCS of soil samples

S. No.	Description	UCS* (kg/cm^2)	% Increase w.r.t. NSS**
1	Natural soil samples	0.51	–
2	Soil samples extracted from lime columns tank	0.85	66%
3	Soil samples extracted from flyash–lime columns tank	0.81	58.82%

*Unconfined compressive strength
**NSS: Natural soil sample

4 Experimental Results and Discussions

4.1 Unconfined Compression Strength Test

Unconfined compression tests are conducted on the natural soil samples, soils samples extracted between the lime columns from Tank 1 and Tank 2 according to the Indian Standards code IS 2720 (Part 10): 1991 [7].

Unconfined compressive strength tests are done on three samples collected from the testing tanks. The samplers were pushed slowly into the treated soil block with aid of hydraulic jack for collecting less/undisturbed soils. The soil samples are collected in between the lime and flyash–lime columns and tested for UCS. The results showed that there is tremendous improvement in the UCS of soil samples treated with lime and flyash–lime columns compared with natural soil samples. The comparison of UCS of soil samples are given in the Table 3.

The Fig. 8 showing the variations of unconfined compressive strengths of the natural soil samples and soil samples extracted from lime columns tank and flyash–lime columns tank. The increase in the UCS of soil samples extracted from the natural ground and Tank 1 is due to pozzolanic reactions occurred between the lime and soil [8]. The strength of the UCS soil sample extracted from Tank 1 is less than soil samples extracted from Tank 2, and it is attributed to the flyash mixed is not having sufficient pozzolanic material as like lime. But, the combination is economical as flyash is available for very lesser cost when compared with lime. The curing period is taken as 28 days, because it can give reliable results than 7 days curing period [9].

5 Conclusions

The following conclusions are drawn from the experiments conducted and results obtained from the experiments.

1. The unconfined compressive strength of the soil specimens obtained from lime columns and lime–flyash columns treated soil tanks are appreciably greater than the natural soil samples.

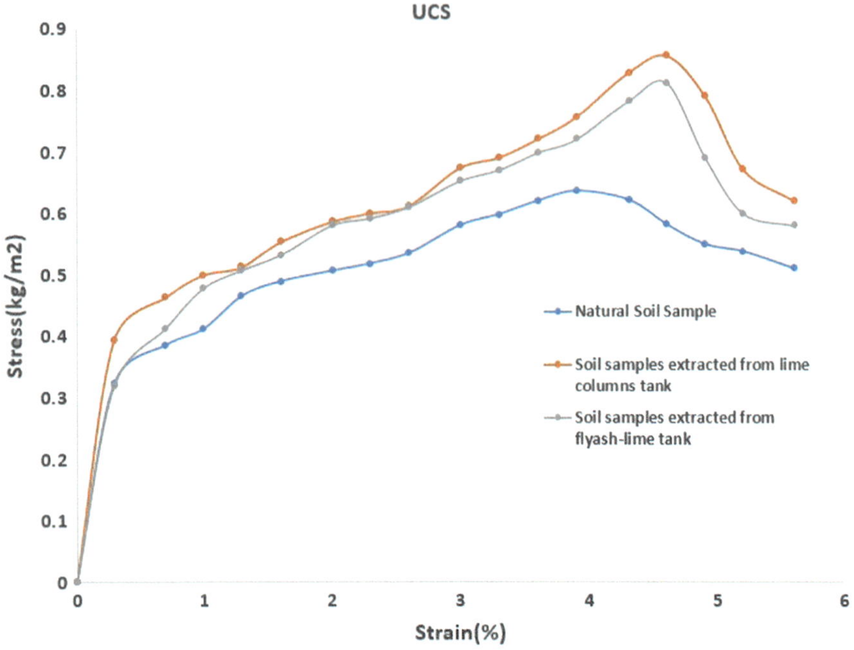

Fig. 8 Variations of unconfined compressive strengths of the soil samples

2. The soil samples extracted from the flyash replaced lime columns tank has the unconfined compressive strength lesser than the soil samples extracted from lime columns treated soil tank and the percentage decrease of strength is very less.
3. The lime–flyash columns treated soils are the better and sustainable solution for stabilizing the expansive soils.

References

1. Bell FG (1996) Lime stabilization of clay minerals and soils. Eng Geol 42(4):223237
2. Bell FG (1989) The enhancement of the properties of clay soils by the addition of cement or lime. In: Proceedings of the 12th international conference on soil mechanics and foundation engineering
3. Broms (1984) Stabilization of soil with lime columns. In: Design handbook. 3rd edn, Lime Column AB, pp 833–855
4. Rajasekaran G (2000) Compressibility behaviour of lime-treated marine clay. Ocean Eng 29(2002):545–559
5. Reddy PP (2012) A study on strength characteristics of flyash, lime and sodium silicate mixtures at their free pouring conditions. Int J Comput Eng Res (ijceronline.com) 2(4)
6. Basumajumdar A, Das AK, Bandyopadhyay N, Maitra S (2005) Some studies on the reaction between fly ash and lime © Indian Academy of Sciences. Bull Mater Sci 28(2):131–136
7. IS 2720-Part 10 (1991) Methods of test for soils—determination of unconfined compressivestrength, Bureau of Indian Standards, New Delhi

8. Prasad PD, Nagarnaik PB, Ajay RG (2016) Effect of fly ash and lime on stabilization of expansive soil. i-manager's J Civil Eng 6(2):8–12
9. Manuel M, Donald TD (1962) Lime and fly ash proportions in soil, lime and fly ash mixtures, and some aspects of soil lime stabilization. Highway research board bulletin: n. pag

Numerical Analysis of Stone Columns in a Layered Soil System

Subhash Kumar Gupta, Soumen Naskar, and Awdhesh Kumar Choudhary

1 Introduction

Stone column is considered to be one of the most efficient and cost-effective methods for improving the weak soil. It reduces the settlements and increases the strength of soft soil. Stone column is a dense aggregate column made of granular material that is embedded in weak soil strata. The granular column consists of compacted gravel or crushed stone which are arranged in the form of a column using a probe vibrator. It is one of the most common ground improvement techniques. Stone column improves the ultimate bearing capacity of the ground. It was earlier known as stone column as only stones were used for filling. Stone columns are mainly failed due to bulging [1, 2], general shear failure [3], and sliding [4]. A long stone column having a length greater than its critical length (i.e. about 4 times the diameter of the column) fails by bulging irrespective of whether it is end bearing or floating. To mitigate this problem, stone columns are of often encased with an appropriate geosynthetic material which prevents their failure by bulging. Encasement of the ordinary stone column adds extra restrain from all the sides which enhances its axial capacity in weak soil. This further leads to a decrease in settlement value and also raises the stress concentration of the stone column resulting in reduction of load value placed above the soil [5–7]. Encasement provides resistance to bulging by cylinder action. As columns bulge under loading, hoop stresses are developed in the encasement. These stresses are resisted by the tensile strength of the geosynthetic. This causes

S. K. Gupta (✉) · S. Naskar · A. K. Choudhary
Department of Civil Engineering, National Institute of Technology, Jamshedpur 831014, India
e-mail: subhashrouniyar@gmail.com

S. Naskar
e-mail: snaskar737@gmail.com

A. K. Choudhary
e-mail: awdhesh.ce@nitjsr.ac.in

B. T. Jose et al. (eds.), *Proceedings of the Indian Geotechnical Conference 2022 Volume 4*, Lecture Notes in Civil Engineering 479,
https://doi.org/10.1007/978-981-97-1753-8_17

high ultimate load carrying capacity and less settlement of composite foundation. The pattern of installation of stone columns can be square or triangular. But densest packing can be achieved only through the equilateral triangular pattern [8].

The above literature review shows that the experimental, analytical and numerical study on the settlement and bearing capacity of the stone column improved soil has been investigated considering surrounding soil to be homogeneous in nature. However, in reality, soils are layered with variable degree of stiffness of different layers. Very few studies have been conducted on the performance of stone column in layered soil [9–11]. In view of this, effect of layering was investigated by performing three-dimensional numerical analyses in PLAXIS^{3D}. Numerical study has been conducted to understand the behaviour of ordinary stone columns as well as encased stone columns in layered soil system.

In the present study, behaviour of stone columns in soft soil has been investigated through a series of numerical analysis. The influence of undrained shear strength of clay bed on load carrying capacity of stone column has been analysed. The influence of layered soil system on the overall behaviour of OSC and ESC has been brought out. Parametric analysis in terms of friction angle of stone aggregates has been performed.

2 Numerical Modelling

Numerical modelling was done with the help of finite element-based software PLAXIS^{3D} to provide better understanding on the behaviour of stone columns in a layered soil system under different test conditions. Initially, a homogeneous geometric soil model having dimension 1000 mm × 1000 mm × 1000 mm (length × width × height) was developed considering Mohr–Coulomb model as shown in Fig. 1. Further, varying depth of two clay layered geometric soil model with same dimensions was prepared as shown in Fig. 2. The model was fully fixed at the bottom (both horizontal and vertical movement was restricted) and partially fixed on the sides (horizontal movement was restricted). A circular-shaped footing having diameter equal to 100 mm was placed at the centre of the group of stone columns which was arranged in triangular pattern having spacing equal to 3 times the diameter of stone column.

Stone columns in layered clay have been used in this study. The material properties of clay and stone column were adopted from literature of [12]. The properties of clay and stone aggregates are presented in Table 1. For encasing the stone column, the properties of geogrid structure element have been adopted from literature of [13] and are presented in Table 2.

The PLAXIS^{3D} software automatically generates a three-dimensional model for this study. Medium meshing has been done in the stone columns in layered soil. The calculation phase was then performed and result of the numerical models has been analysed.

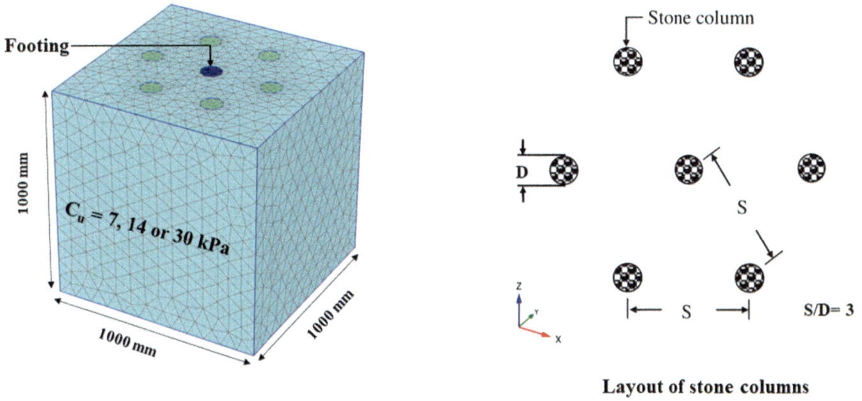

Fig. 1 Typical model geometry of stone columns in homogeneous soil

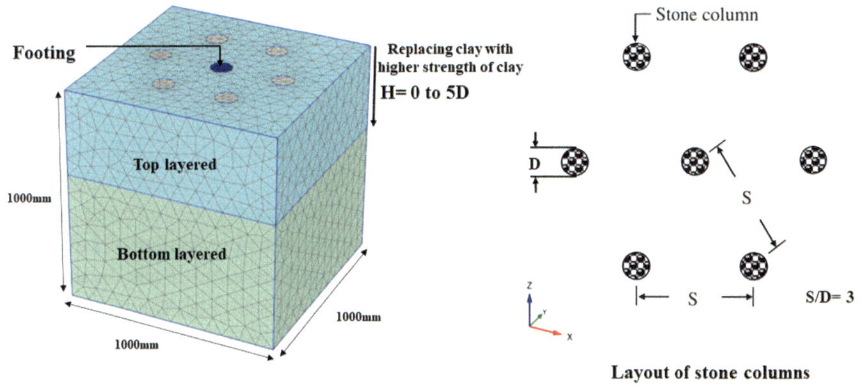

Fig. 2 Typical model geometry of stone columns in layered soil

Table 1 Input parameters for modelling stone columns in soft clay [12]

Material	C_u (kPa)	E (kPa)	μ	$\phi°$	$\Psi°$	γ_{dry} (kN/m^3)	γ_{bulk} (kN/m^3)
Clay	30	5500	0.42	–	–	15.56	19.45
Clay	14	3100	0.45	–	–	14.60	18.98
Clay	07	2150	0.47	–	–	13.60	18.38
Stones	–	55,000	0.30	43	10	16.62	–

Table 2 Properties of encasement materials used [13]

Material	Axial stiffness EA (kN/m)	Shear stiffness GA (kN/m)
Geogrid	210	80.77

3 Results and Discussion

In the present study, the bearing capacity of stone columns (ordinary as well as encased) in layered soil system, under circular footing, has been analysed. For this, numerical models have been developed in PLAXIS^{3D}. However, before detailed investigation and analysis, validation of the proposed models with the experimental studies of ordinary stone column (OSC) reported by Ambily and Gandhi [12] and encased stone column (ESC) by Debnath and Dey [13] have been carried out. The material properties of clay and stone column and dimension of stone column and footing used in the numerical model were adopted from experimental work performed by Ambily and Gandhi [12]. Subsequently, stiffness of encasement material was taken from experimental work performed by Debnath and Dey [13].

The numerical simulation was done, and accordingly, influence of undrained cohesion of clay (C_u) on the performance of stone column in homogeneous soil has been investigated. The influence of layered clay on the load carrying capacity of stone column and the effect of friction angle of stone aggregates (ϕ_a) on the overall behaviour of stone column has been studied.

3.1 Validation of Ordinary Stone Columns

At first, the numerical model was developed using finite element package PLAXIS^{3D} to analyse the load-settlement behaviour of ordinary stone column in clayey soil. The model dimensions and properties of soil and stone column were kept same as reported in the experimental model tests conducted by Ambily and Gandhi [12]. The clay and stones are modelled using Mohr–Coulomb's criterion and drained behaviour was considered for all the materials. The results obtained from numerical analysis are then compared with existing experimental results reported by Ambily and Gandhi [12] as presented in Fig. 3. It can be seen that results from the numerical analysis are in good agreement with the experimental results for different C_u values ranging from 7 to 30 kPa. Hence, it can be said that the present numerical model is capable enough to simulate the behaviour of ordinary stone columns placed in clay beds.

3.2 Validation of Encased Stone Columns

To provide better confinement and reduce bulging, ordinary stone columns are often encased with geosynthetics materials. Therefore, encased stone columns can be provided in place of ordinary stone columns. The numerical model comprising of encased stone column is validated with experimental work reported by Debnath and Dey [13]. The model dimensions, properties of soil, and stone column were kept identical as reported in the experimental model tests by Debnath and Dey [13]. Further,

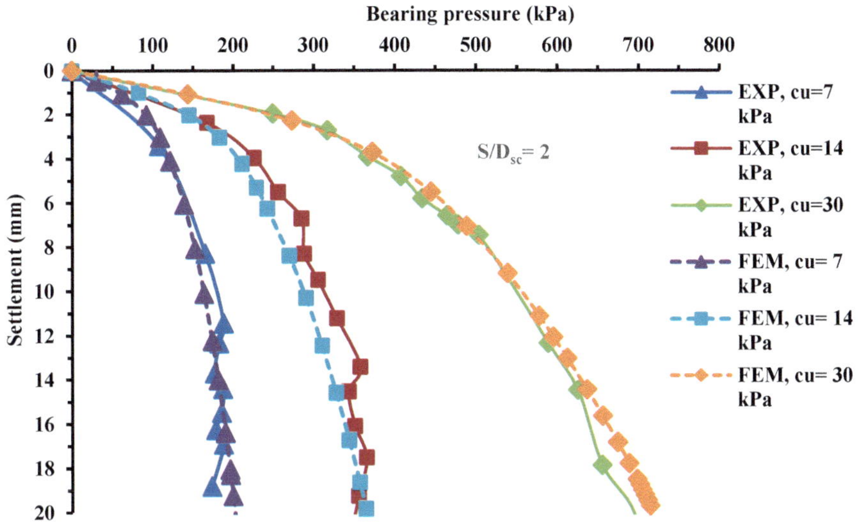

Fig. 3 Validation of OSC

the axial stiffness and shear stiffness values of encasement material were 210 kN/m and 80.77 kN/m respectively. The comparison of numerical results with experimental one for ESC is shown in Fig. 4. It can be noticed that the numerical results are well comparable to the experimental results with marginal error. This error is probably due to inaccuracy in adopting the material properties of geosynthetic used as an encasement material during experimental work by Debnath and Dey [13].

3.3 Behaviour of Stone Columns in Homogeneous Clay

The response of footing placed above unreinforced, ordinary stone column reinforced, and encased stone column reinforced clay bed has been investigated in this section. In view of this, soil model having dimension 1000 mm × 1000 mm × 1000 mm has been created as shown in Fig. 1. Group of stone columns was placed in triangular pattern with spacing being equal to 3 times the diameter of stone column, i.e. 100 mm and a circular footing having diameter equal to the diameter of stone column is placed at the centre of group of stone columns [12]. The value of undrained cohesion of clay, i.e. $C_u = 7$ kPa has been kept the same in all the cases. The bearing pressure vs. settlement curve for unreinforced clay bed, clay bed consisting of OSC and ESC has been presented in Fig. 5. It can be observed that with the provision of OSC in unreinforced clay bed, the footing is able to bear greater loads under same value of settlement. Further when OSC was replaced by ESC, the load carrying capacity of the composite system gets significantly increased. It is because of the

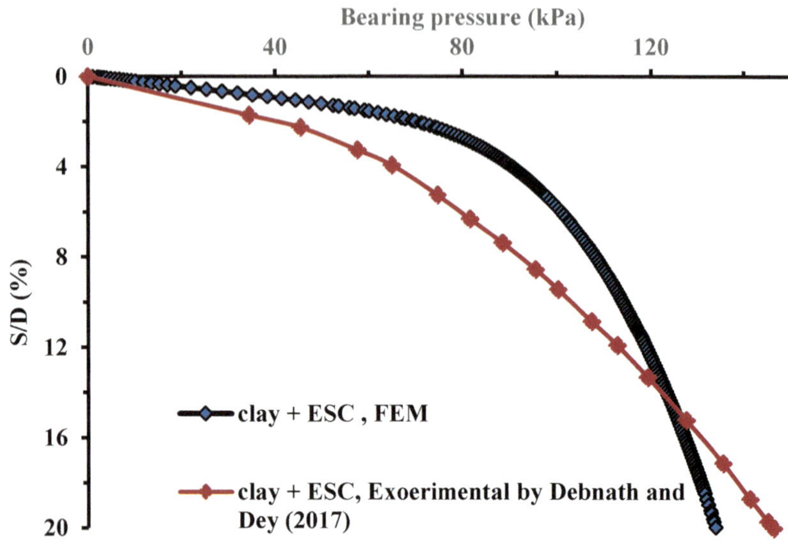

Fig. 4 Validation of ESC

fact the OSC in soil provides lateral confinement to upcoming load and encasement of OSC provides higher resistance to radial bulging of the same.

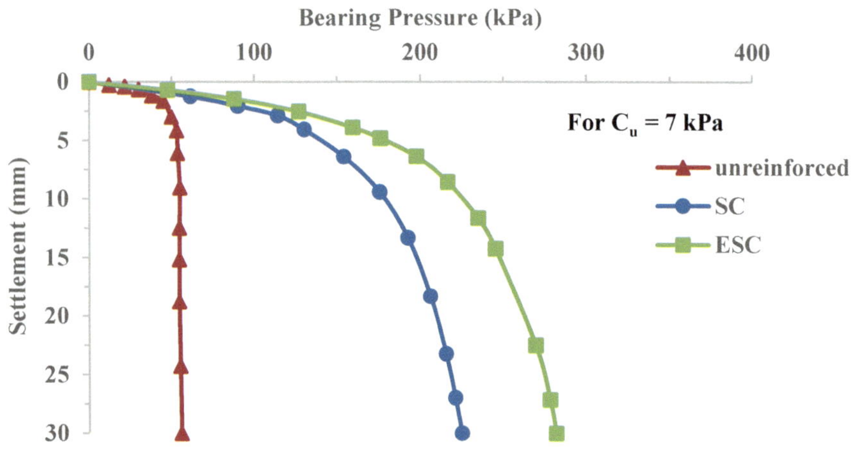

Fig. 5 Bearing pressure-settlement response of stone columns in homogenous clay

3.4 Behaviour of Stone Columns in Layered Clay

The behaviour of stone columns in layered clay has been investigated by changing single layer soil system (homogeneous clay) into a two layered soil system. In view of this soft clay up to depth equal to 'H' is being replaced by clay having higher undrained shear strength as shown in Fig. 2. The model geometry has been kept the same as in Sect. 3.3. In this layered system, undrained shear strength of clay, C_u of the top layer is always kept greater than that of the bottom layer.

Figures 6 and 7 show the bearing pressure vs. settlement curve in homogenous and layered clay for OSC and ESC, respectively. It can be seen that the bearing capacity of OSC and ESC in layered clay increases with an increase in the depth of top layer. However, beyond depth of top layer equal to 4D, the further rate of increment in bearing capacity is found to be negligible for both OSC and ESC. The performance improvement is represented in terms of improvement factor ratio (IFR) which is equal to the ratio of bearing pressure at 30 mm settlement of OSC (or ESC) in layered system to the bearing pressure at 30 mm settlement of homogenous soft clay. The increase in IFR with increase in depth of top layer in case of both OSC and ESC is presented in Tables 3 and 4, respectively. It was found that the bearing capacity of the composite system comprising of OSC increases with an increase in the depth of top layer (H). However, the performance improvement is found to be negligible beyond the depth of top layer equal to 4D as indicated by rate of increase in IFR values. Similar behaviour has been observed in case of ESC as well. However, the performance improvement is found to increase at a marginal rate as can be seen from the Table 4. Hence, it can be said that the optimum depth of upper clay layer to achieve maximum increase in bearing pressure is 4D. The vertical displacement contours of OSC and ESC in layered clay is shown in Figs. 8 and 9. It can be observed that the displacement contours are longer in layered soil system as compared to homogeneous clay system. However, beyond the depth of 4D in layered soil system, the displacement contours are nearly same as that of homogeneous clay. Similar behaviour has been observed in case of ESC as well. But displacement contours were found to be spread more along the column in case of ESC as compared to that of OSC.

3.5 Influence of Friction Angle of Stone Aggregates

The influence of variation in friction angle of stone aggregates for both OSC and ESC in layered soil system having top layer $C_u = 30$ kPa with depth of 4D and bottom layer having $C_u = 14$ kPa is presented in Figs. 10 and 11. From Fig. 10, it can be observed that the bearing capacity of OSC in layered soil system increases with increase in friction angle of stone aggregates. The increment observed is from 663 to 820 kPa when friction angle of stone aggregates is changed from 37 to 46°. This is because of the fact that more frictional resistance has been mobilized to counter the

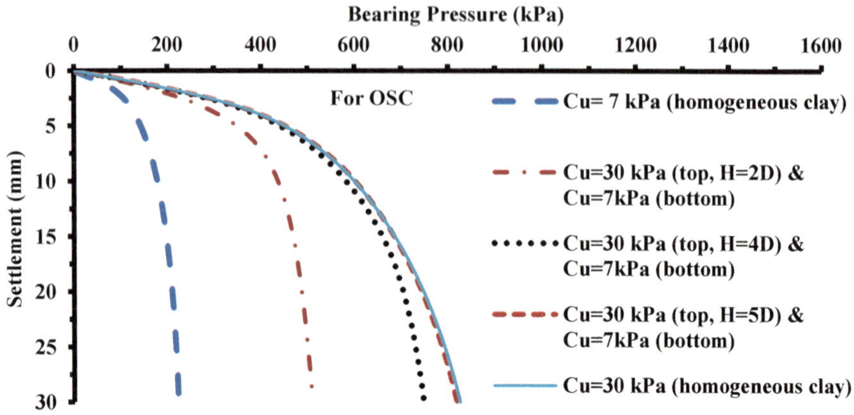

Fig. 6 Bearing pressure-settlement response of OSC in layered clay

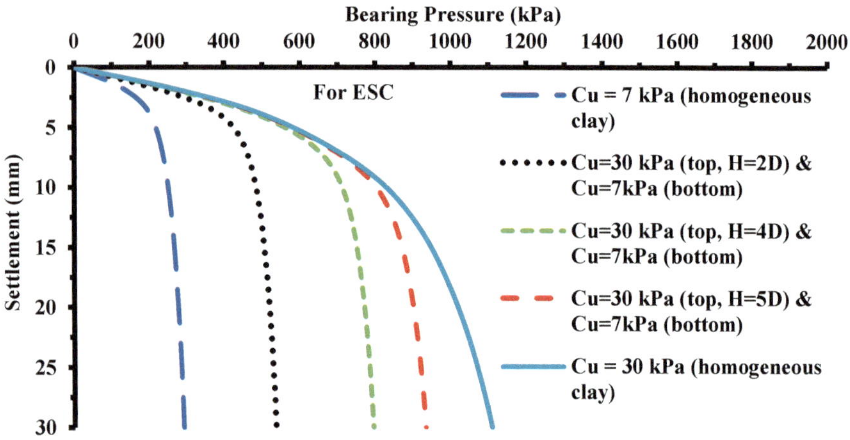

Fig. 7 Bearing pressure-settlement response of ESC in layered clay

Table 3 Rate of increment of IFR with increase in top soil layer depth (H) in OSC

Sl. No.	H/D	Bearing capacity (kN/m^2)	Improvement factor ratio (IFR)	Rate of increment (%)
1	0	225	1.00	–
2	1	362	1.61	61.0
3	2	511	2.27	41.0
4	3	655	2.91	28.0
5	4	749	3.33	14.4
6	5	825	3.66	10.0
7	10	831	3.70	01.0

Table 4 Rate of increment of IFR with increase in top soil layer depth (H) in ESC

Sl. No.	H/D	Bearing capacity (kN/m^2)	Improvement factor ratio (IFR)	Rate of increment (%)
1	0	293	1.00	–
2	1	408	1.39	39.00
3	2	538	1.84	32.4
4	3	681	2.33	26.6
5	4	796	2.72	16.7
6	5	937	3.20	10.3
7	10	1115	3.80	18.7

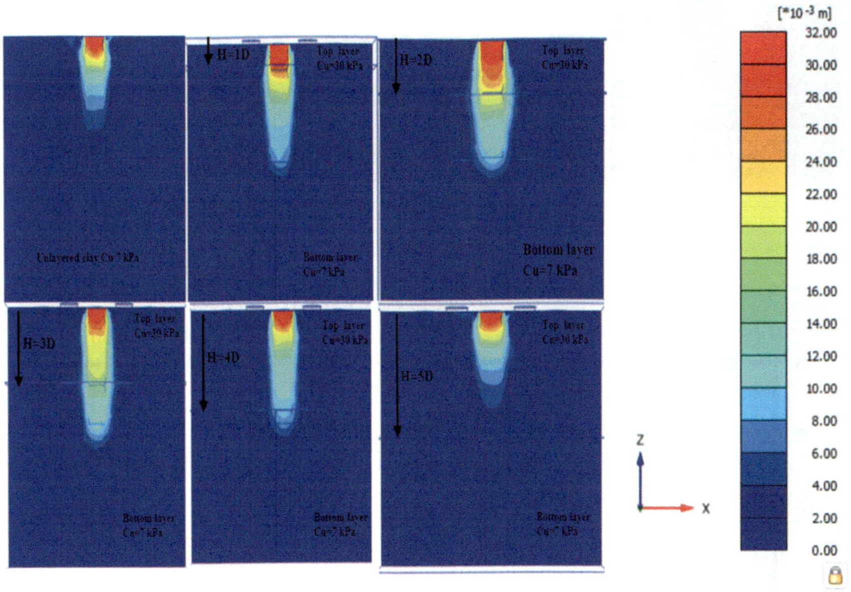

Fig. 8 Vertical displacement contours of OSC in layered clay

upcoming load as the friction angle of stone aggregates is getting increased. Similar behaviour has been observed in case of ESC, but percentage increment is less as compared to OSC case. Even for lesser value of friction angle of stone aggregates, encasement provides more lateral confinement to the column which leads to an increased performance of the composite system.

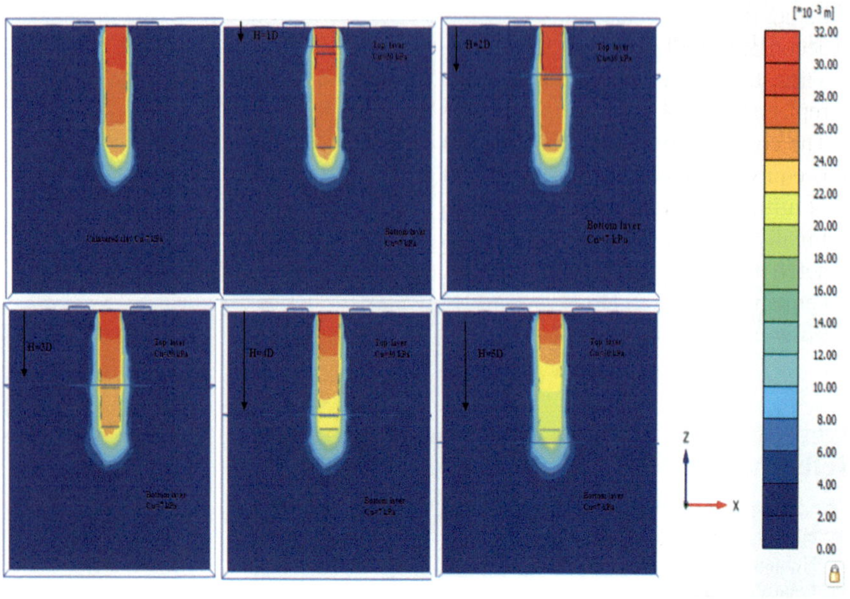

Fig. 9 Vertical displacement contours of ESC in layered clay

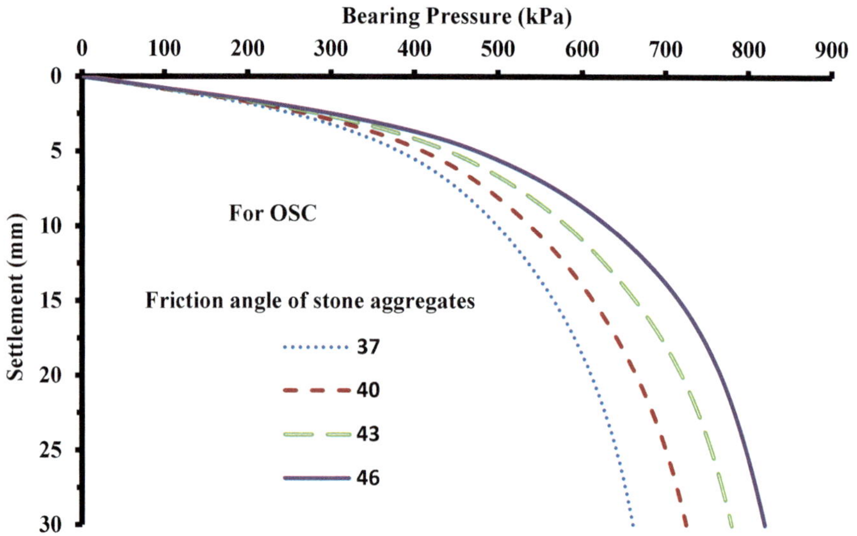

Fig. 10 Bearing pressure-settlement response of OSC for varying friction angle of stone aggregates

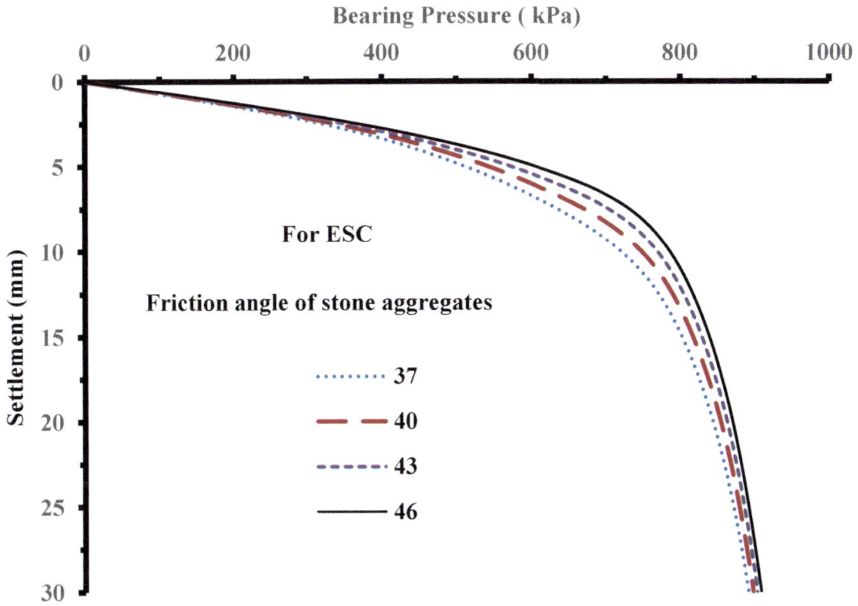

Fig. 11 Bearing pressure-settlement response of ESC for varying friction angle of stone aggregates

4 Conclusion

Numerical simulation of ordinary and encased stone columns have been done using PLAXIS^{3D} to analyse the influence of undrained shear strength of clay and friction angle of stone aggregates in a layered soil system. Based on the various results and analysis, the following conclusions have been brought out:

1. It has been observed that the bearing capacity keep on increasing in all the three cases, namely unreinforced clay, OSC reinforced clay, and ESC reinforced clay. It is because of the fact the OSC in soil provides lateral confinement to upcoming load and encasement of OSC provides higher resistance to radial bulging of the same.
2. In OSC and ESC layered clay system, improvement was observed when the stiffer clay layer was above the weak clay layer. Further, best performance was observed when soft clay is replaced by clay having highest undrained shear strength (C_u = 30 kPa here).
3. For OSC in layered clay system, the bearing capacity increases with increase in the depth of top layer of clay (having higher C_u value than bottom one). Beyond a depth of 4D, the increase in bearing capacity was observed to be negligible. Therefore, it can be said that for optimum performance, a depth equal to 4D can be replaced with stiffer clay.

4. For ESC in layered clay system, the bearing capacity increases with increase in the depth of top layer of clay. Beyond a depth of 4D, the increase in bearing capacity was observed to be significant, but the rate of increment is very less. Therefore, the recommended top layer depth in ESC reinforced layered soil is 4D.
5. In layered clay system, the effect of friction angle of stone aggregates was observed to be significant in OSC, i.e. the bearing capacity increases with increase in friction angle. However, in case of ESC, the effect of friction angle of stone aggregates on bearing capacity of system is very marginal, i.e. very little increase was observed in bearing capacity with increase in friction angle.

References

1. Hughes J, Withers N (1974) Reinforcing of soft cohesive soils with stone columns. Ground Eng 7(3)
2. Hughes JMO, Withers NJ, Greenwood DA (1976) A field trial of reinforcing effect of stone column in soil. In: Proceedings of the ground treatment by deep compaction. Institution of Civil Engineers, London, pp 32–44
3. Madhav MM, Vitkar PP (1978) Strip footing on weak clay stabilized with a granular trench or pile. Can Geotech J 15:605–609
4. Aboshi H, Ichimoto E, Harada K, Emoji M (1979) The com-poser-A method to improve the characteristics of soft clays by inclusion of large diameter sand columns. In: Pro-ceedings of international conference on soil reinforcement, Paris, pp 211–216
5. Malarvizhi SN, Ilamparuthi K (2008) Numerical analysis of encapsulated stone columns. In: The 12th international conference of international association for computer methods and advances in geomechanics (IACMAG) 1–6 October, 2008 Goa, India, pp 3719–3726
6. Deb K, Basudhar PK, Chandra S (2007) Generalized model for geosynthetic-reinforced granular fill-soft soil with stone columns. Int J Geomech 7:266–276
7. Fattah MY, Zabar BS, Hassan HA (2016) Experimental analysis of embankment on ordinary and encased stone columns. Int J Geomech 16(4):1–13
8. IS 15284-Part 1(Stone Columns) (2003) Design and construction for ground improvement-Guidelines. Bureau of Indian Standards, New Delhi
9. Shivashankar R, Babu MRD, Nayak S, Rajathkumar V (2011) Experimental studies on behaviour of stone columns in layered soils. Geotech Geol Eng 29:749–757
10. Mohanty P, Samanta M (2015) Experimental and numerical studies on response of the stone column in layered soil. Int J Geosynth Ground Eng 1:27
11. Nayak S, Balaji M, Preetham HK (2019) A study on the behaviour of stone columns in layered soil system. Transp Infrastruct Geotechnol 7:85–102
12. Ambily A, Gandhi SR (2007) Behavior of stone columns based on experimental and fem analysis. J Geotech Geo-environmental Eng 133(4):405–415
13. Debnath P, Dey AK (2017) Bearing capacity of geogrid reinforced sand over encased stone columns in soft clay. Geotext Geomembr 45:653–664

Stabilization of Red Mud Using Mineral Precursors by Geopolymerization Process

Suresh Prasad Singh, Sourav Sarkar, and Sarthak Rana

1 Introduction

Urbanization, rapid industrialization, and the development of modern infrastructure produce huge amount of waste material which requires large area for disposal and storage. In recent decades, it has been clear that the use of industrial by-products in civil engineering projects has become a major issue. As a result, its use provides a partial answer to both ecological and environmental issues. Red mud is an industrial waste leftover that results from Bayer's bauxite alumina refining process. It is a type of solid waste that is naturally alkaline. On an average, 1.5 tonnes of RM are created from 1 ton of alumina, however due to its chemical composition, high alkalinity, and heavy metals content, its use in the real world is severely limited. Furthermore, it needs a considerable amount of space for storage and disposal, posing a serious environmental danger in terms of pollution of the surrounding land, air, and water, as well as human health. As a result, thorough treatment and research into the usage of this resource are required. The recovery of valuable elements presents in red mud, such as Fe, Ca, Al, Ti, V, Cr, Ni, Na, K, Mn, Zn, and others, has also been explored as part of the treatment processes. RM is also used as an adsorbent to absorb pollutants from the air and water, as well as to enhance fertility of soil. RM is high in Si, Al, Fe, and Ca, which can be utilized to make a geopolymer that can be used to make cement, pottery, concrete, and bricks. Various studies on the use of RM in construction industries have been carried out. According to additional research, red

S. P. Singh (✉) · S. Sarkar · S. Rana
Department of Civil Engineering, National Institute of Technology Rourkela, Rourkela, India
e-mail: spsingh@nitrkl.ac.in

S. Sarkar
e-mail: 220ce1453@nitrkl.ac.in

S. Rana
e-mail: 221ce1183@nitrkl.ac.in

B. T. Jose et al. (eds.), *Proceedings of the Indian Geotechnical Conference 2022 Volume 4*, Lecture Notes in Civil Engineering 479,
https://doi.org/10.1007/978-981-97-1753-8_18

mud can be used as a base material with higher strength by mixing it with various proportions of other industrial wastes rich in alumina and silica to create a new product using the geopolymerization method, which has significant economic and environmental benefits. Despite the fact that RM contains some toxic minerals that harm the environment, neutralization and de-alkalization technology is presently used. RM is mixed with rice husk ash which gives an advantage of using in subgrade material with good strength properties [1]. Also, synthesis and characterization of red mud and rice husk ash-based geopolymer composites are used in various construction materials is performed in various studies [10]. Further, studies state that the strength development of solely ground granulated blast furnace slag geopolymers creates a huge impact on industrial site [3].

A geopolymer is a novel type of green gelation material created by the chemical interaction of natural minerals silica and alumina in the presence of an alkaline activator. At room temperature, the Si–O and Al–O bonds in the raw resources are broken and then rearranged by the alkaline catalysts [2]. Three processes are involved in the geopolymerization reaction mechanisms that are dissolution-recombination and solidification. The SiO_4 and AlO_4 from the first step the alumino-silicate materials are immersed in alkaline solution, then monomers are dehydrated and lead to alumino-silicate oligomers, resulting in the formation of a gel. Alumino-silicate polymeric gel is created by draining excess water from the gel phase. The polymerization of a geopolymer is a difficult process that is influenced by a variety of parameters such as raw material reactivity, Si/Al ratio, alkaline solution concentration, and curing conditions. The mechanical and microstructural properties of geopolymer are affected by this component. RM can be used with other highly active materials such as FA, GBFS, Metakaolin, RHA, and others for improved results. Geopolymer has the advantage of not requiring water for cure. Curing also takes place at normal temperature or below it. In the synthesis and characterization of geopolymer, the curing duration is critical. Further, the microstructure, mechanical, and chemical studies define the factor affecting red mud based geopolymer [11].

2 Materials and Methodology

2.1 Materials

The experimental study includes stabilizing red mud with mineral precursors such as GBFS and rice husk ash (RHA), and assessing the geopolymer-based product formed in terms of physical, chemical, mineralogical, and morphological aspects, as well as comparing it to untreated red mud. The second section compares the behavior of experimental ways to stabilize red mud utilizing various ratios of RM–GBFS and RM–RHA mixes, as well as the addition of alkali activator at various quantities. Red mud (RM) was brought from Vedanta Aluminum Limited, Lanjigarh, Odisha. The red mud was air dried, homogenized, and crushed to powder to increase geopolymerization reaction and reduce compositional error in the final product. The granulated blast furnace slag (GBFS) was obtained from the Rourkela Steel Unit's (RSP) slag granulation plant in Sundargarh, India. It is produced during the pig iron manufacturing process. Silicates, alumino-silicates, and calcium are the primary components. The waste products were dried in the oven to remove the moisture in the raw material. RHA was obtained from Haripriya Agro Industry in M. Rampur, Kalahandi, India. It's also an industrial by-product generated by burning of rice husk to produce electricity, which is a renewable biomass energy source. Because RHA is composed of porous particles, it has less unit weight and a large external surface area. Growing environmental concerns and a demand for long-term biomass energy have led to a practical and cost-effective solution.

2.2 Methodology

In order to synthesis geopolymer, the red mud is mixed with mineral precursor like GBFS, rice husk ash, fly ash, etc. In this research, RM is mixed with GBFS and RHA with varying fractions ranging from 0 to 20% of the dry mass of the mixture at 5% increments. These mixes are termed as RM, RM-S1, RM-S2, RM-S3, RM-S4 and RM-R1, RM-R2, RM-R3, RM-R4 for GBFS and RHA, respectively. Also desired amount of NaOH solution is added in quantity of 1 M, 2 M, and 4 M with these mixtures to form the geopolymer. The unconfined compression strength (UCS) tests were conducted on specimens compacted to their respective maximum dry density (MDD) at optimum moisture content (OMC) and left for 0, 7, 14, and 28 days curing at constant temp of 27 ± 2 °C. The physical, chemical, morphological, and mineralogical properties of cured specimens were evaluated. This includes the grain size distribution (GSD), specific gravity (SG), liquid limit (LL), plastic limit (PL), dry density (Y_d), optimum moisture content (OMC), and unconfined compressive strength (UCS). All these tests were conducted as per the provisions of Indian Standards code of practice. The pH values are determined using an electronic pH meter.

3 Results and Discussion

3.1 Index and Engineering Properties

The specific gravity of the RM, GGBS, RHA and RM–GGBS, RM–RHA mix are studied according to IS-2720 part-3 (1980) [5]. The GGBS has the specific gravity ($\boldsymbol{G_s}$) value of 2.84, which is higher than conventional earth material. However, the specific gravity of RM is found to be 3.24. This is found to be higher than that of GGBS (2.84) and RHA (2.16) due to the presence of higher percentage of iron. The

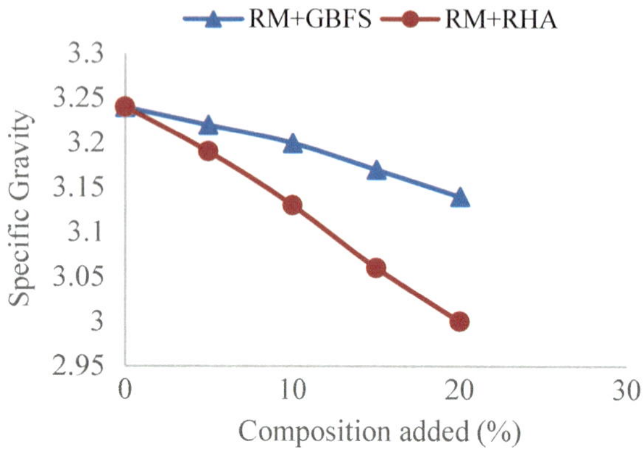

Fig. 1 Specific gravity of RM–GBFS/RHA mixes

specific gravity of RM mix with 5, 10, 15, and 20% GGBS or RHA is found to decrease with increase in GGBS and RHA as shown in Fig. 1.

The PH of red mud and RM mixed with different fractions of GBFS and RHA is presented in Fig. 2. It has been seen that after adding GBFS and RHA, the pH of the mixture is reduced because of the low pH value of GBFS and RHA.

The plasticity properties of red mud show that it has a low plasticity, with a liquid limit of 29% (< 50%) and plastic limit of 25% and plasticity Index was found to be 4%. However, slag and RHA are found to be non-plastic material. The mixture of RM, GBFS, and RHA with different fraction and its LL and PL are listed, and the variation with GBFS and RHA is shown in Fig. 3a, b, respectively. These tests were conducted as per IS-2720 part-5 (1985) [7].

The grain size distribution for RM, RHA, and GBFS is made as per IS-2720 part-4 (1985) [6]. The gradation curve of these materials is presented in Fig. 4. From the grain size distribution and plasticity properties, RM is categorized as low compressible silt (ML) as per IS-1498 (1970) [8].

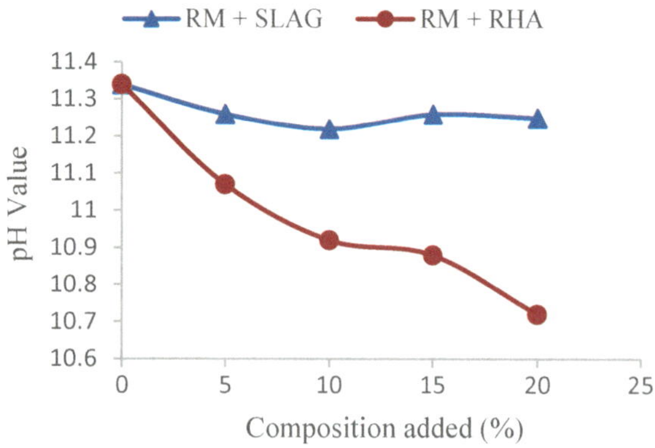

Fig. 2 pH value of RM–GBFS/RHA mixes

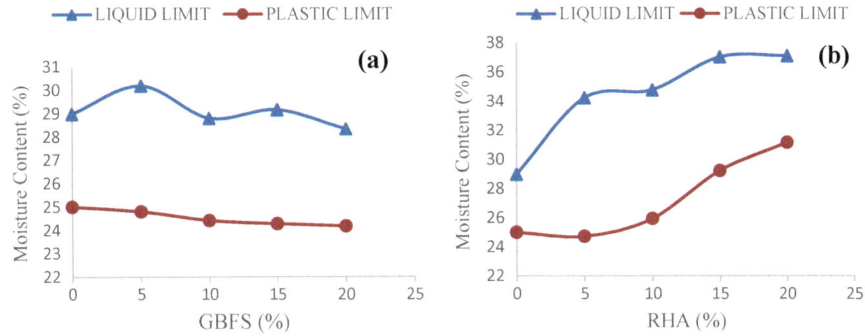

Fig. 3 LL and PL of RM in addition with **a** GBFS and **b** RHA

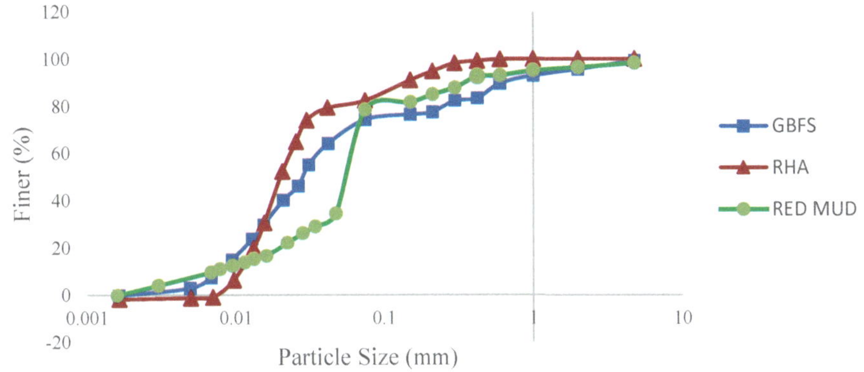

Fig. 4 Grain size distribution curve of RM, GBFS, and RHA

The differential free swell (DFS) value of raw materials that is RM, GBFS, and RHA is presented in Table 1. It is observed that DFS values of these materials are negative in nature. Further, the DFS values of RM–GBFS and RM–RHA mixes are also negative. Materials like fly ash, rock powder, and pure kaolinite minerals occupy a higher sediment volume in non-polar solvents like kerosene than in water and thus observed to have negative DFS values. This shows that there is no inter-particle attraction in the substances and that they are dispersive in nature. Figure 5a, b shows the values and change in DFS of RM–GBFS and RM–RHA mixes, respectively.

Compaction tests were conducted on RM–GBFS and RM–RHA mixes using compactive efforts of 595 kJ/m^3 and 2674 kJ/m^3 as per IS-2720 part-7 (1980) [9] and the compaction curves were obtained. For each combination of RM–GBFS and RM–RHA, the OMC and MDD values were determined from the compaction curves. Figure 6 shows how MDD and OMC values change with GBFS content and RHA content. The MDD increases and the OMC falls in RM–GBFS mixtures as the proportion of slag added to red mud increases. Addition of GBFS results in change in gradation of the mix and reduces the plasticity values. This causes a well packing

Table 1 Index and engineering properties of raw materials

Physical properties	Red mud	GBFS	RHA
Color	Red	Gray	Black
Specific gravity	3.24	2.84	2.16
pH	10.43	10.24	7.12
Liquid limit (%)	29	Non-plastic	Non-plastic
Plastic limit (%)	25		
Coefficient of uniformity (Cu)	11.88	1	–
Coefficient of curvature (Cc)	0.345	6	–
Optimum moisture content (OMC)	27.3%	19%	–
Maximum dry density (MDD)	1.65 g/cc	16.9 g/cc	–
DFS value (%)			

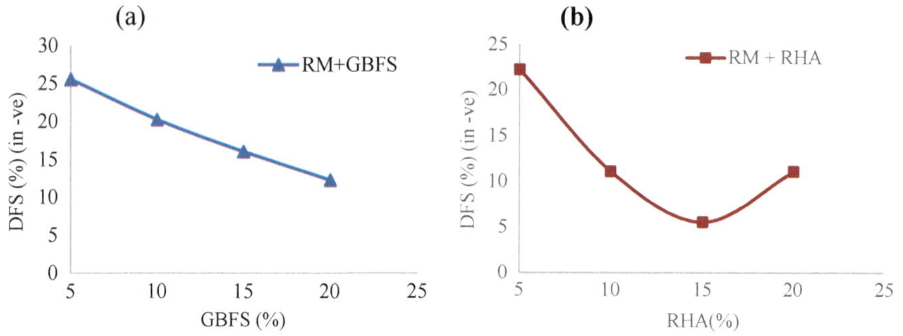

Fig. 5 DFS values of **a** RM–GBFS and **b** RM–RHA mixes

of particles during dynamic loading, thus increasing the dry density. The decrease in OMC could be related to the loss of adsorbed water when red mud is replaced by slag. As the amount of RHA added to RM is raised, the MDD drops and the moisture content increases in the red mud RHA combination. The rise in OMC with RHA is due to the inclusion of RHA, which reduces the amount silt sized fraction and increases clay and colloidal sized particles with higher specific surface areas, requiring more water to lubricate the particle surface. This also implies that additional water would be required for greater compaction of the soil and RHA mixture. Figures 7 and 8 show the moisture content and dry density relationship of RM–GBFS and RM–RHA mixture.

Unconfined compressive strength (UCS) tests were carried out on RM–RHA and RM–GBFS mixes compacted to their respective MDD at OMC. These tests were conducted as per IS-2720 part-10 (1991) [4] code provisions with a loading rate of 1.25 mm/min. UCS test specimens were wax coated and cured at an average temperature of 30 °C for 0, 7, 14, and 28 days. For each test conditions, three identical specimens were tested and the average value is reported in Fig. 9. Again, an extra

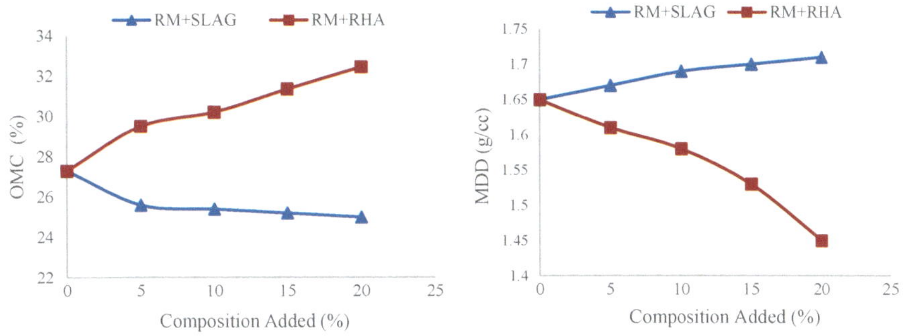

Fig. 6 Variation in OMC and MDD with addition of GBFS and RHA

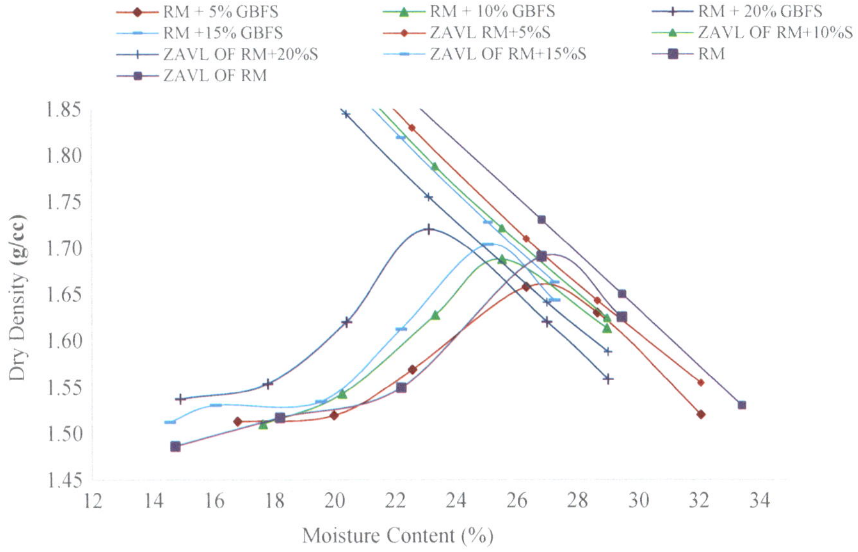

Fig. 7 Compaction curves of RM–GBFS mixes

additional dose of 1 M, 2 M, and 4 M NaOH solution was added to the RM–RHA and RM–GBFS mixes to accelerate the geopolymerization reaction. The UCS of these specimens was also determined after curing periods of 0, 7, 14, and 28 days. These results are presented in Figs. 10, 11, and 12, respectively.

Figure 13a–c shows the XRD patterns of the RM, RHA, and GBFS, respectively, and Fig. 14d, e shows cured RM–GBFS and RM–RHA geopolymers. The presence of hematite and calcite in RM is detected with sharp peaks. The lack of apparent broad humps in the RM pattern implies that the amorphous phases are either absent or present in small amounts. For geopolymerization, the RM offers primarily NaOH and Al in the form of amorphous Al_2O_3 or dissolved $NaAlO_2$, but little Silica. In Fig. 13b as seen in XRD plot of RHA, at 10–35°, the RHA pattern has one sharp

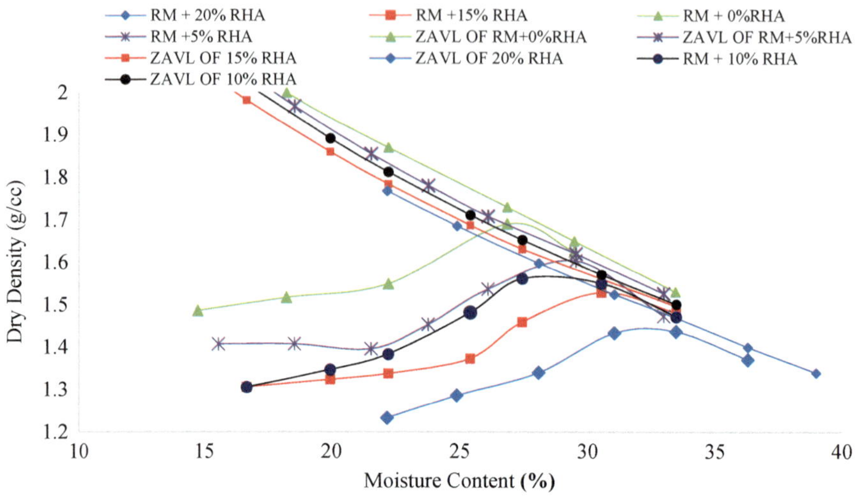

Fig. 8 Compaction curves of RM–RHA mixes

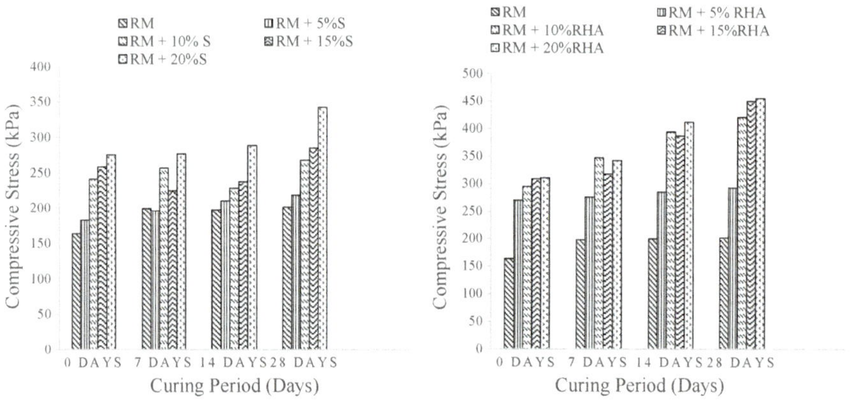

Fig. 9 UCS of RM–GBFS and RM–RHA mixes at different curing periods

peak overlaid on a broad hump. Cristobalite, a high-temperature silica polymer, is responsible for the steep peak. Amorphous silica, a significant component of RHA, is responsible for the very broad hump. The RHA contains mostly amorphous silica with cristobalite and trace crystalline quartz. According to its chemical makeup, the XRD pattern of GBFS powder is shown in Fig. 13c. It has a board hump that ranges from 30 to 40°. The presence of an amorphous (glassy) phase accounts for this hump. The XRD pattern of the GBFS powder matches that of prior studies conducted by other authors. A quantitative XRD investigation previously determined that 97.7% of this GBFS is amorphous in nature, with minor crystalline content, with one sharp

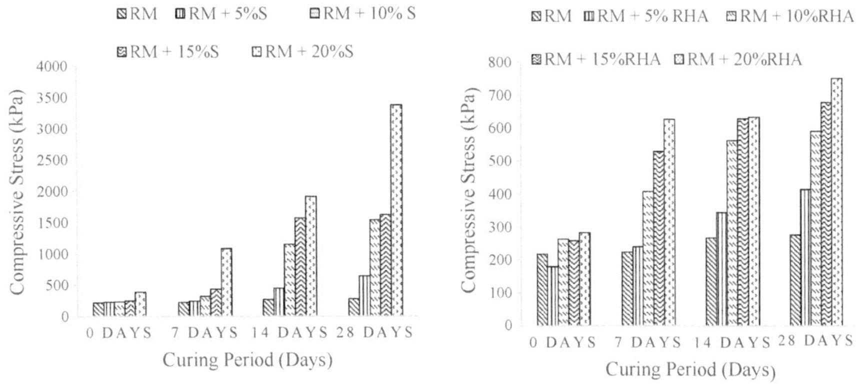

Fig. 10 UCS of RM–GBFS/RHA mixes with an addition of 1 M NaOH solution at different curing period

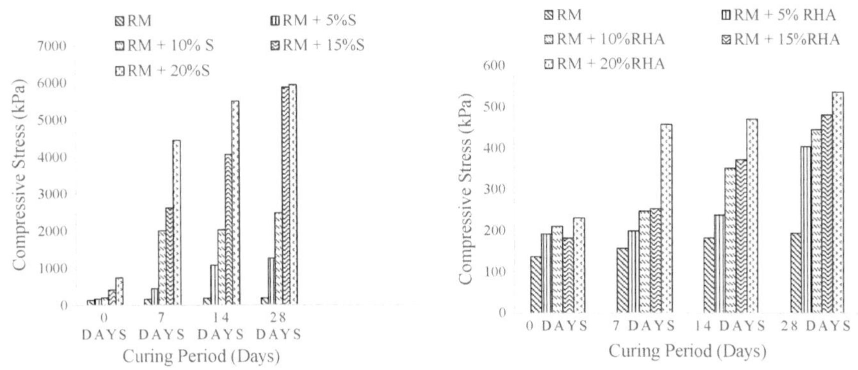

Fig. 11 Variation in UCS of RM with GBFS and RHA and an addition of 2 M NaOH with curing period

peak superimposed over a broad hump at 10–35° diffraction angles. Cristobalite, a high-temperature silica polymer, is responsible for the steep peak.

Table 2 shows the real concentration, which is based on the average concentration of three specimens. The dangerous heavy metals designated by the EPA for toxicity characteristics (As, Cr, Pb, Ni, and Hg) are high in the red mud within tolerable values. It has been observed that stabilizing red mud with activated GBFS can stop Hg from leaching. Other heavy metal leaching, on the other hand, increases after stabilization but remains within permitted limits. According to the study report, the chromium in the red mud is present as Cr^{3+}, which is not a health danger. The silent increase in pH caused by the use of NaOH as an activator may be to blame for the increased leaching. Although the heavy metal concentration is within the permitted thresholds for toxicity characteristics, it is beyond the WHO's recommended drinking water

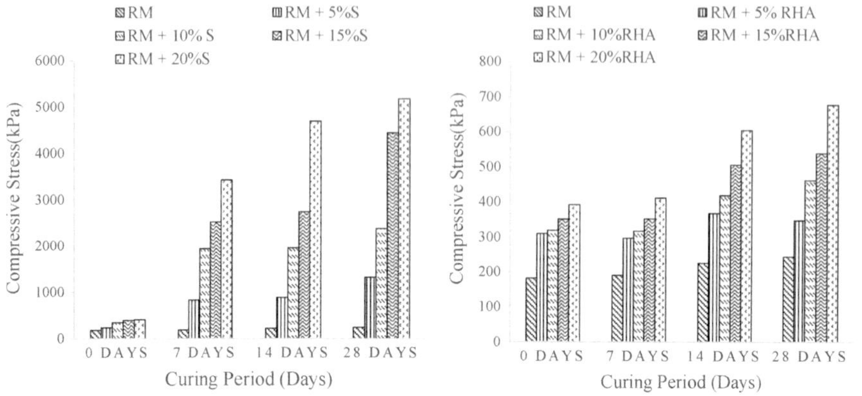

Fig. 12 Variation in UCS of RM with GBFS and RHA and an addition of 4 M NaOH with curing period

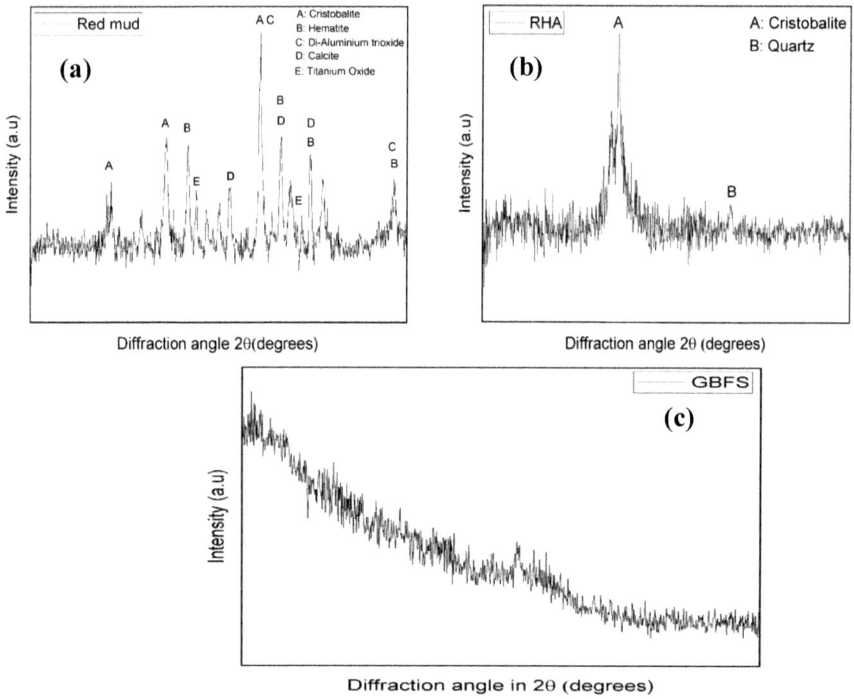

Fig. 13 XRD micrograph of **a** red mud, **b** RHA, and **c** GBFS

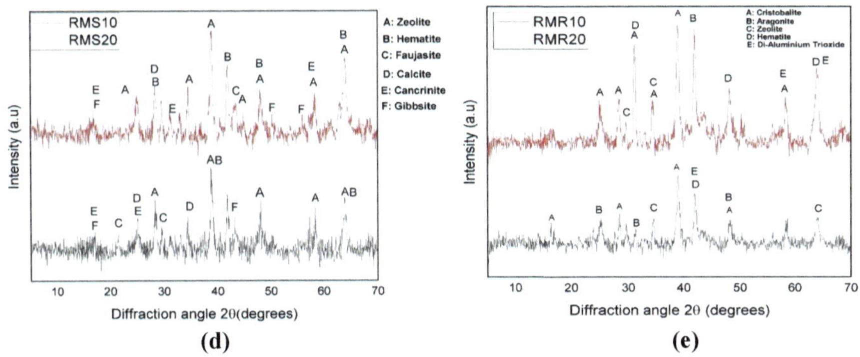

Fig. 14 XRD micrographs of **d** GBFS stabilized RM and **e** RHA stabilized RM

Table 2 Heavy metal concentrations in the leachate (ppm)

		RM		GBFS		RHA
	Mean	SD	Mean	SD	Mean	SD
Na	2	0.388	1.256	0.0159	1.309	0.0166
K	6.573	0.1015	55.19	0.148	255.1	1.39
Ca	0.013	0.0048	27.37	0.088	0.148	0.0135
Fe	2.686	0.0284	0.207	0.087	0.142	0.045

standards. As a result, the water quality of the surrounding drinking water source must be investigated anytime this material is utilized as a geotechnical material.

4 Conclusions

This work uses geopolymerization to convert three industrial wastes, red mud (RM), granulated blast furnace slag (GBFS), and rice husk ash (RHA) into valuable construction materials, resulting in a new type of RM–GBFS and RM–RHA geopolymer composites. NaOH was the sole non-waste substance employed in this study. The degree of geopolymerization and its influence on mechanical, chemical, and microstructural properties were determined using a variety of synthesis parameters, including GBFS/RM and RHA/RM ratios, alkalinity condition, and curing period. The following conclusions can be taken from the experimental findings.

- From result it has seen that there is a change in liquid limit and plastic limit of RM upon addition of RHA or GBFS. Graph shows that by adding GBFS there is a slight decrease in LL and PL in case of GBFS and an increase in LL and PL in case of RHA.

- The OMC decreases and MDD increases by adding GBFS to RM however by adding RHA it was found that OMC increases and MDD decreases.
- RM, GBFS, RHA, and their mixes exhibit negative differential free swell indicating that they are dispersive in nature. The dispersive percentage of RM is found to be 54.5% by double hydrometer method and also by crumb test it was found that after adding GBFS and RHA the depressiveness reduces.
- Addition of GBFS and RHA in RM shows an increase in UCS up to 20% by the quantity of RM–GBFS and RM–RHA mixture. Maximum UCS value is recorded in the mix composition containing 20% GBFS or RHA at curing period of 28 days. Additional doses of alkali showed a tremendous improvement in UCS value up to 4 M NaOH concentration.
- At high pressure, more amorphous silicate gel polymers and zeolite structures are formed, according to XRD data. This data shows that the geopolymer reaction is more powerful. The geopolymers have good mechanical characteristics as their porosity decreases and their structure densifies.
- The toxic heavy metals (As, Cr, Pb, Ni, Zn, and Hg) as identified by EPA for toxicity characteristics was found to be increased after stabilization but within permissible limit.

References

1. Deewal et al (2018) Influence of rice husk ash source variability on road subgrade properties. Niger J Technol (NIJOTECH)
2. Duxson P, Fernandez-Jimenez A, Provis JL, Lukey GC, Palomo A, Van Deventer JSJ (2007) Geopolymer technology: the current state of the art. J Mater Sci 42(9):2917–2933
3. Aziz IHA, Abdullah MMAB, Salleh MAAM, Chaiprapa J, Azimi EA, Sandu AV (2020) Strength development of solely ground granulated blast furnace slag Geopolymers. Constr Build Mater. 250, 118720
4. IS 2720 part 10 (1991) Methods of test for soils, Part 10: Determination of unconfined compressive strength. Bureau of Indian Standards, New Delhi
5. IS: 2720 (part 3/sec 1) (1980) Determination of specific gravity. Bureau of Indian Standards, New Delhi
6. IS: 2720 (part 4) (1985) Grain size distribution of soil. Bureau of Indian Standards, New Delhi
7. IS: 2720 (part 5) (1985) Determination of liquid limit, plastic limit and shrinkage limit. Bureau of Indian Standards, New Delhi
8. IS: 1498 (1970) Classification of soil. Bureau of Indian Standards, New Delhi
9. IS: 2720 (part 7) (1980) Determination of water content-dry density relation using Light compaction. Bureau of Indian Standards, New Delhi
10. He J, Jie J, Zhang J, Yu Y, Zhang G (2013) Synthesis and characterization of red mud and rice husk ash based geopolymer composites. Cem Concr Compos.https://doi.org/10.1016/j.cemconcomp.2012.11.010
11. Manjusha M, Yashida N (2021) Factors affecting the mechanical properties and microstructure of geopolymers from red mud and granite waste powder. Ceram Int. 10.1016

Assessment of Desiccation Cracks in Soil Fused with Biochar and Dolochar Using Digital Image Processing

Prasanna Kumar Behera, G. Suneel Kumar, Vamsi Alla, Arnab Ghosh, and Rabi Narayan Behera

1 Introduction

Soil cracking is a complicated phenomenon that is impacted by several characteristics such as soil mineralogy, temperature fluctuations, and soil layer thickness. Soil cracking patterns are influenced by the alternating soaking and drying of soils, as well as the physiochemical qualities present in soil. Prior research has shown that soil and crack geomorphology can have a considerable impact on the intrinsic qualities of the soil. Desiccation fissures developed on the soil's top surface can affect the soil's characteristics linked to geotechnical, hydraulic, and environmental applications [11]. Changes in moisture content, stiffness, and tensile strength of the soil can cause cracking patterns to form. Cracks form on the soil surface because of the drying process caused by the evaporation of water from the exposed soil surface, which is mostly caused by volumetric shrinkage of the soil. When the threshold condition is achieved, desiccation fractures appear on the soil's surface. The tensile strains created during water evaporation are also connected to crack growth. When the tensile stresses created surpass the soil's natural tensile strength, fractures can grow and degrade the soil [14].

Soil cracking originates in weaker soil zones and can cause stability concerns in dams, tailing ponds, earthen embankments, landfills, liners, and other structures. The increased hydraulic conductivity caused by crack development can substantially impair the effectiveness of contamination barriers. Moisture infiltration caused by cracking has a significant impact on the stability of natural slopes and vertical cuts, as well as the carrying capacity of the foundation. Crack propagation may be efficiently prevented by enhancing soil strength and treating the soil with appropriate additions such as lime, cement, and sand. Izzo & Miletic [7] investigated the effect

P. K. Behera · G. Suneel Kumar · V. Alla · A. Ghosh · R. N. Behera (✉)
National Institute of Technology Rourkela, Rourkela 769008, India
e-mail: beherarabin@nitrkl.ac.in

B. T. Jose et al. (eds.), *Proceedings of the Indian Geotechnical Conference 2022 Volume 4*, Lecture Notes in Civil Engineering 479,
https://doi.org/10.1007/978-981-97-1753-8_19

of recycled carpet fibers and a gelatin-based bioplastic on desiccation cracking in cohesive soils. Kalkan [8] examined the suitability of silica fume as a stabilization material to reduce the development of desiccation cracks in compacted clayey liner and cover systems. Water evaporation from an exposed soil surface generally leads to volumetric shrinkage of the soil [9]. Desiccation fissures form in the soil and appear on the soil surface after the threshold condition is achieved. Some of the criteria that aid in understanding crack connectivity include the morphological properties of the soil-crack pattern, such as crack area density, crack length, and crack width. The assessment and study of these factors can aid in understanding the variance in crack pattern propagation and characterization of soil cracking behavior, as well as its effect on geotechnical and environmental applications. Imaging methods may be used to efficiently study fracture forms and propagation. Furthermore, numerical analysis can be performed to investigate the fracture pattern in soil.

2 Materials and Methodology

2.1 Materials

There are two types of materials taken for analysis, such as biochar and dolochar along with the locally available soft soil. The soil used for the purpose of testing was taken from the shores of the nearby river by using means of local transportation and was stored carefully. Before conducting various tests, the soil lumps were crushed and required soil size is obtained using appropriate sieves, then it was oven dried for 24 hours, and finally, it was used for various experiments.

Biochar. The biochar collected is a compound which is rich in carbon content and produced by the pyrolysis of biomass at a relatively elevated temperature under limited availability of oxygen. Pyrolysis is a typical process for manufacturing biochar in an oxygen-controlled environment, such as turning biomass into oil or to produce hydrogen, which enhances the economic effectiveness of extracting and transportation of the biomass to the facilities where it will be converted into various forms. "Slow pyrolysis" is a process involving the breakdown of biomass by heating over a considerable length of time (generally 3–5 days) to create enormous volumes of gas biochar, and oil. To prepare the biochar, air-dried wheat straw was heated at a rate of 7 °C/min in a heating furnace until the temperature of carbonization grew up to 400 °C. Then, it was cleaned and sieved through appropriate sieve sizes before conducting all the experiments. Initially, biochar was characterized to determine its index properties. Before adding it to the soil, it was kept in the oven for at least 24 hours, and then it was taken out and cooled outside for half an hour before starting any kinds of experiments.

Dolochar. The dolochar samples were broken down into pieces having very small dimensions and then sieved by passing them through a 4.75 mm IS sieve. Dolochar is a sponge iron by-product with no clay component. The dolochar is non-expansive and non-plastic in nature. Because dolochar is a by-product of sponge iron, its specific gravity is higher than Expansive soil. The collected samples were oven dried for 24 hours initially before commencement of any experiments. To access the capability of the taken additives for reducing the desiccation cracks, three different proportions were chosen for biochar and dolochar individually. For biochar, 2.5, 5, and 10% compositions were taken for preparation of the samples, whereas for dolochar, 5, 10, and 20% were taken as the appropriate proportions. As the soil taken was of intermediate compressibility, bentonite was also added along with biochar and dolochar separately to visualize its effects on enhancing the cracking resistance of soil.

Bentonite. The bentonite used in this study was sodium (Na^+) bentonite. It was stored in an oven for 24 hours for complete removal of any sort of moisture from it, and then it was used as a binding material in this specific study.

Sample preparation. The soil taken was of medium compressibility with IS classification of MI (intermediate compressible silty soil). Hence, to convert the soil into a more expansive form to facilitate the cracks and to see the effects of the additives such as biochar and dolochar, bentonite was also added to the soil. Then finally, biochar and dolochar were added to the bentonite-amended soil, so that the samples were prepared with the mixtures of three types of materials, i.e., soil, bentonite, and biochar/dolochar. Hence, a total of eight samples were prepared for all the tests that are conducted, i.e., Base soil (S), Soil + 10% Bentonite (S10B), Soil + 10% Bentonite + 2.5% biochar (S10B2.5BI), Soil + 10% Bentonite + 5% biochar (S10B5BI), Soil + 10% Bentonite + 10% biochar (S10B10BI), Soil + 10% Bentonite + 5% dolochar (S10B5D), Soil + 10% Bentonite + 10% dolochar (S10B10D), Soil + 10% Bentonite + 20% dolochar (S10B20D).

2.2 Methodology

Primary Tests. Desiccation tests were conducted as the primary tests. The bentonite-amended samples after adding the biochar and dolochar were mixed thoroughly and prepared for desiccation cracking tests. For accessing the crack resistance of biochar and dolochar, the soil samples were prepared with adding biochar, dolochar separately along with addition of bentonite as well.

Initially, the soil was passed through a 425-μ IS sieve and oven dried for 24 hours at 105 °C. All the requisite additives were also prepared in similar manner. Then, the soil and additives mixture were placed in between specially prepared rubber containers, filled thoroughly with the help of spatula and given pressure to prepare the samples having same thickness for all the samples. For desiccation cracking, two factors are predominant, i.e., the thickness of soil sample and the amount of water added. Depending on the thickness of the sample and the amount of water added to the soil, the time required for the desiccation cracks to appear varies between a few hours to a few days to months, even years. In order to facilitate the development of cracks in the soil sample, the thickness of the samples was kept very less so that cracks can be generated within the due period. The amount of water should also be taken into consideration since it is another predominant factor for desiccation cracking. In our case, the samples were prepared at the liquid limit so that a uniform thickness of the soil sample could be generated, thereby making the top surface of the soil smooth which would be convenient while image acquisition and processing would be easier to conduct.

After preparation of the samples, they were kept in laboratory condition for sufficient period for the cracks to develop on the surface of the soil samples. Generally, a period of 24–72 hours is sufficient for the cracks to develop, depending on the amount of water added and the thickness of the samples.

Secondary Tests. A certain set of tests were conducted with the bentonite-amended soil added with biochar and dolochar to determine the effect of those additives on engineering properties of soil. The grain size distribution of the soil and the additives were determined by both dry sieving and hydrometer analysis according to IS: 2720 (Part-4): 1985 [6]. The specific gravity of the soil as well as the additives were determined in accordance with IS: 2720 (Part-3): 1980 [2]. The Atterberg limits of the soil-bentonite mixture along with the prepared samples were determined according to IS: 2720 (Part 5): 1985 [5].

Compaction Characteristics. The standard compaction test was performed on the sample in accordance with IS code: 2720 (part 7): 1980 [3] for light compaction.

Free Swell Index. The free swell index (FSI) was performed to determine the potential of a sample to swell, i.e., swelling index of the sample as per IS code: 2720 (Part XL): 1977 [4]. Two oven dried specimens of 50 g each passing through a 425 μ IS sieve were taken and poured into a 100 ml capacity glass cylinder filled with distilled water and another with kerosene oil up to the 100 ml mark. The entrapped air was removed by shaking or stirring with a glass rod, and the sample was allowed to suspend for at least 24 hours. The reading of the final volume of sample in each cylinder was noted, and free swell index was calculated.

Unconfined Compressive Strength. UCS of a specimen is defined as the strength corresponding to either the failure stage or the maximal vertical strain (ε) equal to 20% of the original height, whichever arrives first. The specimens were sheared at a strain rate of 1.2 mm/min in this investigation. IS code: 2720 (part 10): 1991 [1] was used to conduct the UCS test. A typical compression testing equipment was used to perform the UCS test on samples. The tested specimens had a height of 72 mm and a diameter of 36 mm.

Image Acquisition. The images of the top surfaces of the cracks were taken using a digital image capturing device. The captured images are taken with a pixel size of 64 mega pixels, i.e., 9248 $\times$ 6936 number of pixels were present in the image. Since the height of capturing the images influences the size of the pixels, the images were taken at three constant heights, i.e., 15, 21.5, and 23.5 cm. Also to check the gradual

Fig. 1 Images of the samples taken for analysis

development of the cracks, the pictures were taken at a regular interval, i.e., 3, 6, and then finally at 24 hours, since there were eight number of samples present, which are S, S10B, S10B2.5BI, S10B5BI, S10B5D, S10B10D, and S10B20D. Hence, a total of 81 number of pictures were taken for analysis. Images taken at a height of 15 cm and after 3 h from the time of preparation of the samples are shown in Fig. 1.

Image Processing and Alterations. After taking the pictures, they were transferred into a computer for processing. Peters & Ranson [10] used digital imaging techniques to calculate the stresses with in the boundary of an object. Sutton et al. [12] used an improved digital correlation method to obtaine the full-field in-plane deformations of an object. Digital image correlation is an image processing technique used to visualize geotechnical failure processes [13]. Initially, the background of the images was removed, and a circular portion was cut in the middle of the samples. It was done to remove the effect of the boundaries. Since when the soil gets dried up, some of the soil layers get attached to the boundaries of the container in which they were kept and by which there is a presence of a greater number of cracks in the peripheries of the soil samples as compared to the central positions. The pictures captured at different heights had different number of pixels present in them, which is why it was also necessary to keep the area taken into consideration constant for all the pictures taken at a constant height. Hence, three types of pre-sets were prepared, for pictures taken at 15 cm, the area taken was having greater number of pixels as compared to the pictures taken at a height of 23.5 cm. Hence to keep the uniformity and integrity of the samples intact, different pre-sets were used for pictures taken at different

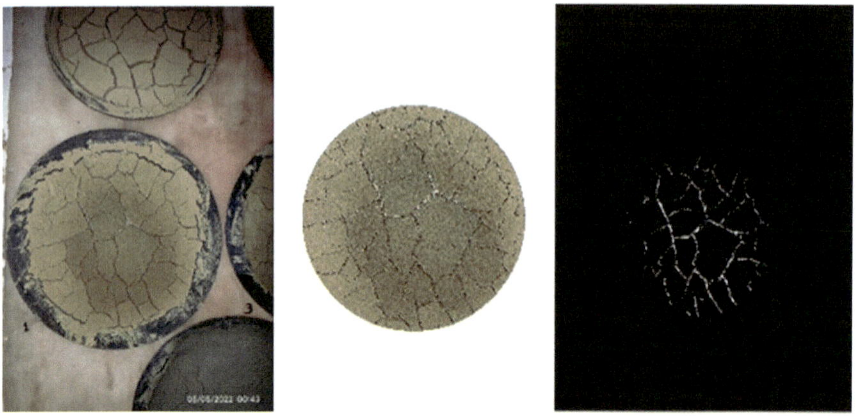

Fig. 2 Background removal and quantification of cracks using MATLAB digital image processing

heights. After separating the background from the image, this image was processed using "MATLAB-2020."

Data Extraction The images are analyzed using different commands such as imread, uigetfile, and rgb2gray via MATLAB software. Later, the image was converted into grayscale from the original RGB format which contains a series of colors and data, which interacts with the cracks while processing and might give erroneous results. By turning the image into grayscale, the cracks and rest of the soil were distinctly visible. A visual representation of the changes in the top surface of the soil after altering it through MATLAB to identify the patterns of the cracks in soil is shown in Fig. 2.

Masking with a Threshold. The noise present in the image was reduced using the masking feature present in MATLAB. For this, a low threshold and high threshold was defined, and the function "imfill" was used to create the mask. Then, the function "bwareafilt" was used to further enhance the mask using the lower and higher threshold values.

Data Interpretation. The area and width of the cracks are calculated from the converted pixelated image. In the image, only white colored skeletal lines are present which are easy to calculate areas for, which are later analyzed for the cracking characteristics of different specimens.

To quantify the crack resisting characteristics of the additives, certain parameters were defined. The first parameter was "total area" of the cracks which is basically the addition of all the areas of the cracks generated in the specimen. However, the total area can't give any information regarding the number of cracks generated in the specimen, which is why another parameter was taken into consideration, known as "average area". This new parameter was able to quantify the number of cracks as well. But we couldn't compare specimens having different top surface area with the help of either of those parameters, as pictures taken from different heights cannot

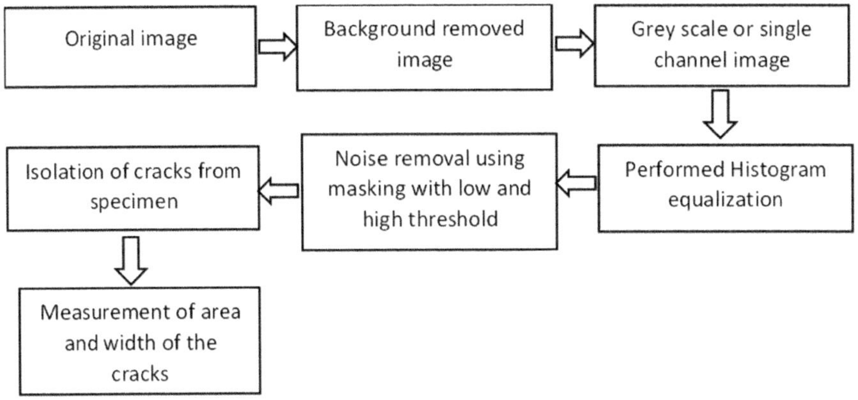

Fig. 3 Schematic diagram of quantification of cracks generated using MATLAB digital image processing

be compared extensively as their area will be different. Which is why, to normalize the area values in a single scale, another parameter was defined which is known as "area ratio" or "cracking index" or "desiccation ratio." It is defined as the ratio of total area of the cracks generated to the total top surface area of the specimen.

This cracking index has been widely used in the past as well to quantify the cracking characteristics. A higher value of cracking index signifies a greater number of crack generation and lower efficiency of crack resistance for a given additive. The entire process of extraction of useful crack related data from raw images is explained using a flowchart in Fig. 3.

3 Results and Discussions

3.1 *Effect on Free Swell Index*

The variation of differential free swell index (DFS) with percentage composition of biochar and dolochar is shown in Fig. 4. The DFS value for biochar reduces up to 5%, and then it remains constant which shows that addition of biochar reduces the swelling. However, no observable trend was obtained in case of addition of dolochar.

According to Yang et al. [15] the rate of evaporation curve of biochar is a steep curve which may be divided into three stages, i.e., rapid, steady, and slow. Biochar decreases the initial rate of evaporation when the biochar percentage is at 5%. Hence, the water holding capacity of biochar increases resulting lesser compression and thereby reducing the value of DFS. However, after the biochar percentage is increased beyond 5%, the initial rate of evaporation increases again, resulting increment in DFS value.

Fig. 4 Variation in DFS with % composition of biochar and dolochar

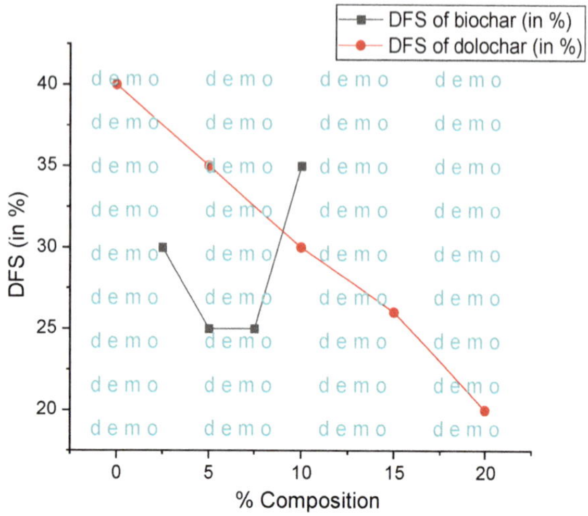

In case of dolochar, the fractal dimension of the soil sample decreases as dolochar concentration increases, while the water content of the soil sample increases. Because of the dolochar's wide pores and specific surface area, stable agglomerates of dolochar with soil particles are easily formed, thereby reducing the volume of the dolochar amended soil samples with increase in percentage of dolochar, which in turn reduces the DFS value gradually.

3.2 Changes in Cracking Characteristics

One of the first parameters defined in the analysis was the total area of the cracks. It was the simplest and basic characteristics defined. It was defined as the area of all the cracks generated in a certain specimen. The cracking resistance of the additives was assessed from this parameter. The variation of total area of the cracks with the percentage composition of biochar and dolochar is shown in Fig. 5.

From Fig. 5, it can be observed that for S10B5BI and S10B10D, the values were lesser as compared to other values, and then the values of total area were least for the optimum percentage composition, i.e., S10B5BI10D.

The values of total area of cracks averaged over the no of cracks generated are termed as average area of cracks. In addition to the total area of cracks, the average area also considers the number of cracks generated in a particular specimen, since if the number of cracks is higher, the average area will be lower, and vice versa. However, if the total area is also decreasing, the average area will reduce as well. The variation of average area of the cracks with different percentage compositions of biochar and dolochar is presented in Fig. 6.

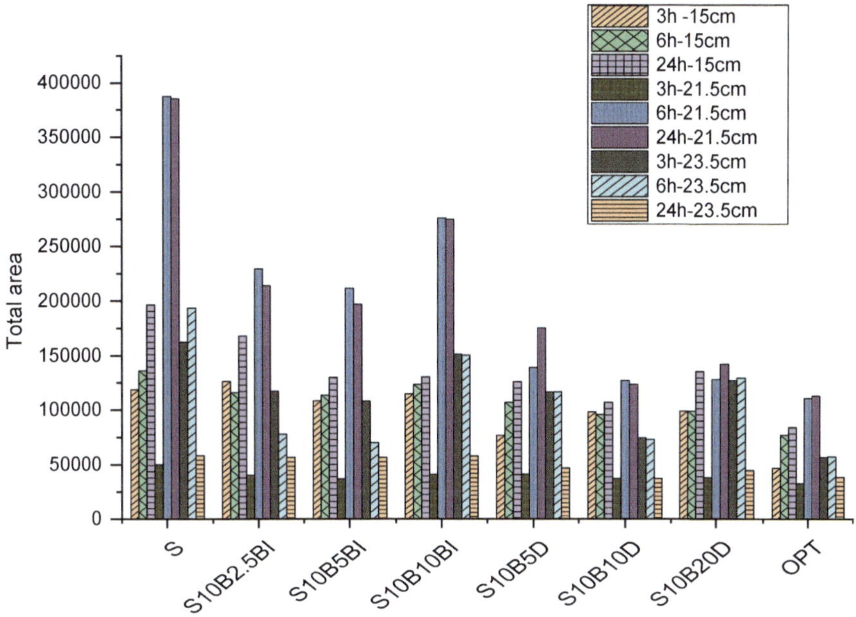

Fig. 5 Variation of total area of cracks for the compositions

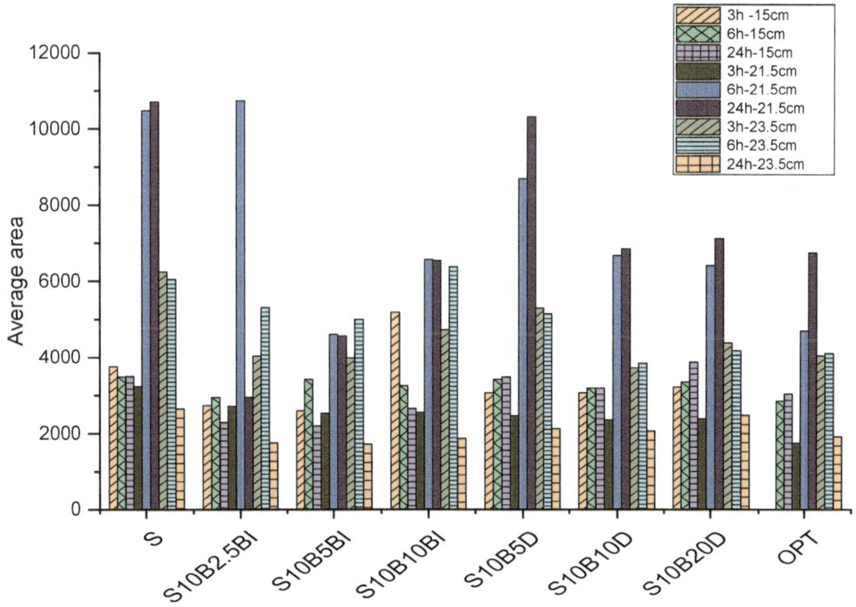

Fig. 6 Variation of average area of cracks for the composition

From Fig. 6, it was concluded again that the average area of cracks was minimum for the samples S10B5BI and S10B10D in most cases of the specimens. Average area of cracks is maximum for S10B10BI during the initial period since there are introduction of very minute cracks but in a very large number. This increases the average area of the cracks, but the practical significance of this curve is not of much importance.

The cracking index is defined as the ratio of area of the cracks to the total top surface area of the specimen. The top surface area for the samples taken from a height of 15 cm, 21.5 cm, and 23.5 cm was 18,895,914, 10,446,273, and 6,701,206 pixels, respectively, since the diameter of the specimens taken was 4905, 3647, and 2921 pixels.

The variations of cracking index with different compositions of biochar and dolochar are shown in Fig. 7. For maximum cases, again the area ratio is minimum for the samples S10B5BI and S10B10BI, which are considered as the optimum samples in our test. The area ratio is decreasing with the addition of biochar and dolochar up to 5% and 10% respectively, and beyond which it keeps on increasing. The comparison between area ratios of different samples over time are shown below for 15 cm and 21.5 cm of heights.

The variation of average width of the cracks generated are shown in Fig. 8. It can be observed that the minimum width of cracks was observed for the samples S10B5BI, S10B10D, and S10B5BI10D, which again proves the optimum dosage of biochar and dolochar for mitigation of cracks.

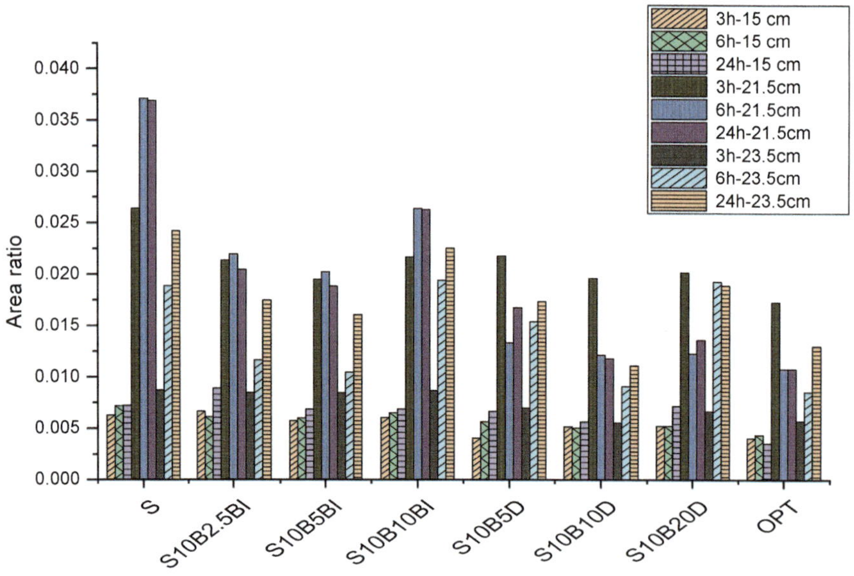

Fig. 7 Variation of cracking index for the compositions

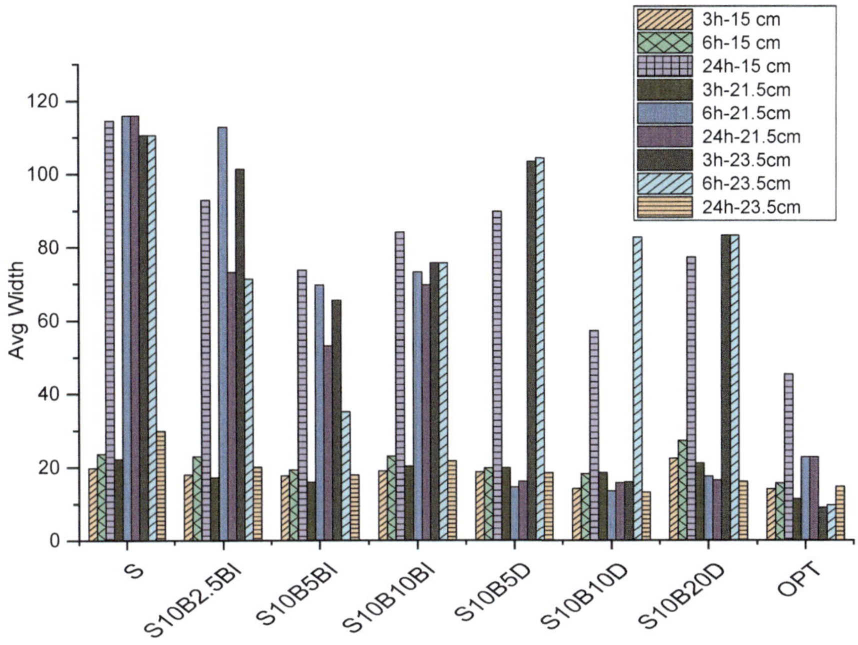

Fig. 8 Variation of average width of cracks for the compositions

4 Conclusions

From the study, following are some of the conclusions that are derived:

- The differential free swell values decrease by addition of biochar up to 5%, and then it keeps steady up to 7.5% and then starts on increasing. There is no trend followed by addition of dolochar in bentonite-amended soil.
- By addition of biochar in bentonite added soil, the fractal dimensions reduced significantly. The total area, average area, and area ratio (cracking index) reduced with the increase in biochar content up to 5% beyond which they started to increase. Similar trends were found with the addition of dolochar as well; however for dolochar, the optimum percentage at which the minimum number of cracks were formed was found to be 10%.
- From the quantitative analysis of cracks, it was found that with the addition of biochar, the total area, average area, area ratio, and average width are reduced by 33%, 37%, 5%, and 35% respectively at its optimum content, which is 5%. Similarly, by addition of dolochar the total area, average area, area ratio, and the average width are reduced by 45%, 28%, 21%, and 50% respectively at its optimum content which is at 10% dolochar. By adding both the optimum percentages of biochar and dolochar in the soil, the reduction in total area, average area, area ratio, and the average width were 57%, 135, 50%, and 69%, respectively.

References

1. IS 2720 (Part 10) (1991) Indian standard methods of test for soils determination of unconfined compressive strength (second revision)
2. IS 2720-3-1 (1980) Methods of test for soils, Part 3: Determination of specific gravity
3. IS: 2720 (Part VII) (1980) Indian Standard methods of test for soils part vii determination of water content-dry density relation using light compaction (Second Revision)
4. IS: 2720 (Part 40) (1985) Methods of test for soils: determination of free swell index of soil
5. IS: 2720 (Part 5) (1985) Indian Standard. Method of test for soils. Part 5 determination of liquid and plastic limit. (Second Revision)
6. IS-2720 (Part 4) (1985) (Reaffirmed- May 2015) Methods of test for soils: Grain size analysis
7. Izzo MZ, Miletić M (2019) Sustainable improvement of the crack resistance of cohesive soils. Sustainability 11(20):5806
8. Kalkan E (2009) Influence of silica fume on the desiccation cracks of compacted clayey soils. Appl Clay Sci 43(3–4):296–302
9. Kumar PBR, Sharma RS (2004) Effect of fly ash on engineering properties of expansive soils. J Geotech Geoenvironmental Eng 130(7):764–767
10. Peters WH, Ranson WF (1982) Digital imaging techniques in experimental stress analysis. Opt Eng 21(3):213427. https://doi.org/10.1117/12.7972925
11. Qin Y, He H, Ou X, Bao T (2019) Experimental study on darkening water-rich mud tailings for accelerating desiccation. J Clean Prod 240:118235
12. Sutton MA, Wolters WJ, Peters WH, Ranson WF, McNeill SF (1983) Determination of displacements using an improved digital correlation method. Image Vis Comput 1(3):133–139. https://doi.org/10.1016/0262-8856(83)90064-1
13. Take WA (2015) Thirty-Sixth Canadian geotechnical colloquium: advances in visualization of geotechnical processes through digital image correlation. Can Geotech J 52(9):1199–1220
14. Tang CS, Shi B, Liu C, Suo WB, Gao L (2011) Experimental characterization of shrinkage and desiccation cracking in thin clay layer. Appl Clay Sci 52(1–2):69–77
15. Yang B, Xu K, Zhang Z (2020) Mitigating evaporation and desiccation cracks in soil with the sustainable material biochar. Soil Sci Soc Am J 84:461–471

Behavior of Soft Clay Treated with Vibro Stone Columns Under Earthen Embankment Loads—Field Study

Anurag Chafale, B. Govind Raj, and Madan Kumar Annam

1 Introduction

The construction of embankments over problematic soils such as soft, compressible ground is unavoidable due to lack of suitable land for infrastructure and other developments. The geotechnical engineering tasks consist of ensuring the global stability, controlling surface settlements, and enhancing bearing capacity. Various methods such as Prefabricated Vertical Drains (PVDs), concrete piles, stone columns, and geosynthetic reinforcement have been developed for economical and safe construction of high embankments resting over soft ground [1, 11, 12].

Eventually, the final choice from variety of techniques depends upon the subsoil conditions, project cost, and time of completion including long-term construction advantages. Stabilizing soft ground by installation of vibro stone columns has numerous benefits such as improved bearing capacity, accelerated consolidation, decrease horizontal deformations, increased slope stability, and mitigating liquefaction [2, 4–6, 8, 10, 13]. Especially, the degree of improvement of in situ soft clays depends on two factors, when stone columns are used. The first one is introducing stiffer aggregate columns in the soft soil, and the second is consolidation of the surrounding soil during the loading process.

Review of embankments built over treated soft ground using vibro stone columns is presented in this paper. Earthen embankment of 10 m high was constructed over 13-m-deep soft marine clay treated with vibro stone columns. Behavior of embankment during construction stage is monitored through instrumentation such as settlement

A. Chafale (✉) · B. Govind Raj
Keller Ground Engineering India Pvt. Ltd., Mumbai 400059, India
e-mail: anuragchafale@kellerindia.com

M. K. Annam
Keller Ground Engineering India Pvt. Ltd., Chennai 600024, India

B. T. Jose et al. (eds.), *Proceedings of the Indian Geotechnical Conference 2022 Volume 4*, Lecture Notes in Civil Engineering 479,
https://doi.org/10.1007/978-981-97-1753-8_20

markers and inclinometers. Interpretation of the observed data is carried out to understand performance of the embankment in service condition (observational and analytical methods). Further, improvement between intercolumn stiffness was studied by comparing pre- and post-penetration testing results. A careful watch over soil investigation, design process, and post construction monitoring over the entire process is thoroughly monitored to assess reliable prediction of embankment performance.

2 Project Environment

The project site is situated in West Coast of India. Anticipated load intensity due to 10-m-tall earthen embankment was about 200 kPa and expected unimproved surface settlement was more than a meter. This made practicing engineer's attention over global stability. In addition, the construction site falls under Seismic Zone-IV having Peak Ground Acceleration (PGA) of 0.24 g and earthquake magnitude of 7.5. These conditions posed a question of cyclic softening in the event of earthquake.

2.1 Subsoil Conditions

About 6 eCPTs and 3 boreholes data were available to understand the in situ subsoil conditions. Soil investigation was carried out up to 30 m below Existing Ground Level (EGL). The subsoil comprising of 1-m-thick filled soil followed by very soft to firm silty clay up to the depth of 13 m below EGL. There is occasional presence of silt lenses between 4 and 7 m deep. This layer is further underlain by dense to very dense silty sand followed by hard clayey silt up to termination of borehole. Ground water table was encountered almost at surface (say 1 m below EGL) during the time of investigation. SPT-N vs depth and cone resistance (q_c) and friction ratio (R_f) vs depth are presented in Fig. 1.

2.2 Geotechnical Concerns

Since the subsoil conditions below the proposed embankment being soft clay, it will undergo excessive long-term settlements and will experience global stability. Further, project location falls under Seismic Zone-IV; hence, it will prone to liquefaction/ cyclic softening in the event of earthquake.

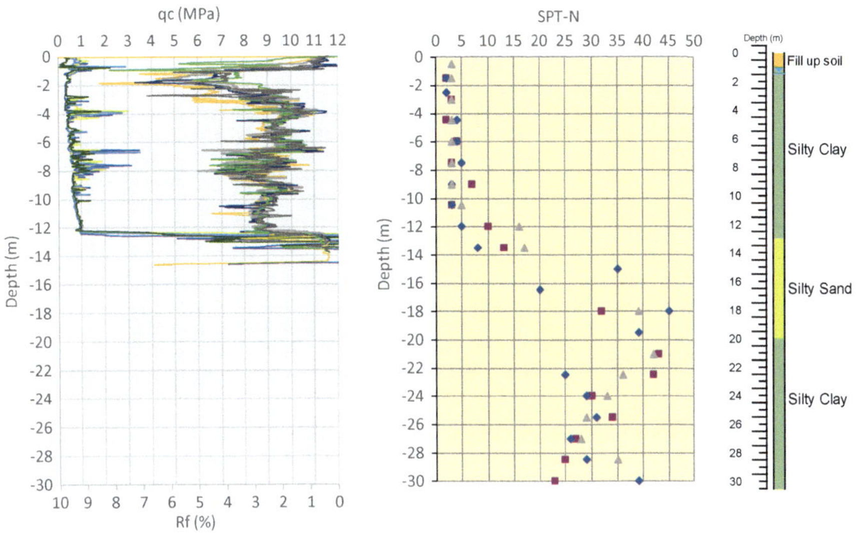

Fig. 1 Cone resistance (q_c), friction ratio (R_f) and SPT-N versus depth

3 Ground Improvement Works

The subsoil within the project site comprising of soft in nature results in high settlements, low bearing capacity, and inadequate global stability. This condition demands appropriate ground improvement technique. Further, rate of settlements needs to be accelerated, so that permissible settlements in service condition can be controlled within the performance requirements. Ground improvement technique using vibro stone columns was chosen in this project addressing geotechnical challenges, satisfying project schedule and cost economics.

Typical arrangement of vibro stone columns is illustrated in Fig. 2. Advantage of slow rate of embankment filling was considered to accelerate the consolidation and to elude most of the excessive settlements. Ground improvement design is carried out according to Priebe's design methodology [9] using "Keller Improvement Designer" (KID) software. Time rate consolidation analysis of treated ground is done using radial drainage consolidation theory [7]. Vibro stone columns with Area Replacement Ratio (ARR) of 20% and treatment depth of 13 m below EGL were finalized. Further, 1-m-thick granular blanket was laid above cut off level of vibro stone columns to form connectivity between columns for pore pressure dissipation when embankment loading is done.

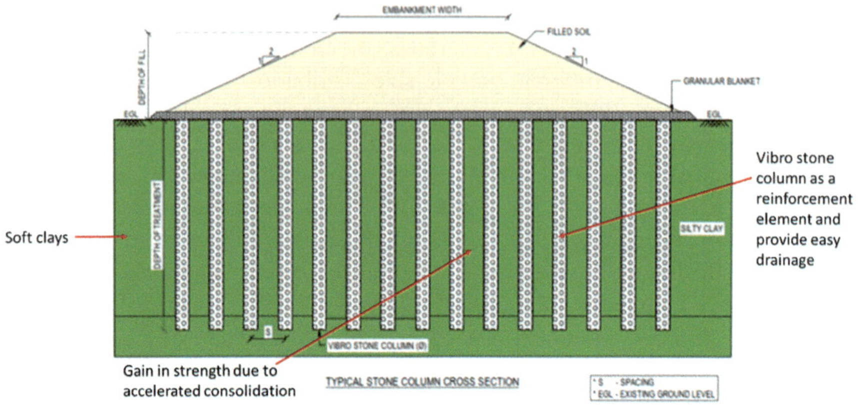

Fig. 2 Typ. cross section illustrating ground improvement scheme

4 Construction and Monitoring

A selected stretch of embankment (10 m fill plus 1 m buffer to match the top level after construction stage settlement) was built over the treated ground in 2 months duration. Slow rate of filling (say 0.4 m to 0.5 m/day) was maintained during construction of embankment. After reaching the desired top level of embankment, it was held for sufficient rest period to satisfy 95% of consolidation. Loading of embankment is monitored in such a way that majority of settlements are completed during construction leaving serviceability settlements which are well within the permissible limits. Asaoka et al. 1978 were considered for interpretation of degree of consolidation [3]. Six settlement plates and two inclinometers were installed to monitor settlements and lateral movements, daily. Key observations are listed below.

- Recorded settlements are in the range of 579–619 mm.
- Rate of settlement was gradually decreased and stabilized after 70 days of waiting period.
- 95 to 97% of consolidation was achieved within 90 days of rest period.
- The balance long-term settlements, approx. 30–35 mm will occur during service life of embankment.

Embankment loading and observed vs predicted settlements are presented in Fig. 3. Degree of consolidation using Asaoka's method is presented in Fig. 4.

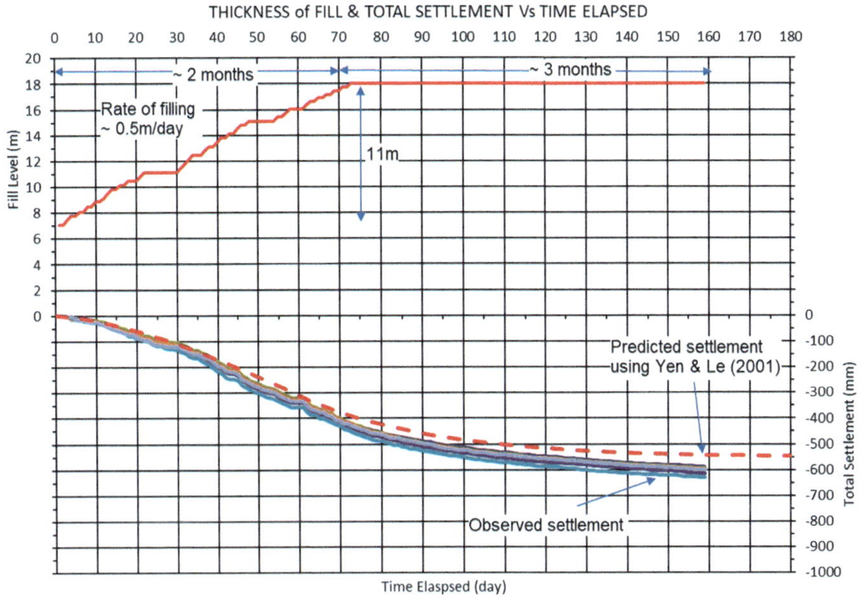

Fig. 3 Plot showing embankment height and observed vs predicated settlements

5 Post-treatment Soil Investigation

Efficacy of improved ground was checked by conducting 5 post-eCPTs and 2 boreholes. The exploration depth of post-treatment BHs and eCPTs were limited up to the treatment depth. Post-improvement results indicate considerable improvement in SPT-N and q_c values when compared to in situ conditions. It shall be noted that improvement in the subsoil stiffness (constrained modulus) as the result of the consolidation of clay layer is in line with design assumption. The results of post-SPT-N and eCPTs against depth are plotted in Fig. 5.

6 Conclusions

Ground improvement using vibro stone columns successfully demonstrated the performance criteria of 10-m-high embankment in the present case. Embankment loading acts like preloading which increases the effective stress and reduces the compressibility of weak ground through consolidation. Proposed ground improvement scheme is not only addressed the service stage settlements but also accelerated the rate of settlements. Around 95% consolidation achieved within 90 days of rest period via effective radial drainage. Pre- and post-soil investigation results clearly indicate improvement of shear strength and soil stiffness of the treated ground. This

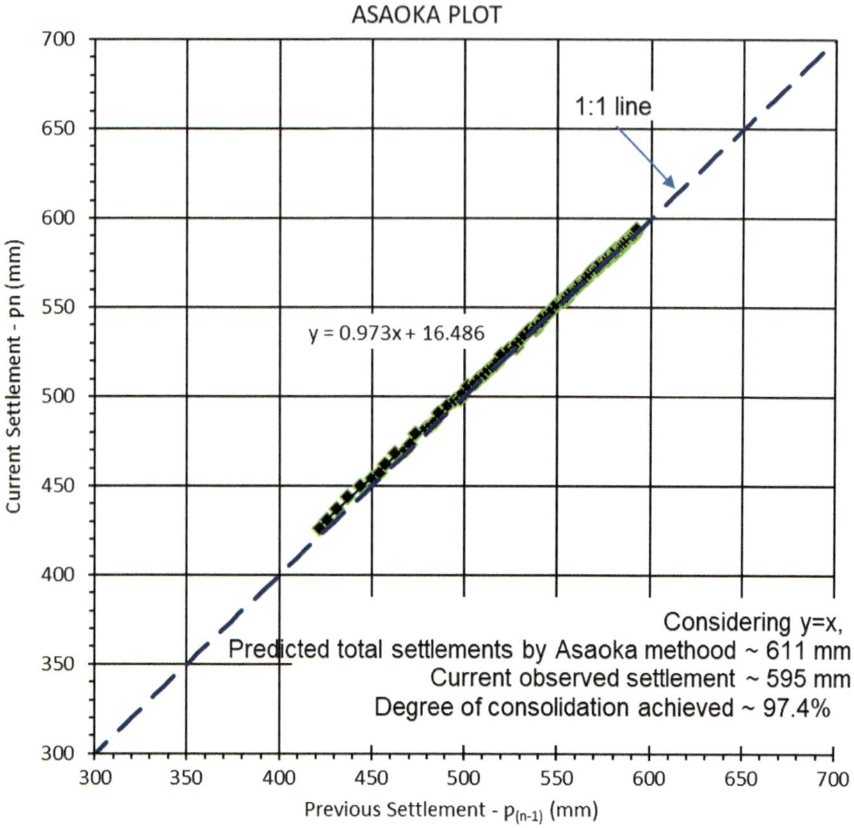

Fig. 4 Estimation of total settlement and degree of consolidation [3]

implies more than 1/10 reduction in the estimated service stage settlement compared to unimproved case. The design approach used has been verified and substantiated in comparison with field monitoring and in situ test results.

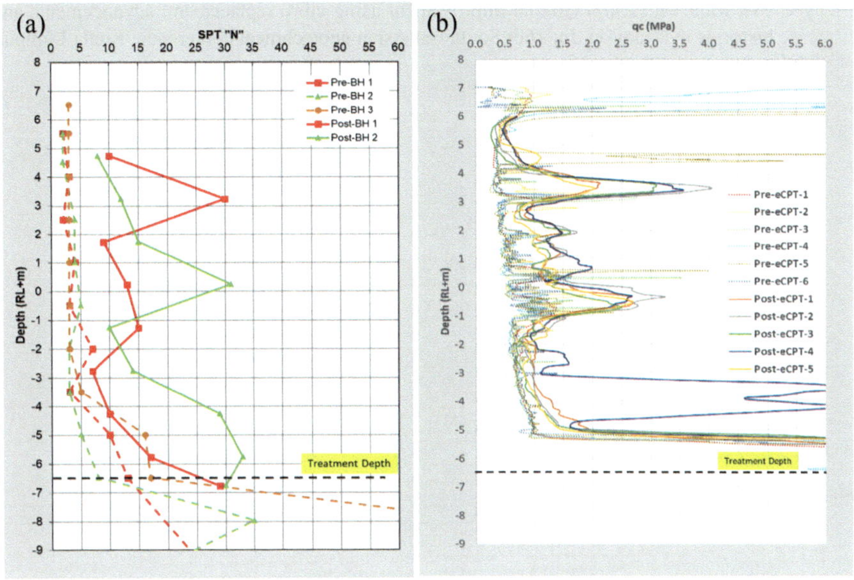

Fig. 5 Comparison of **a** pre- and post-SPT-N and **b** pre- and post-cone tip resistance (q_c)

References

1. Abhishek SV, Madhav MR (2016) Embankments on soft ground: an overview. In: Foundation and soft ground engineering conference, Thu Dau Mot University, pp 27–44
2. Almeida MSS, Lima BT, Riccio M, Jud H, Almeida MCF, Roza F (2014) Stone columns field test: monitoring data and numerical analyses. Geotech Eng J SEAGS AGSSEA 45:103–112
3. Asaoka A (1978) Observational procedure of settlement prediction. Soil Found 18(4):89–101
4. Basack S, Indraratna B, Rujikiatkamjorn C (2006) Analysis of behavior of stone column stabilized soft ground supporting transport infrastructure. Advances in Transportation Geotechnics. In: The 3rd international conference on transportation geotechnics, vol 143, pp 347–354
5. Chafale A, Annam MK (2022) A review on ground improvement with surcharge in addressing liquefaction mitigation. In: Dynamics of soil and modelling of geotechnical problems. Lecture notes in civil engineering, vol 186. Springer, Singapore, pp 367–375
6. Chafale A, Annam MK (2021) Mitigation of liquefaction in silty soils using Vibro stone columns with surcharge. In: Proceedings of Indian geotechnical conference, pp 1–4
7. Han J, Ye S (2001) Simplified method for consolidation rate of stone column reinforced foundations. J Geotech Geoenvironmental Eng 127(7):597–603
8. Poorooshasb HB, Meyerhof GG (1996) Consolidation settlement of rafts supported by stone columns. Geotech Eng 27(2):83–92
9. Priebe HJ (2005) Design of vibro replacement. Ground Eng 38(1):1–16
10. Raju VR, Sondermann W (2005) Ground improvement using deep vibro techniques. Ground Improv-Case Histories 3:601–638
11. Rowe RK, Hinchberger SD (2003) Geosynthetic reinforced embankments on soft clay foundation: predicting reinforcement strains at failure. Geotext Geomembr 151–175
12. Shen S, Chai J, Hong Z, Cai F (2005) Analysis of field performance of embankments on soft clay deposit with and without PVD-improvement. Geotext Geomembr 463–485

13. Yee YW, Raju VR (2007) Ground improvement using vibro replacement: advancements and case histories in Malaysia. In: 16th Southeast Asian geotechnical conference, Kuala Lumpur, Malaysia

Control of Shrinkage Characteristics of Expansive Soil by Using Low Compressible Soils

R. K. Bharti, H. Prakash, S. Noor, M. Gupta, and R. Chitra

1 Introduction

The entire southern peninsula of India to the south of the Narmada River, which is spread over eight states of the country, is known as the Deccan Plateau. The pedogenesis of expansive soils is caused by minerals as well as high temperature and low rainfall, which is the main reason of the formation of an expansive type of soils. The Deccan Traps is the basalt lava plateau filled mainly with mica and iron ore minerals that are responsible for the formation of black cotton soil and other types of expansive soils. Expansive soils are widely distributed in most parts of India, and about 20% of the land is covered by black cotton soils. Maharashtra, Tamil Nadu, Andhra Pradesh, Telangana, Kerala, Madhya Pradesh, Rajasthan, and Gujarat are the most well-known regions of expensive soil deposits in India.

There are three basic particles available in soils as clay, silt, and sand. Clay here is usually referred to as range in expansive soils if the clay is rich in hydrophilic minerals such as smectite group minerals, montmorillonite, and kaolinite. It is also known as highly compressible soil. Due to its highly moisture-sensitive property, the change in soil volume increases significantly when exposed to water and contracts drastically when water is released. Due to the shrink and swell characteristics of expansive soils, these create the great problem for the stability of a structure. Therefore, any project in the field of civil engineering such as the construction of embankments, retaining walls, high-rise buildings, etc., should be seriously evaluated or analyzed for satisfactory performance. Many researchers have suggested a lot of solutions to encounter this problem, but most of them are difficult to execute in the field and economically expensive. Some researchers have suggested a mechanical technique like expulsion of air voids by compaction, stone or sand columns, cement or steel fiber

R. K. Bharti (✉) · H. Prakash · S. Noor · M. Gupta · R. Chitra
Central Soil and Materials Research Station, New Delhi, Delhi 110016, India
e-mail: rk.bharti@nic.in

B. T. Jose et al. (eds.), *Proceedings of the Indian Geotechnical Conference 2022 Volume 4*, Lecture Notes in Civil Engineering 479,
https://doi.org/10.1007/978-981-97-1753-8_21

grouting, soil stabilization by mixing suitable soil techniques and chemical treatment, etc. Most of these techniques aim to change the physical and basic index properties of the expansive soil which controls the shrinkage and swelling behavior. To alter the behavior of expansive soil, soil mixing is one of the suitable methods to treat the expansive soil. By using the low compressibility soil, the liquid limit, plasticity index, shrinkage limit, shrinkage index as well as free swelling index of the problematic soil can be improved. Seed et al. [10] defined that a well-defined relationship can be established between the percentage of clay size and soil activity. Van Der Merwe [16] proposed a method based on prediction of heave from the plasticity index and percentage of clay fraction of soils to describe potential severity of volume change for clay soil. McCormack and Wilding [6] explained the phenomenon of swelling and found that expansive soils continue to swell at moisture contents above 1/3-bar tension and continue to shrink at moisture contents below 15-bar tension and below the shrinkage limit. Thomas [15] related the soil shrink-swell with soil properties and to partition variability into map unit components. Taboada [13] distinguished between soils with different magnitudes of swelling, as well as consequences on soil structural behavior. Chertkov [2] presented that clay shrinkage curve, under the influence of a silt–sand admixture as well as inter/intra aggregate structure, is transformed to the soil shrinkage curve.

In the current study, an attempt has been made to control the shrink and swell behavior of the expansive soil by adding low compressibility soils in different proportions such as 10, 15, 20, 25, 30, and 35% by weight. The test results show that by adding low compressibility clay in the expansive soil, the shrinkage limit and FSI are can be controlled or improved up to the satisfactory level.

2 Material

The soil samples taken for improvement of highly expansive soil are comprised percentage of clay content of 72.7% and fine particles 96.8% by weight. According to the Unified Soil Classification System, the sample is classified as High Plasticity Clay (CH). The percentage of different particles (clay, silt, and sand) as determined by the mechanical analysis method as per IS: 2720 (part-4) is presented in Table 1. The engineering and index properties of the soil like Atterberg limit, volume shrinkage, shrinkage limit, and free swell index are also presented in Table 1. The shrinkage limit and free swell index of the tested soil indicate that soil has high shrink and swell characteristics.

3 Methodology

The following laboratory tests are carried out on the mixed soil samples taken for improvement as per the respective Bureau of Indian Standards:

Table 1 Properties of fine-grained soil taken for improvement

0.002 mm and less	72.70
0.002–0.075 mm	26.10
0.075–0.425 mm	0.50
0.425–2.0 mm	0.60
2.0–4.75 mm	0.10
4.75 mm and above	0.00
Liquid limit	103.30
Plastic limit	39.00
Plastic index	64.00
Specific gravity	2.37
Shrinkage index	59.87
Shrinkage limit (%)	4.43
Shrinkage ratio	2.11
Volumetric shrinkage (%)	1.49
Free swelling index (%)	203.0

- Specific Gravity IS: 2720 (Part-3)
- Mechanical Analysis IS: 2720 (Part-4)
- Atterberg's Limit: Liquid Limit and Plastic Limit IS: 2720 (Part-5)
- Shrinkage Factor IS: 2720 (Part-6)
- Free Swell Index (FSI) IS: 2720 (Part-40).

Classification of all the soil samples is as per IS: 1498.

3.1 Shrinkage Factor Tests

To understand the volume change potential of expansive soil, there are numerous direct and indirect methods available in the literatures. However, many methods for determining the shrinkage factors, namely shrinkage index (I_s), shrinkage ratio (R) and volumetric shrinkage (V_S) are found unsatisfactory. Expansive soils are highly retentive of moisture and undergo to large volume change with the small change in water content. The value of shrinkage limit is very low in comparison with liquid limit and plastic limit. The liquid limit is the minimum percentage of water content at which the soil begins to flow, while the plastic limit is the minimum water content at which the soil behaves like plastic without elastic rebound, volume change, and cracking.

In this present study, a method as described in the IS: 2720 (Part XX) has been chosen to determine the volume change behavior of the expansive soil with the change in moisture content. The remolded soil has been used to determine the shrinkage limit. The low compressible soil is mixed with expansive soil in different proportions of

10, 15, 20, 25, 30, and 35% by weight and shrinkage characteristics is determined of each mix by the method as described in IS: 2720 (Part-5).

The shrink index (I_s) was calculated using the following equation as given in the IS: 2720 (Part-6).

$$\omega_s = \omega - ((V - V_0)/\omega_0) \times 100, \quad (1)$$

where ω_s is the shrinkage limit in percent, ω is moisture content of wet soil pat, V volume of wet soil pat in ml, V_0 volume of dry soil pat in ml, and ω_0 weight of oven-dry soil pat in g.

3.2 Free Swell Index (FSI)

A geotechnical engineer always identified the damage possibilities to the structure due to swelling of expansive soil. For such identification, there are numerous methods to understand the swelling characteristics of expansive soil, and free swell index is one of them. The swelling potential depends on several factors such as dry density, water content, structure load, and chemical composition of soil. But the more critical condition that has been adopted by the many standards in laboratory investigation is free swelling potential change, i.e., swelling without any external interference. The following formula is used to calculate the free swelling index.

$$\text{Free swell index, percent} = (V_d - V_k)/V_k \times 100, \quad (2)$$

where V_d is the volume of the soil specimen read from the graduated cylinder containing distilled water, and V_k is the volume of the soil specimen read from the graduated cylinder containing kerosene.

In the present study, the free swelling index test is performed in accordance with the IS: 2720 (Part-40) on the expansive soil as well as mixed soil samples.

4 Results

The test results of the shrinkage factor test of expansive soil and the mix. of expansive soil with low compressible soil in proportion of 10, 15, 20, 25, 30, and 35% by weight are presented in Table 2. The mechanical analysis test results show that clay particle size (less than 0.002 mm) is decreasing with the addition of low compressible soil and particle greater than clay sizes (0.002 mm) are increasing significantly.

The graph presented in Figs. 1, 2 and 3 shows the relation of shrinkage limit vs. clay percentage, shrinkage index vs. clay percentage, and free swelling index vs. clay percentage. It is evident from the Graph 1 that as the percentage of low compressible soil increases, the shrinkage limit increases (Shrinkage characteristics is decreasing).

Table 2 Volume change potentials of mixed soil

% of mixing clay		0	10	15	20	25	30	35
0.002 mm and less		72.70	67.90	63.40	59.60	58.90	54.75	52.31
0.002–0.075 mm		26.10	26.50	32.10	31.80	30.10	29.75	28.65
0.075–0.425 mm		0.50	2.20	2.20	4.30	5.23	7.50	8.49
0.425–2.0 mm		0.60	2.30	0.90	4.10	5.34	6.25	8.87
2.0–4.75 mm		0.10	1.10	1.40	0.20	0.43	1.75	1.68
4.75 mm and above		0	0	0	0	0	0	0
Atterberg limits	LL	103.50	86.30	82.10	79.20	75.20	71.95	68.49
	PL	39.0	40.50	37.50	35.80	34.00	32.95	31.48
	PI	64.50	45.80	44.60	43.40	41.20	39.00	37.01
Specific gravity		2.72	2.69	2.67	2.67	2.66	2.66	2.65
Shrinkage index		60.07	38.37	34.30	31.86	27.21	23.44	20.26
Shrinkage limit (%)		4.43	7.43	10.30	11.54	13.99	15.56	16.76
Free swelling index (%)		205.0	165.0	133.0	114.0	86.0	76.0	65.0
Degree of expansion (%)		Very high	High	High	High	Medium	Medium	Medium

While the Graph 2 indicates that as the percentage of low compressible soil increases, shrinkage index decreases. Graph 3 indicates that with increase of low compressible soil, the free swelling index is decreasing.

It can be observed from the above graphs, that the effect of mixing the low compressible soil in expansive soil is appreciable up to 20% quantity low compressible soil. Beyond that, the effect is not much prominent. So, it can be concluded that there is an optimum quantity of low compressible soil that can mixed with expansive soil to improve the shrink-swell characteristics of expansive soil. In the present study, the optimum quantity of low compressible soil is 20%.

The study also shows that shrinkage limit of the expansive soil changes from 4.43 to 16.76% which indicates an increase of 278.33% in shrinkage limit with addition of 35% low compressible soil. The shrinkage index changes from 60.06 to 20.26 which indicates a 66.27% reduction in shrinkage index. The free swell index changes from 205.0 to 65.0% which indicates a reduction of 68.29% in free swelling characteristics of soil.

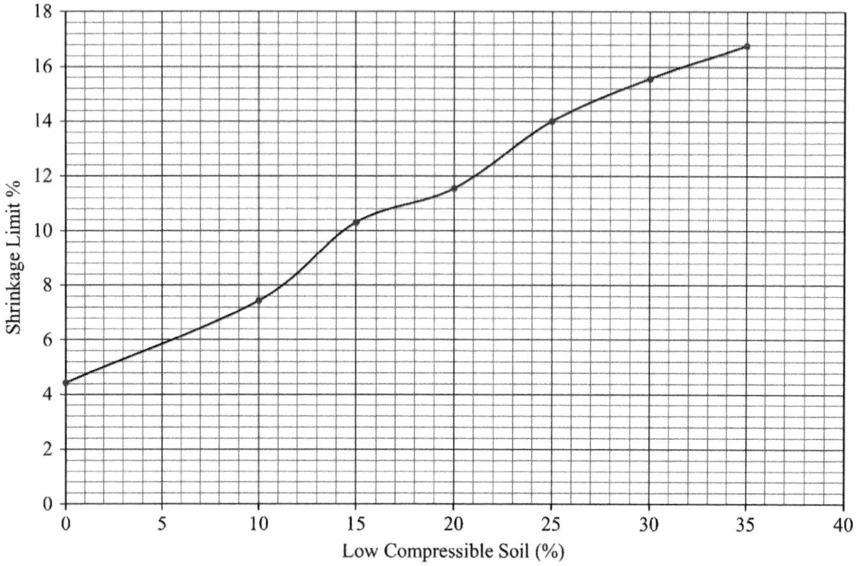

Fig. 1 Shrinkage limit versus low compressible soil (%)

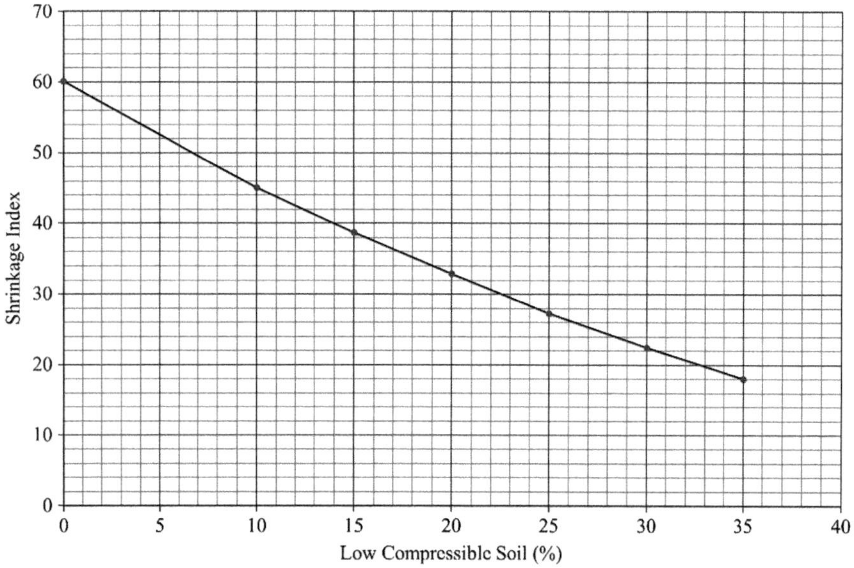

Fig. 2 Shrinkage index versus low compressible soil (%)

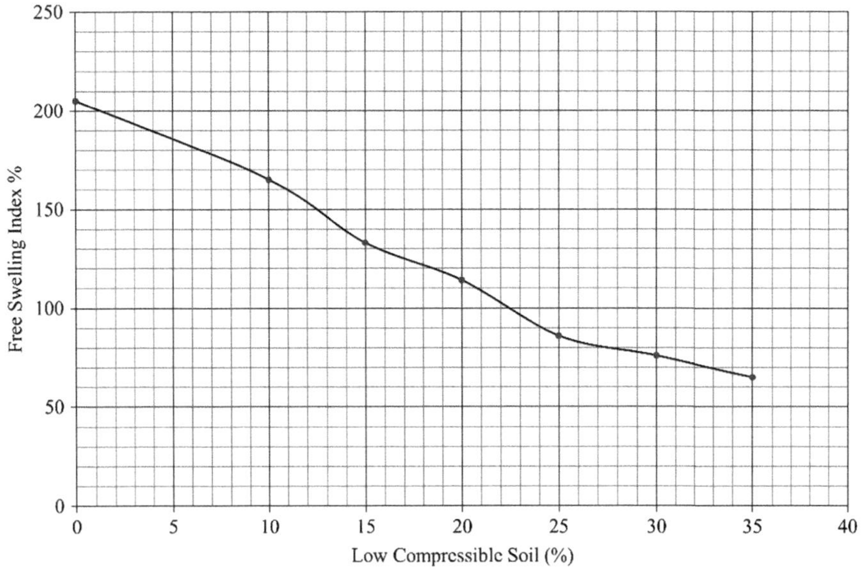

Fig. 3 Free swelling index versus low compressible soil (%)

5 Conclusion

In conclusion, this study has shown that shrink-swell characteristics of the expansive soil can be controlled by mixing the low compressible soil and the optimum quantity of low compressible soil is 20%. Strong correlations were found between the shrinkage limit, the shrinkage index and the free swell index with the percentage of low compressible soil. As the percentage of low compressibility soil increases in expansive soil, the shrinkage limits curve exponentially increases, whereas the shrinkage index and the free swelling index decrease exponentially.

It is also concluded that the shrinkage limit and the free swelling index are on the satisfactory level and indicate that adding low compressible soil can decrease the expansive potential of soils, thereby improving their compressibility characteristics. Therefore, the properties of highly expansive soil can be altered with the help of low compressive soil mix.

References

1. Abduljauwad SN, Al-Sulaimani GJ (1993) Determination of swell potential of Al-Qatif clay. Geotech Test J 16(4):469–484
2. Chertkov VY (2007) The reference shrinkage curve at higher than critical soil clay content. Soil Sci Soc Am J 71(3):73–92

3. Dakshanamanthy V, Raman V (1973) A simple method of identifying an expansive soil. Soils Found Jpn Soc Soil Mech Found Eng 13(1):97–104
4. Golait YS, Kishore MP (1990) A new approach for free swell index and evaluation of swelling potential for construction sites. Indian Geotech J V/ Session VIII/ Art. 2:485–489
5. Lambe TW, Whitman RV (1979) Soil mechanics. Wiley, New York
6. McCormack DE, Wilding LP (1975) Soil properties influencing swelling in canfield and Geeburg soils. Soil Sci Soc Am J 39(3):496–502
7. Mitchell PW (1986) The design of raft footing on expansive soil. The Institution of Engineers, Australia, pp 328–335
8. Nayak NV, Christensen RW (1971) Swelling characteristics of compacted expansive soils. Clays Clay Miner 19:251–261
9. Seddon KD (1992) Reactive soils. In: Peck, Nelson, Olds, Seddon (eds) Engineering geology of Melbourne. Balkema, Rotterdam, pp 33–37
10. Seed HB, Woodward RJ Jr, Lundgren R (1962) Prediction of swelling potential for compacted clays. Soil Mech Found Div Am Soc Civ Eng 88(SM3):53–87
11. Skempton AW (1953) The colloidal activity of clays. In: Proceedings of the 3rd international conference on soil mechanics and foundation engineering, vol I. ICOSOMEF, Zurich, Switzerland, pp 57–61
12. Sridharan A, Rao SM, Murthy NS (1985) Free swell index of soils: a need for redefinition. Indian Geotech J 15(2):94–99
13. Taboada AM (2003) Soil shrinkage characteristics in swelling soils. In: Lecture given at the College on Soil Physics, Trieste, Italy
14. Terzaghi K, Peck R (1967) Soil mechanics in engineering practice, 2nd edn. Wiley, New York
15. Thomas PJ, Baker JC, Zelazny LW (2000) An expansive soil index for predicting shrink–swell potential. Soil Sci Soc Am J 64(1):268–274
16. Van Der Merwe DH (1964) The prediction of heave from plasticity index and percentage clay fraction of soils. Civ Eng S Afr 6(6):103–107
17. Young GS, Parmar M (1999) Shrink-Swell testing in the Sydney region. In: Proceedings 8th Australian New Zealand conference on Geomechanics, pp 221–224, Hobart

Effect of Microbial Enzyme on Fly Ash and Assessment of Compressive Strength

Joyprakash Naskar and Anil Kumar Sharma

1 Introduction

Sustainable building material demand is increasing day by day. Some traditional construction materials like sand, cement, and aggregates have been used in previous decades. Today, the number of thermal power plants is increasing day by day to meet the demand for power requirements. Hence, fly ash generation is also high. Recycling and reuse of fly ash is an economical and eco-friendly technique to conserve natural resources for sustainable development. The use of fly ash as a building material is recognized worldwide [1]. To prevent the adverse environmental impact of fly ash, recycling rates are increasing to minimize its negative impact and promote better sustainable development [2, 3].

In recent years, biocementation, which is based on calcite precipitation using microbes, has been recognized worldwide as a novel consolidation technique, used in cement concrete mortars, expansive soil, and sand consolidation to improve engineering properties and compressive strength [4–6]. A novel method that uses urease-producing microbes to produce carbonate has been introduced and examined by prior studies to address the remediation of metal leaching from CFA. Microbial-induced carbonate precipitation.

(MICP) is a well-known name for this method [5, 7–9]. By using MICP to immobilize cadmium in soils, Kumari et al. (2014) observed a 90% conversion of soluble cadmium to carbonate-bound cadmium [10]. Microbial-induced calcium carbonate precipitation is a mineralization process leading to CaCO3 precipitation to fill the pore structure of fly ash and bind them. The microbe reaction process induces calcite precipitation via urea hydrolysis by urease enzyme as follows:

- Urea gets hydrolyzed by the urease enzyme and produces carbamate and ammonia:

J. Naskar · A. K. Sharma (✉)
Department of Civil Engineering, National Institute of Technology Patna, Patna 800005, India
e-mail: aks.ce@nitp.ac.in

B. T. Jose et al. (eds.), *Proceedings of the Indian Geotechnical Conference 2022 Volume 4*, Lecture Notes in Civil Engineering 479,
https://doi.org/10.1007/978-981-97-1753-8_22

$$CO(NH_2)_2 + H_2O \rightarrow NH_2COOH + NH_3 \quad (1)$$

- After hydrolysis of carbamate, 1 mol of ammonia and carbonic acid are formed:

$$NH_2COOH + H_2O \rightarrow NH_3 + H_2CO_3 \quad (2)$$

- Carbonic acid dissociates into one mole of bicarbonate and a hydrogen ion:

$$H_2CO_3 \rightarrow H^+ + HCO3^- \quad (3)$$

- Ammonia dissociates into 2 mol of ammonium ions and hydroxide ions:

$$2NH_3 + 2H_2O \rightarrow 2NH^+ + 2OH^- \quad (4)$$

- Due to the presence of ammonium ion, pH increases, and bicarbonate ions dissociate and carbonate ions are formed and a combined reaction occurs:

$$HCO_3 + +H^+ + 2\,NH_4^+ + 2OH^- \rightarrow CO_3^{2-} + 2NH_4^+ + 2H_2O \quad (5)$$

- Bacterial cell walls are negatively charged. Due to this, bacterial cell walls attract $Ca2^+$ and cations and stick to the cell surface; further $Ca2^+$ reacts with CaCO3 and precipitation of calcium carbonate occurs at the bacterial cell surface as a nucleation site and the size of CaCO3 mineral increases:

$$Cell + Ca^{2+} \rightarrow \left(Cell - Ca^{2+}\right) \quad (6)$$

$$\left(Cell - Ca^{2+}\right) + CO_3^{2-} \rightarrow (Cell - CaCO_3) \quad (7)$$

This study focuses on improving the compressive strength of fly ash by using bacterial solutions and cementation solutions. Nutrients, calcium, and urea have been given as ingredients for bacterial growth and bio cementation.

2 Materials

2.1 Source of Bacteria

Bacteria, namely *Sporosarcina pasteurii* (MTCC 1761), were procured from the Microbial Type Culture Collection and Gene Bank (MTCC), Chandigarh, India.

Table 1 Nutrient source

Ingredients	Gram/Litre
Tryptone	10.00
Yeast extract	5.00
Sodium chloride	10.00

Table 2 Biocementation composition

Mineral	Molarity
Urea	0.5
Calcium chloride	0.5

2.2 Nutrients for Bacterial Growth

LB (Luria Bertani) Broth was used as a nutrient for bacterial species as given in Table 1.

The nutrients were mixed with distilled water and autoclaved for 15 min at 121 °C.

2.3 Biocementation Medium Preparation

The biocementation solution was prepared, and pH was adjusted to 7.5 with 1 N sodium hydroxide as given in Table 2.

2.4 Source of Fly Ash

A fly ash sample was collected from the National Thermal Power Corporation located at Barh in Bihar, India. Samples were transported with the utmost precaution for studying various properties of fly ash samples.

3 Methodology

3.1 Microbial Solution Preparation

Frozen *Sporosarcina pasteurii* was inoculated in LB (Luria Bertani) broth nutrient medium and placed in an incubator at 37 °C for 24 h. The bacterial culture was then sub-cultured in 250 ml of LB broth. Biological solution was checked at Optical

Density (OD) at 600 nm wave length for determining cell concentration. In this study, an OD600nm of 0.8 was used, and cell concentration was taken as 6.34107 cells/ml. The bacterial cell concentration in growth media was measured by the expression [11].

$$Y = 8.59 \times 10^7 \times Z^{1.3627} \tag{8}$$

Note: (Z = OD600nm reading, Y = cell concentration per millilitre).

3.2 Cementation Solution Preparation

Urea and calcium chloride were mixed in distilled water used as a cementation media preparation. Table 2 shows the detailed composition of cementation media. In addition, 3gm of LB broth per litre was added for the supply of nutrients to bacteria [12]. Nutrients are added for the multiplication of bacteria and a higher ureolysis rate.

3.3 Physical Properties of Fly Ash

On the collected fly ash sample, various tests such as specific gravity, gradation, and standard proctor test were performed in accordance with the IS-2720 [13–17]. Unconfined compressive strength was conducted on an untreated fly ash sample and a MICP-treated fly ash sample. Table 3 gives different physical and engineering properties of a fly ash sample.

Table 3 Physical and engineering properties of fly ash

Properties	Value
Specific gravity (G)	2.01
% Finer than 75 μm	40.04
Sand (%)	59.96
Silt (%)	40.04
Coefficient of uniformity (Cu)	7.86
Coefficient of curvature (Cc)	1.21
Optimum moisture content (%)	31.47
Maximum dry unit weight (kN/m^3)	11.22
Unconfined compressive strength, untreated (kPa)	--
Unconfined compressive strength, MICP treated (kPa)	422
pH	7.85

Fig. 1 PVC moulds for specimen preparation

3.4 Specimen Preparation

A fly ash column was prepared in a cylindrical PVC mould of 40 mm in diameter and 90 mm in length to maintain a ratio of greater than 2 (height: diameter). Moulds were prefabricated and formed at the National Institute of Technology Patna as shown in Fig. 1. Twelve numbers of oven dried fly columns were prepared by adding the bacterial solution at OMC. PVC moulds were fitted with silicon pipe at both ends and made ready for treatment.

$$(NH_2)_2CO + 2H_2O \rightarrow CO_2 + H_2O + NH_3 \quad (9)$$

3.5 MICP Treatment Process

Figure 2 shows the adopted method of MICP treatment in this study. A fly ash column was placed in a vertical position with clamps and stands. One end of a silicon pipe was fitted with a peristaltic pump to maintain a constant rate of flow of bacterial media and cementation solution. The other end of the silicon pipe is placed in a container to collect effluent. At the time of column preparation at OMC, two pore volumes of cementation solution were passed through the fly ash column and retained for 10 h. After that, one pore volume of bacteria medium was given to the sample and retained for two hours. After that, two pore volumes of cementation solution were given to the fly ash sample again and retained for 12 h. This cycle continues for 28 days,

Fig. 2 MICP treatment process through peristaltic pump

after which the samples are cured for seven days. Thereafter, samples were dried as prepared for the unconfined compressive strength test.

4 Result and Discussion

4.1 Urease Test

The hydrolysis of bacteria is confirmed by a urease test. In this process, after hydrolysis, urea produces ammonia and carbon dioxide. Ammonia increases the basicity of the medium. Phenol red reagent is used as an indicator. It changes colour from light orange (pH 6.8) to magenta (pH 8.1). Urease enzyme producing bacteria turn the medium pink within 24 h [18]. 10 ml of cementation solution and 5 ml of bacterial solution were mixed in a test tube and placed in an incubator at 37 °C for 24 h. Colour change is detected, and it confirms that it is a urease enzyme producing bacteria and it is efficient for hydrolysis of urea. Figures 3 and 4 show the confirmation of urease activity of *Sporosarcina pasteurii*.

$$(NH_2)_2CO + 2H_2O \rightarrow CO_2 + H_2O + NH_3 \quad (10)$$

4.2 Compressive Strength Test

Unconfined compressive strength (UCS) [19] tests were carried out to determine the compressive strength of untreated and treated cylindrical samples of fly ash. However, it must be highlighted that the UCS value of untreated samples was very negligible. In

Fig. 3 Urease activity test at time *T*0

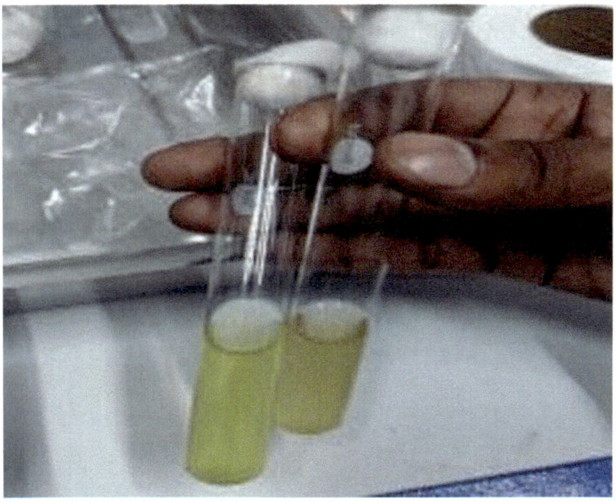

Fig. 4 Urease activity test result after 24 h

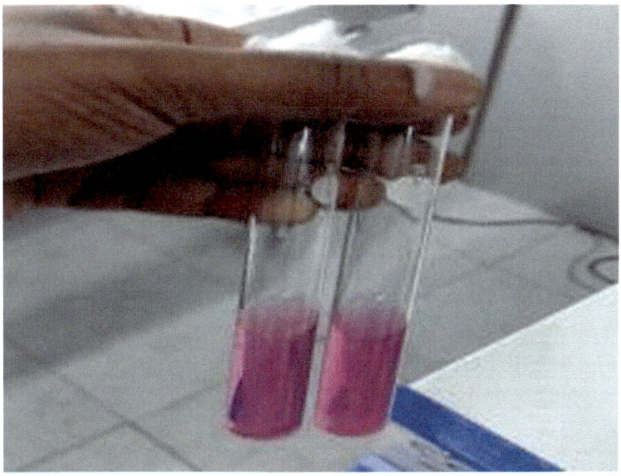

a similar study, bio-treated fly ash samples developed strength of about 33.6 kPa [20]. As shear strength becomes a governing design criterion for geotechnical engineering constructions, numerous studies on the application of MICP to improve the shear strength of soil and related materials have been done [21, 22]. Almost all of these studies revealed an increase in UCS ranging from 36 to 12,400 kPa. The stress–strain behaviour of fly ash samples treated with MICP for seven days is depicted in Fig. 5. Similar improvement in UCS values has been observed by other researchers. In case of fine sand, the UCS values of samples stimulated for both aerobic and anaerobic bacteria growth increased by 254–283% [23]. In an experimental investigation of the MICP treatment of calcareous foundations on an artificially reclaimed island, the UCS reached around 821 kPa after nine MICP treatments [24]. The treated fly ash samples attained a compressive strength of 422 kPa. As observed, it shows a significant increase in strength in comparison to untreated samples, for which the

UCS value was negligible. Figure 6 shows the tested samples after the UCS test. Therefore, it may be concluded that MICP treatment is also effective in stabilizing fly ash.

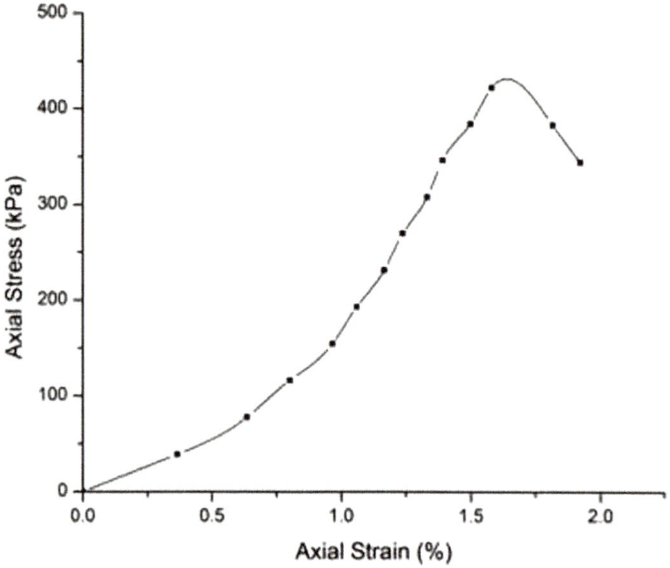

Fig. 5 Stress–strain curve for MICP treated fly ash after 28 days

Fig. 6 MICP-treated fly ash sample after UCS test

5 Conclusion

Based on this experimental study, it can be concluded that microbial-induced calcite precipitation (MICP) has been found to be effective in improving the UCS of fly ash. The unconfined compressive strength of fly ash increased after 28 days of MICP treatment and the UCS value is 422 kPa. That is quite impressive at OD600 of 0.8 bacterial concentrations and 0.5 M of biocementation solution concentration. Further investigation is need to establish the effect of MICP with other engineering properties such as compressibility and permeability characteristics.

References

1. Su P, Zhang J, Li Y (2019) Solidification/stabilization of stainless steel pickling residue with aluminum potassium sulfate amended fly ash. J Clean Prod 234:400–409
2. Bhattacharyya P, Reddy KJ, Attili V (2011) Solubility and fractionation of different metals in fly ash of powder river Basin coal. Water Air Soil Pollut 220(1):327–337
3. Cetin B, Aydilek AH, Li L (2012) Experimental and numerical analysis of metal leaching from fly ash-amended highway bases. Waste Manage 32(5):965–978
4. Seifan M, Samani AK, Berenjian A (2016) Induced calcium carbonate precipitation using Bacillus species. Appl Microbiol Biotechnol 100(23):9895–9906
5. Xu H, Zheng H, Wang JN, Ding XQ, Chen P (2019) Laboratory method of microbial induced solidification/stabilization for municipal solid waste incineration fly ash. MethodsX 6:1036–1043
6. Li M, Fang C, Kawasaki S, Achal V (2018) Fly ash incorporated with biocement to improve strength of expansive soil. Sci Rep 8(1):1–7
7. Zhang J, Wen K, Li L (2021) Bio-modification of coal fly ash using urease-producing bacteria. Fuel 286:119386
8. Chen P, Zheng H, Xu H, Gao YX, Ding XQ, Ma ML (2019) Microbial induced solidification and stabilization of municipal solid waste incineration fly ash with high alkalinity and heavy metal toxicity. PLoS ONE 14(10):e0223900
9. Kadapure SA, Kulkarni GS, Prakash KB (2017) A laboratory investigation on the production of sustainable bacteria-blended fly ash concrete. Arab J Sci Eng 42(3):1039–1048
10. Kumari D, Pan X, Lee DJ, Achal V (2014) Immobilization of cadmium in soil by microbially induced carbonate precipitation with Exiguobacterium undae at low temperature. Int Biodeterioration and Biodegradation 94:98–102
11. Ramachandran SK, Ramakrishnan V, Bang SS (2001) Remediation of concrete using microorganisms. ACI Mater J-Amer Concrete Instit 98(1):3–9
12. Cheng L, Shahin MA, Cord-Ruwisch R, Addis M, Hartanto T, Elms C (2014) Soil stabilisation by microbial-induced calcite precipitation (MICP): investigation into some physical and environmental aspects. In: 7th international congress on environmental geotechnics. Engineers Australia Melbourne, Australia, vol 64(12). pp 1105–1112
13. IS : 1498—1970, Classification And Identification of Soils for General Engineering Purposes, (First Rev.), Bureau of Indian Standard (BIS)
14. IS : 2720 (Part 1)—1983, Preparation of dry soil samples for various tests, (Second Rev.), Bureau of Indian Standard (BIS)
15. IS : 2720 (Part VII)—1980, Determination of water content-dry density relation using light compaction, (Second Rev.), Bureau of Indian Standard (BIS)
16. IS : 2720 (Part 13)—1986, Direct Shear Test, (Second Rev.), Bureau of Indian Standard (BIS)

17. IS:2720 (Part III/Sec 1)—1980, Determination of Specific Gravity, (First Rev.), Bureau of Indian Standard (BIS)
18. Morgan A, Babu D, Reiz B, Whittal R, Suh LY, Siraki AG (2019) Caution for the routine use of phenol red–it is more than just a pH indicator. Chemico-Biol Interact 310:108739
19. IS 2720 (Part 10):1991, Determination of Unconfined Compressive Strength, Bureau of Indian Standard (BIS)
20. Yadav A, Vineeth Reddy K, Khan M, Kalyan Kumar G, Bandhu A (2020) Bio-treatment of fly ash. In: Advances in computer methods and geomechanics. Springer, Singapore, pp 505–517
21. Fatehi H, Abtahi SM, Hashemolhosseini H, Hejazi SM (2018) A novel study on using protein based biopolymers in soil strengthening. Constr Build Mater 167:813–821
22. Chen R, Ramey D, Weiland E, Lee I, Zhang L (2016) Experimental investigation on biopolymer strengthening of mine tailings. J Geotech Geoenviron Eng 142(12):06016017
23. Pakbaz MS, Kolahi A, Ghezelbash G (2022) Assessment of microbial induced calcite precipitation (MICP) in fine sand using native microbes under both aerobic and anaerobic conditions. KSCE J Civ Eng 26(3):1051–1065
24. Liu HL, Ma GL, Xiao Y, Ding XM, Fang XW (2019) In situ experimental research on calcareous foundation stabilization using MICP technique on the reclaimed coral reef islands. Chinese Ground Improvement 1(1):26–31

Stabilization and Durability Analysis of Medium Compressibility Clay Using Rubber Fibers

K. L. Venkateshwara Reddy and Suresh Prasad Singh

1 Introduction

Soil is a dynamic entity in construction. Soil in the site should be able to bear the load but in adverse conditions soil fails in shear, bearing, etc., and hence failure of the structure occurs. To avoid these failures, investigation of the soil is mandatory. It is observed that sometimes the soil doesn't meet the engineering requirements, thereby arises the need for soil stabilization, which improves the soil properties by mechanical and chemical actions.

Considering loads of tons of waste rubber tires produced yearly across the country, which not only poses the problem for disposal but also increases to environmental contamination and health risks, as majority of the waste rubber tires are illegally dumped in landfills. It was observed that if rubber tires are subjected to fire conditions, they burn intensely releasing toxic gases into the atmosphere. Further rubber tires are highly durable and biologically non-degradable and hence can act as effective stabilizing material. Thus, consumption of such refuse and industrial wastes and their secondary products as alternatives to construction materials may effectively subsidize to environmental preservation and minimization of their adversative effects on the environment. The properties of clay including liquid limit, plastic limit, plasticity index, dry density, optimum moisture content, unconfined compressive strength, and CBR tests were performed for the different trails. In this paper, rubber fibers with different proportions were used to find their effect on stabilization of clay.

Bekhiti et al. [1] reinforced samples with rubber fibers content of 0, 0.5, 1, and 2% and cement contents of 5, 7.5, and 10% and found maximum UCS and ductility behavior at a rubber fibers content of 2%. Buritatum et al. [2] incorporated rubber fibers at contents of 0, 1, 2.5 and 10% for the reinforcement of the clay and observed

K. L. V. Reddy (✉) · S. P. Singh
Department of Civil Engineering, National Institute of Technology, Rourkela 769008, India
e-mail: 219CE1031@nitrkl.ac.in

B. T. Jose et al. (eds.), *Proceedings of the Indian Geotechnical Conference 2022 Volume 4*, Lecture Notes in Civil Engineering 479,
https://doi.org/10.1007/978-981-97-1753-8_23

when mixed with rubber fibers, the swelling rate decreased to 9.75% at a fiber content of 10%. Kumar et al. [4] used lime and elastic tire crushed powder and found that by putting 20% of lime and 12% of crushed tire with clay, the compressive strength was enhanced. Signes et al. [5] found addition of rubber particles to the expansive clay soil exponentially reduced swelling and optimum rubber content to prevent its swelling was 3%. Wang et al. [6] studied the effects of crumb rubber on the frost resistance of cement-soil for different rubber contents (by weight) of 0, 5, 10, 15 and 20% and found higher growth rate of the compressive strength of the rubber cement-soil under a negative temperature. Yadav et al. [7] focused on inclusion of waste rubber tire fibers on geotechnical properties of cemented clay and found optimum proportion for UCS and CBR was 2.5%. Yadav et al. [8] used shredded rubber tire passing through 75 micron I.S Sieve and found addition of 4% crumb rubber to the clay soil gave optimum moisture content and shear strength value increased compared to the normal soil.

2 Materials and Methodology

2.1 Materials

Clay. The soil used was collected from a site at NIT, Rourkela. The soil passing through 4.75 mm sieve was collected and stored at room temperature in lab. Geotechnical properties of the soil were analyzed in the laboratory, and it was classified as Medium Compressibility Clay as per IS 1498 (1970). Wet sieve analysis of clay was done as per IS: 2720 (Part 4): 1985. The clay was then tested for liquid limit, plastic limit, and specific gravity using density bottle.

The geotechnical properties of the clay obtained are tabulated (Table 1):

Rubber Fibers. Rubber has elastic properties and induces ductility to soil. Rubber fibers extracted from tires were collected for experimental purpose. The size of rubber fibers used in the study was of 3mm x 15mm. These rubber fibers were added to clay

Table 1 Geotechnical properties of clay

Sl. No.	Properties	Value
1	Natural moisture content	5–7%
2	Specific gravity	2.67
3	Liquid limit	40%
4	Plastic limit	19%
5	Plasticity index	21%
6	Optimum moisture content (OMC)	13%
7	Maximum dry density (MDD)	1.9 g/cc
8	Free swell index	25%

Table 2 Properties of rubber fibers

Sl. No.	Properties	Value
1	Specific gravity	1.26
2	Water absorption capacity	4%

in proportions of 0.5, 1, 1.5, 2 and 2.5% by dry weight of clay, and the samples were tested.

The properties of rubber fibers used are tabulated (Table 2):

2.2 Methodology Adopted

To evaluate the efficiency of rubber fibers as a stabilizer in clay soil, sequence of tests was conducted by varying the content of rubber fibers that were added to clay soil in proportions by dry weight of the clay taken. The tests performed and Indian Standard codes adhered to are as follows:

- Light compaction test—IS: 2720 (Part 7)—1983
- Heavy compaction test—IS: 2720 (Part 8)—1983
- Unconfined compressive strength (UCS) test—IS: 2720 (Part 10)—1991
- California bearing ratio (CBR) test—IS: 2720 (Part 16)—1987
- Dispersion test
- Durability test—IS: 4332 (Part 4)—1968.

3 Results and Discussions

3.1 Compaction Test

Light Compaction Test

See Fig. 1.

Heavy Compaction Test

It was inferred that with the addition of rubber fibers, OMC and MDD for both light and heavy compaction test decreases. This was due to low specific gravity and low water absorption capacity of rubber fibers in comparative to the virgin clay (Figs. 1, 2 and 3).

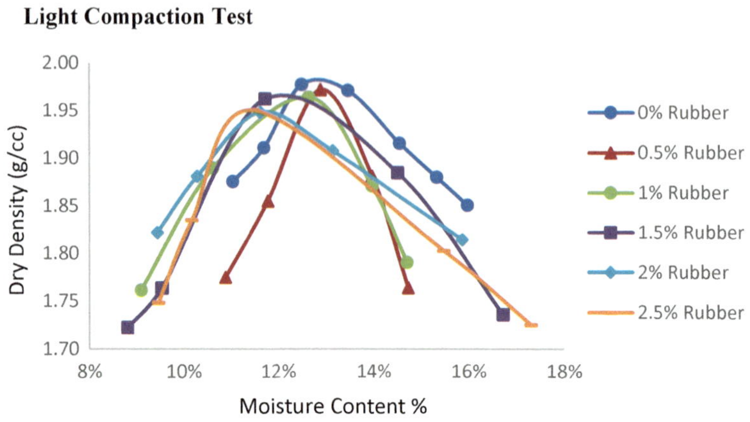

Fig. 1 OMC and MDD of light compaction test for different proportions of rubber fibers

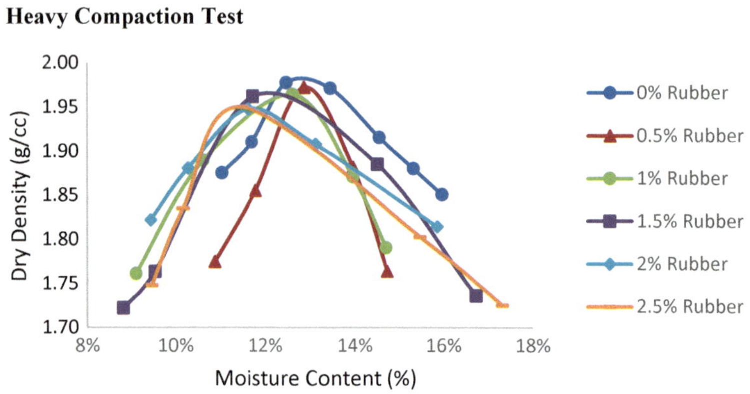

Fig. 2 OMC and MDD of heavy compaction test for different proportions of rubber fibers

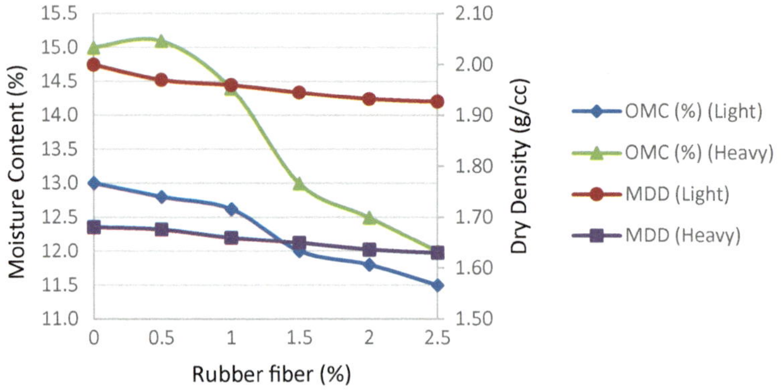

Fig. 3 Comparison of OMC and MDD for light and heavy compaction test

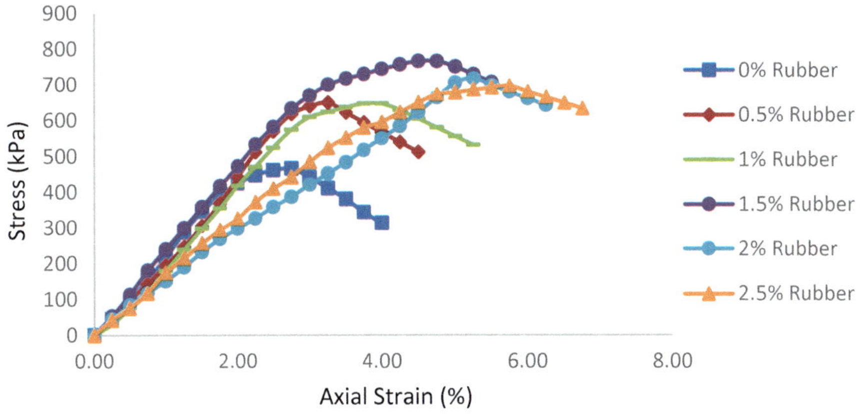

Fig. 4 UCS test for different proportions of rubber fibers

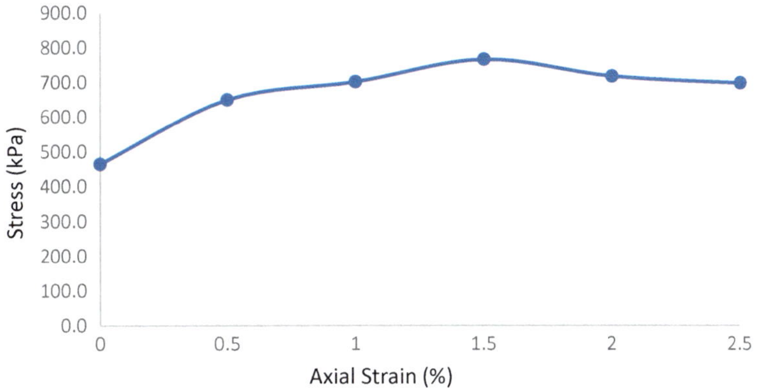

Fig. 5 Variation of UCS with rubber fibers

3.2 Unconfined Compressive Strength Test

It was inferred that with the addition of rubber fibers, UCS increased initially up to 1.5% addition and thereby decreased with further addition of rubber fibers (Figs. 4 and 5).

3.3 California Bearing Ratio Test

Un-Soaked CBR Test

See Fig. 6.

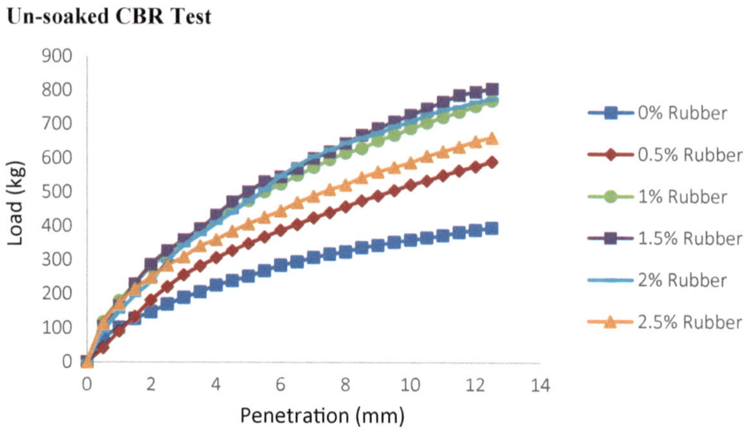

Fig. 6 Un-soaked CBR test for different proportions of rubber fibers

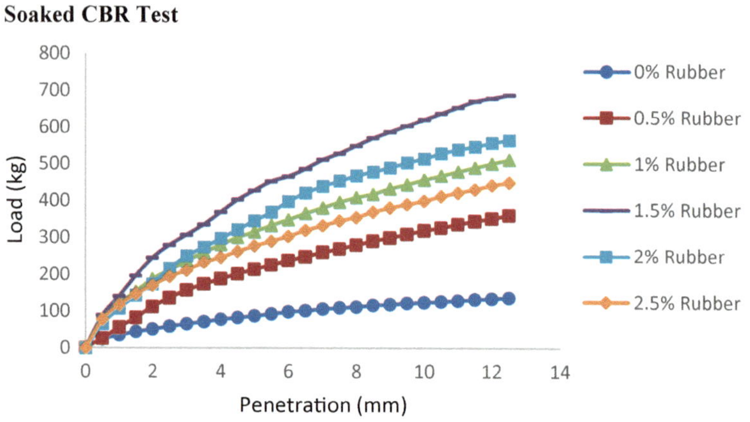

Fig. 7 Soaked CBR test for different proportions of rubber fibers

Soaked CBR Test

It was inferred that with the addition of rubber fibers, CBR value for both un-soaked and soaked samples increased initially up to 1.5% addition and thereby decreased with further addition of rubber fibers (Figs. 7 and 8).

3.4 *Dispersion Test*

This test is helpful in indicating the potential for dispersion of soils. Dispersive soils are prone for soil erosion easily; hence, soils with least dispersion are recommended.

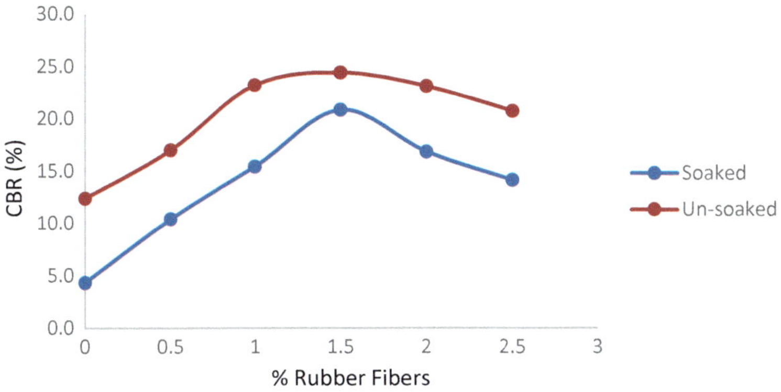

Fig. 8 Comparison of un-soaked and soaked CBR values

It was inferred from the test that dispersion of clay with addition of rubber fibers reduced compared to that of virgin clay. It is obvious that the rubber fibers act as reinforcement when added to clay, thereby holding the clay particles intact, resulting in reduced dispersion.

3.5 *Durability Test*

Wetting and Drying Test

The wetting and drying test was performed for the clay samples to check the durability of clay samples subjected to fluctuations in moisture content. After each wetting and drying cycle, the weight loss for the samples was calculated and durability analysis was performed.

The wetting and drying test was performed for the clay samples, and it was inferred that both virgin clay samples and clay samples with rubber fibers failed in the first cycle. Hence, there was no significant change when rubber fibers were added to the clay.

Freezing and Thawing Test

The freezing and thawing test was performed for the clay samples to check the durability of clay samples subjected to fluctuations in temperature. After each freezing and thawing cycle, the weight loss for the samples was calculated, and durability analysis was performed (Fig. 9).

The freezing and thawing test was performed for the clay samples, and it was inferred that the weight loss for 1.5% rubber clay was less when compared to virgin clay in each cycle. Thus, addition of 1.5% rubber fibers to clay can increase the durability.

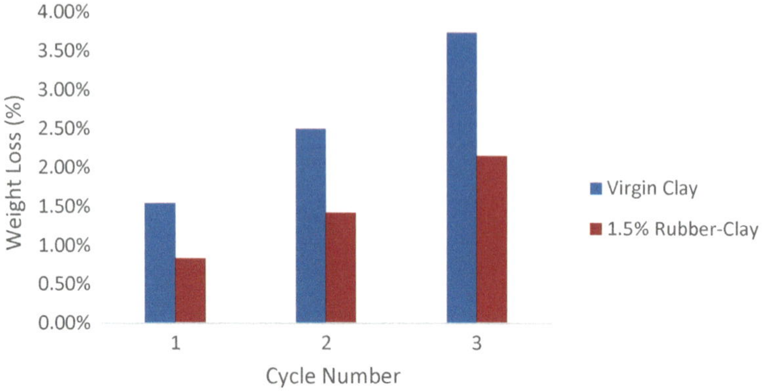

Fig. 9 Weight loss % for freezing and thawing cycles

4 Conclusion

- It was inferred that with the addition of rubber fiber, OMC and MDD decreased. This was due to low specific gravity and low water absorption capacity of rubber fibers in comparative to the virgin clay.
- It was inferred that with the addition of rubber fiber, UCS increased initially up to 1.5% addition and thereby decreased with further addition of rubber fibers. Hence, the optimum percentage of rubber fiber for maximum UCS was 1.5%.
- It was inferred that with the addition of rubber fiber, CBR value increased initially up to 1.5% addition and thereby decreased with further addition of rubber fibers. Hence, the optimum percentage of rubber fiber for maximum CBR was 1.5%.
- It was inferred that with the addition of rubber fiber, there was no notable change in wetting and drying test. However, the weight loss of clay in freezing and thawing test was reduced with addition of rubber fibers which can increase the durability of clay.
- Addition of rubber fibers to the medium plasticity clay improved its strength properties by enhancing the UCS and CBR values at optimal percentage. Hence, this method can be used in embankment applications where higher strength (CBR) is a primary requirement.

References

1. Melik B, Trouzine H, Rabehi M (2019) Influence of waste tire rubber fibers on swelling behavior, unconfined compressive strength and ductility of cement stabilized bentonite clay soil. Construct Build Mater 208:304–313

2. Apinun B, Horpibulsuk S, Udomchai A, Suddeepong A, Takaikaew T, Vichitcholchai N, Horpibulsuk J, Arulrajah A (2021) Durability improvement of cement stabilized pavement base using natural rubber latex. Transport Geotech 28:100518
3. IS 2720—Methods of test for soil (1983) Bureau of Indian Standards, New Delhi
4. Mukesh K (2021) Clay soil stabilization by utilizing secondary lime and rubber tire powder. Mater Today: Proc 37:3471–3474
5. Carlos Hidalgo S, Garzón-Roca J, Martínez Fernández P, Elvira Garrido de la Torre M, Franco RI (2016) Swelling potential reduction of Spanish argillaceous marlstone facies tap soil through the addition of crumb rubber particles from scrap tires. Appl Clay Sci 132:768–773
6. Fengchi W, Ping X, Zhou J, Kang T (2019) Effects of crumb rubber on the frost resistance of cement-soil. Construct Build Mater 223:120–132
7. Yadav JS, Tiwari SK (2017) Effect of waste rubber fibers on the geotechnical properties of clay stabilized with cement. Appl Clay Sci 149:97–110
8. Yadav JS, Tiwari SK (2017) A study on the potential utilization of crumb rubber in cement treated soft clay. J Build Eng 9:177–191

Improvement of Soft Ground Using Nano-Chemical Stabilization

Regi P. Mohan and K. Rangaswamy

1 Introduction

1.1 Background

Low-strength, highly compressible soils with clay or silt on-site make the soil unsuitable for construction activities and pose a problem to civil engineers. It is uneconomical to get the entire soil deposits replaced. The engineering properties of these problematic soils need to be altered to achieve desired values. Ground improvement techniques, including compaction, dewatering, treatment with admixtures, reinforcement techniques, etc., are available to treat the soft clay grounds.

An economical solution to stabilize soils with low strength and high deformation is chemical soil stabilization, in which admixtures are used to improve the properties of soils. Most traditional additives, like lime, fly ash, rice husk ash, etc., require high energy demand for their production, which contributes to global warming [4, 15]. Besides, these have a high cost of excessive maintenance, poor soil structure, leaching away during Indian monsoons, secondary chemical pollution, and detrimental environmental impacts associated with construction [2]. All the above factors have tended to reduce the application of stabilization with traditional chemicals over the past decade.

At present days, science has advanced far better from the olden days. From an economic point of view, many alternative admixtures are being used nowadays to quicken the process of soil improvement and to avoid delays in construction activities. Thus, many more non-traditional additives are developed from the advanced search for new materials. Research is being carried out in this area using alternative

R. P. Mohan (✉) · K. Rangaswamy
Department of Civil Engineering, National Institute of Technology Calicut, Kozhikode, Kerala 673601, India
e-mail: regimohan2007@gmail.com

B. T. Jose et al. (eds.), *Proceedings of the Indian Geotechnical Conference 2022 Volume 4*, Lecture Notes in Civil Engineering 479,
https://doi.org/10.1007/978-981-97-1753-8_24

stabilizing agents such as bio-enzymes, nano-chemicals, nano-materials, and ionic polymers, to understand their behaviour in soil [3, 14, 17].

1.2 Stabilization Using Nano-Materials

Many studies are being conducted on different nano-materials, nano-chemical compounds, etc., to find their effectiveness as soil stabilizers. Unlike traditional materials such as lime and cement, the nano-materials are permanent, non-toxic, biologically and chemically inert, and have excellent durability. The studies conducted by [9, 16] indicated that adding nano-materials to soil leads to an improvement in the strength, Atterberg's limits and a decrease in the permeability properties of soil. Due to its high specific surface area, even a tiny amount of nano-material addition could cause a significant change in the physical and chemical properties of soil [17]. A study on the effect of different nano-materials (nano-clay, nano-MgO, and nano-alumina) on engineering characteristics showed that an increase in nano-clay content increases Atterberg's limits due to the higher specific surface area of nanoparticles. However, the increase in nano-MgO and nano-alumina decreased Atterberg's limits [12]. California Bearing Ratio (CBR) strength values increased considerably with nano-chemical addition, and the design of pavement resulted in a reduced thickness of about 25% lower than that with untreated soil due to reactions of clay soil with nano-chemical [8, 10, 11, 13]. Thus, nano-material stabilization will be of great significance in sustainable ground improvement. From all these studies, it is observed that nanotechnology-based stabilization makes the soil slightly less plastic and improves the strength characteristics of the clay soil to a higher degree. Thus, the main focus of this study is to minimize the use of environmentally detrimental traditional stabilizers like cement, lime, etc.

2 The Objective of the Study

This paper discusses the influence of an organosilane-based nano-chemical addition on the geotechnical engineering properties of a low-strength soil for different curing periods and varying dosages of nano-chemicals. The scope of this paper is limited to studying the variation in the unconfined compressive strength characteristics of a weak soil stabilized with nano-chemical and 1% cement after 7 and 14 days of curing period.

3 Materials and Methods

3.1 Materials

The natural untreated clay soil used in the study has been procured locally from Pantheerankavu, Kozhikode, Kerala, India, at a depth of 1.5–2.0m from the ground level. The native soil used in the present study is termed virgin soil (VS). Characterization tests were conducted in the laboratory according to Indian Standard specifications, and Table 1 summarizes the characteristics of virgin soil used in the study.

The nanotechnology-based organosilane nano-chemical used in the present study is terrasil, manufactured by Zydex industries in Vadodara, Gujarat, India. It is a concentrated liquid in golden brown colour and is water-soluble and designated as a nano-chemical (NC) in the present study. The chemical composition of the nano-chemical used in the study, as provided by the manufacturer, is listed in Table 2.

Based on economic considerations and to minimize environmental pollution, cement usage in the present study is restricted to 1% by weight of soil to act as a binder. Several published papers proved that the strength of nano-material stabilized clay soils is significantly more enhanced in the presence of even a tiny amount of cement content instead the shear strength of clay soils stabilized by nano-chemicals/materials alone. Nano-chemical can act as an accelerator in the hydration of cement,

Table 1 Geotechnical characteristics of the virgin soil

Property	Value
Silt size fractions (%)	60.0
Clay size fractions (%)	22.0
Liquid limit (%)	78.8
Plastic limit (%)	44.1
Plasticity index (%)	34.7
Shrinkage limit (%)	27.5
IS soil classification	MH
Maximum dry density (kN/m^3)	12.6
Optimum moisture content (%)	39.1
Unconfined compressive strength (kPa)	50.0

Table 2 Chemical composition of nano-chemical as provided by the manufacturer [18]

Chemical compound	Value (%)
Hydroxyalkyl-alkoxy-alkylsil	65–70
Benzyl alcohol	25–27
Ethylene glycol	3–5

which imparts strength development due to accelerated pozzolanic reactions in nano-chemical-treated soil in the presence of cement binder.

3.2 Sample Preparation and Testing

The treated soil samples were prepared as follows: initially, the binder cement at 1% by weight of soil is introduced into the air-dried pure clay and mixed in a dry state to distribute uniformly. Later, the specific dosage of nano-chemical is injected into the OMC of the soil to obtain the nano-chemical solution. Nano-chemical solution has been sprayed directly on the dry soil–cement mixture and mixed thoroughly to spread uniformly. The present study chooses the trial dosages of nano-chemicals in the range of 0.03% to 0.05% by the weight of clay soil. Table 3 tabulates the composition of different soil mixes with designations used in the present study.

Unconfined Compressive Strength Test. The present study has also conducted the standard proctor compaction (SPCT) tests according to the Indian Standard light compaction test [5], to examine the influence of nano-chemicals on the variation of compacting density and optimum moisture content of clay soil. The samples were cured for three days before conducting the SPCT tests [15], and it has found that there were slight changes in the maximum dry density (MDD) and OMC since the replacement of soil is very minimal with 1% cement and up to 0.05% nano-chemical by the weight of soil. Hence, the treated soil specimens for the UCS testing have been prepared at a maximum dry density of virgin soil for comparison with the parent clay soil and to maintain uniformity.

Unconfined compression tests were carried out on the nano-chemical-treated soil specimens with 1% cement binder after being cured at 7 and 14 days to examine the influence of nano-chemical on UCS strength variation of clay soil. The curing period allows the chemical reactions to take place and examine the effect of curing on strength improvement. UCS test specimens were stored under humid conditions with a covered curing technique to prevent moisture loss and cure the samples. Three treated soil specimens at particular nano-chemical dosage have been cast in test moulds of constant volume (38 mm diameter and 76 mm height) at compaction density of virgin soil and cured for the specified curing period. UCS testing has been

Table 3 Composition of soil mixes used in the study

Mix	Designation
VS	Virgin Soil
VS + C1	Virgin Soil + 1% Binder
VS + C1 + NC1	Virgin Soil + 1% Binder + 0.03% NC
VS + C1 + NC2	Virgin Soil + 1% Binder + 0.04% NC
VS + C1 + NC3	Virgin Soil + 1% Binder + 0.045% NC
VS + C1 + NC4	Virgin Soil + 1% Binder + 0.05% NC

conducted at a strain rate of 1.5 mm/min according to Indian Standards [6]. The average unconfined compression strength for each nano-chemical soil mixture had reported after conducting three UCS tests. The optimum dosage of nano-chemical is determined that gives the maximum unconfined compression strength of treated clay soil.

Scanning Electron Microscopy: The sample surface morphology had investigated using Field Emission Scanning Electron Microscopy (FESEM). FESEM image analysis was carried out to peer into the morphological details of untreated and nano-chemical-treated samples at optimum dosage after seven days of curing. The image analysis had conducted on Hitachi SU 6600, Japan, at the SEM centre of the Nano-science Department, National Institute of Technology Calicut, Kerala, India.

4 Results and Discussion

A series of UCS tests had performed on various nano-chemical soil mixtures to determine the effect of nano-chemical on UCS of clay soil at a curing period of 7 and 14 days.

4.1 *Strength Characterization*

Unconfined Compressive Strength. The effect of nano-chemical on UCS of clay soil at 7 and 14 days of curing is tabulated in Table 4. It presents the percentage increase in strength with nano-chemical dosage and curing times. It shows the soil mixture contains 0.045% nano-chemical along with 1% cement binder, providing maximum strength improvement of about 184–253% increase over the strength of clay soil. The UC strength has improved significantly with the increased curing period. The UC strength of treated soil mixtures has been found to increase by about 72–253% to the strength of clay with an increase in the curing period from 7 to 14 days. Hence, the nano-chemical with 0.045% is the optimum dosage that gives the maximum UC strength of clay soil.

Table 4 Effect of nano-chemical addition on UC strength

Soil mix	7-day UC strength (kPa)	% Increase in strength	14-day UC strength (kPa)	% Increase in strength
VS	50.0	–	50.0	–
VS + C1 + NC1	86.3	72.6	115.4	130.8
VS + C1 + NC2	123.9	147.8	166.6	233.2
VS + C1 + NC3	142.2	184.4	176.6	253.2
VS + C1 + NC4	106.2	112.4	141.9	183.8

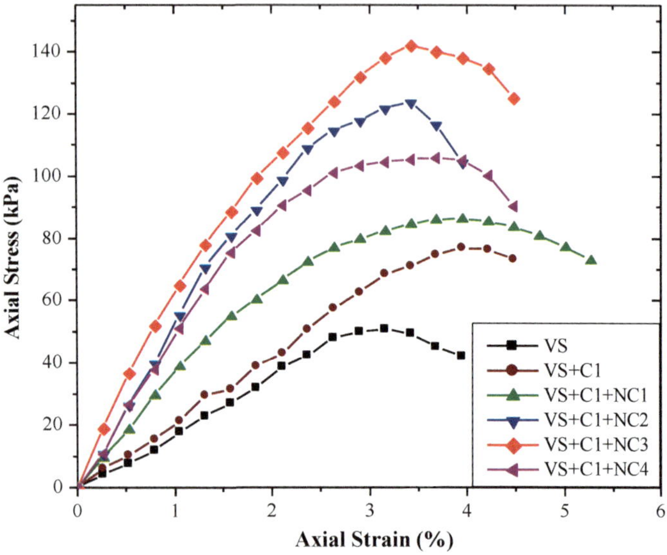

Fig. 1 Stress–strain relations of soil–cement nano-chemical combinations at 7 days of curing

The stress–strain relations of various soil mixtures obtained under UCS testing at 7 and 14 days of curing are shown in Figs. 1 and 2, respectively. The results show that the nature of the stress–strain response is similar for all the soil mixtures. Initially, the relation is linear, and then the stresses increase with strain in nonlinear until it reaches a failure point, i.e. maximum compression strength (UCS), and decreases. It can see that the failure stress (UCS) depends on the dosage of the nano-chemical, and it increases with an increase in nano-chemical up to a particular value and then decreases. It also finds that the flexibility, i.e. increase of strain at failure of soil mixtures, increases by adding nano-chemicals up to the optimal dosage into the clay soil. The strength improvement of nano-chemical-treated clay in the presence of 1% cement binder is more significant than the strength of clay soil untreated with 1% cement. The results comply with published papers that investigated the effect of nano-chemicals on the strength of clay soils [7, 8].

4.2 Microstructural Characterization

Scanning Electron Microscopy: Fig. 3 shows the FESEM images of virgin clay soil, and Fig. 4 shows the FESEM image of the nano-chemical-treated soil mixture (VS + C1 + NC3) containing the optimum dosage of nano-chemical (0.045%) after seven days of curing.

It can be seen from the figures that there is a significant change in the soil fabric of the treated soil mixture prepared at the optimum dosage of nano-chemical compared

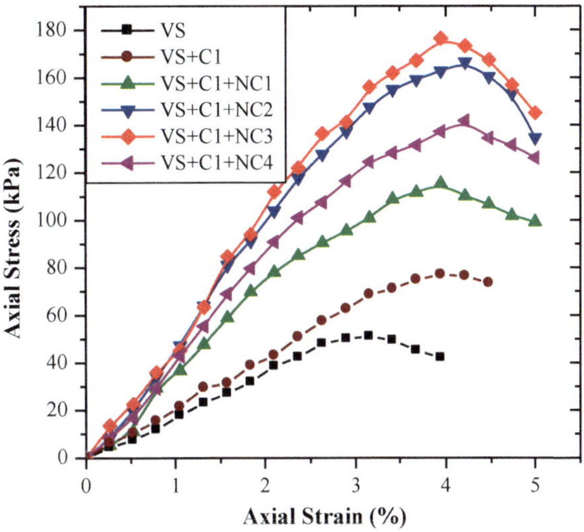

Fig. 2 Stress–strain relations of soil–cement nano-chemical combinations at 14 days of curing

Fig. 3 FESEM image of untreated soil after 7 days of curing

to virgin soil. In nano-chemical-treated soil, the particles are bound together and are more closely packed or agglomerated, with a noticeable reduction in void spaces. It occurs because van der Waals forces develop strongly between the nanoparticles. Thus, addition of nano-chemical considerably reduced the porosity as it fills the micro- and nano-pores between soil particles.

Fig. 4 FESEM image of the treated soil mixture (VS + C1 + NC3) after 7 days of curing

5 Mechanism of Stabilization

The main ingredient of nano-chemical used in the present study is hydroxyalkyl-alkoxy-alkyl silanes. The considerable improvement in unconfined compressive strength is due to the strong reaction of nano-chemical with silanol groups of soil and the resulting siloxane bondage [1], which forms a breathable in-situ membrane. The nano-chemical reacts with the soil particles and makes the surfaces waterproof, stiffening the soil and thus increasing the bond between soil particles [18]. Beyond optimum dosage, the strength decreases due to the loosening of the interaction between soil and nano-chemical in the presence of excess in-active nano-chemical.

6 Conclusions

A novel technique that can be applied as an alternative to environmentally harmful traditional soil stabilization additives is explained in this paper. A comprehensive experimental study was conducted on the soil before and after stabilizing with 1% cement and a nanotechnology-based organosilane compound to determine the effect of nano-chemical addition on the UC strength characteristics of the clay soil. The mechanism and microstructural characterization of nano-chemical-treated soil has presented to support the strength improvement of clay soil with the addition of optimum dosage of nano-chemical and 1% cement binder.

The salient conclusions drawn from the study are as follows:

- The UC strength of clay soil, along with 1% cement as a binder, increased with an increase in the nano-chemical dosage up to 0.045% of the weight of clay soil and is about 253% higher than the UC strength of untreated virgin soil at a 14-day curing period.

- The UC strength of nano-chemical stabilized soil after a 14-day curing period reveals a higher strength improvement than at seven days curing period. It is due to the improved chemical interactions of nano-chemical to flocculate and agglomerate the soil particles, thereby stiffening the soil with curing time.
- The FESEM images of nano-chemical-treated soil showed that interparticle voids had reduced considerably, and the structure had changed to flocculated with the addition of nano-chemical.

In summary, treating weak subgrade soil with an organo-silane-based nano-chemical at a dosage of 0.045% of the weight of clay soil along with 1% cement content is the best combination to achieve higher UC strength. The shear strength achieved by this stabilization method is perfectly sufficient in a way that locally available field soil deposits could enhance the shear strength and can be used in pavement construction works. Thus, typical weak soils unsuitable for construction activities could be stabilized by adding organo-silane-based nano-chemical compound.

References

1. Arkles B, Steinmetz JR, Zazyczny J, Mehta P (1992) Factors contributing to the stability of alkoxysilanes in an aqueous solution. J Adhes Sci Technol 6(1):91–104. https://doi.org/10.1163/156856192X00133
2. Buazar F (2019) Impact of biocompatible nano silica on green stabilization of subgrade soil. Sci Rep 9(1):1–9. https://doi.org/10.1038/s41598-019-51663-2
3. Carroll D, Starkey HC (1971) Reactivity of clay minerals with acids and alkalies. Clays and Clay Mineral 19(5):321–333. https://doi.org/10.1346/CCMN.1971.0190508
4. Ewa DE, Egbe EA, Akeke GA (2016) Effects of nano-chemical on geotechnical properties of Ogoja subgrade. J Res Inform Civil Eng 13(1):820–831
5. IS 2720-Part 7 (1980) Methods of test for soils- laboratory determination of water content- dry density relation using light compaction, Bureau of Indian Standards, New Delhi
6. IS 2720-Part 10 (1991) Methods of test for soils- laboratory determination of unconfined compressive strength, Bureau of Indian Standards, New Delhi
7. Lekha BM, Goutham S, Ravi Shankar AU (2013) Laboratory investigation of soil stabilized with nano-chemical. In: Proceedings of Indian geotechnical conference, Roorkee, pp 1–7
8. Mrudul UV, Damodariya SM (2016) Laboratory investigation of soil stabilized using terrasil. Int J Scient Res Developm 3(12):537–539
9. Noll MR, Bartlett C, Dochat TM (1992) In situ permeability reduction and chemical fixation using colloidal silica. In: National outdoor action conference, Las Vegas, pp 443–457
10. Patel AN, Mishra CB, Pancholi VV (2015) Scientifically surveying the usage of terrasil chemical for soil stabilization. Int J Res Advent Technol 3(6):77–84
11. Prakash MM, Shravan K (2015) Effect of nanochemicals on properties of the black cotton soil. In: National conference on sustainable infrastructure development 2015, Chandigarh
12. Priyadharshini R, Arumairaj PD (2015) Improvement of bearing capacity of soft clay using nanomaterials. Int J Scient Res 4(6):218–221
13. Roshni S, Jeyapriya SP (2017) Experimental study on the use of nanochemical and cement in the modification of subgrade. Int J Sci Eng Res 8(3):1868–1872
14. Sol-Sanchez M, Castro J, Urena CG, Azanon JM (2016) Stabilization of clayey and marly soils using industrial wastes: ph and laser granulometry indicators. Eng Geol 200:10–17. https://doi.org/10.1016/j.enggeo.2015.11.008

15. Taha MR, Khan TA, Jawad IT, Firoozi AA (2013) Recent experimental studies in soil stabilization with bio enzymes-a review. Electron J Geotech Eng 18:3881–3894
16. Yonekura R, Miwa M (1993) Fundamental properties of sodium silicate-based grout. In: Geotechnical conference, Singapore, pp 439–444
17. Zhang G (2007) Soil nanoparticles, and their influence on engineering properties of soils. Geotech Special Publicat (ASCE) 1–13. https://doi.org/10.1061/40917(236)37
18. ZYDEX industries home page. https://zydexindustries.com/soil-stabilization/. Last Accessed 7 Oct 2022

Numerical Modeling on Behavior of Annular Stone Column in Layered Soils

Panuganti Aditya, Boya Manikanta Reddy, and M. Heeralal

1 Introduction

At present, the ascending order of growth in the population leads to huge construction in every area of land. Whichever soil type it may be, there will be no free land available. In the meantime, low shear strength and high compressible soils pose high settlements, which leads to failure of the structure. Engineering applications, which are the remedy for those types of complications for the failure of structures, in particular, geotechnical engineers made a possible solution for the improvement of soil by the application of stone columns. Many past studies dealing with stone columns refer to multiple methods of trials in the field before execution [1]. The stone column is designed with past performing knowledge and no proper guidelines or codes are available. By drilling a borehole in diameter, TH-02–1431 granulated stones are poured into the borehole. Ordinary and encased stone columns were used in the ground modifications. The stone columns, which are placed under moving loads, can give the deformation behavior with analytical calculations. In the clayey soils, experimental results with the making of models by the implementation of stone columns, proved the un-drained shear strength and young's modulus can be

P. Aditya (✉)
Master of Technology, Department of Civil Engineering, VNR Vignana Jyothi Institute of Engineering and Technology, Hyderabad 500090, India
e-mail: 1999adithya369@gmail.com

B. M. Reddy
Department of Civil Engineering, VNR Vignana Jyothi Institute of Engineering and Technology, Hyderabad 500090, India

M. Heeralal
Department of Civil Engineering, National Institute of Technology, Warangal 506004, India
e-mail: mhl@nitw.ac.in

B. T. Jose et al. (eds.), *Proceedings of the Indian Geotechnical Conference 2022 Volume 4*, Lecture Notes in Civil Engineering 479,
https://doi.org/10.1007/978-981-97-1753-8_25

acceptable [3]. The small area of group columns under the pad footings, the settlement, and confinement analysis were done by using 3D models in PLAXIS. The author concludes that along the total length of the column, small areas performed well compared to large areas in spacing of columns [2]. By using the same software, another author performed settlement and bulging analysis with field data by changing the length of the column. They reported that the L/D ratio at 8.3 had an impact on the settlement performance of the column with optimum bulging and stress ratio [4].

In another research under transport infrastructure, the retarding of settlements and pore water pressure were analyzed by using analytical and numerical models with effects of arching, clogging, and smear zone in soft soils [6]. Stone column behavior in c-soils was observed in a study on embankment height and the presence of fines in embankment soil. Numerical models were performed using FLAC software [8]. Improved train critical speed, length of the stone column under cyclic loads with dynamic amplification factor, and parametric analysis performed with 3D numerical models [9]. A numerical study was carried out to find out the critical length of the encasement provided for stone columns. The study reveals the encasement length should be slightly lower compared to the stone column length [10]. In this present manuscript, the modification of stone columns with the provision of geo-reinforcement will be described. Annular stone columns are modified stone columns, which have an internal and external diameter with a hollow at the center of the column. In the area of circular thickness, stone aggregates are placed, and the remaining area will be filled with soil. This type of stone column is going to be analyzed by using numerical models in the geotechnical software. The model is performed in 3D using finite element analysis in PLAXIS software. The deformation behavior and load-carrying capacity of a solitary annular stone column are investigated with numerical models. These models are performed in layered soils and two-layered systems were considered. Complete numerical models created in a 3D application were used for this study's model test in order to fully capture the potential of the novel approach. The paper does not include constructability for experiments or installations.

2 Methodology

To ascertain the loading capacity of an annular stone column under a given displacement, the finite element analysis approach is applied. PLAXIS 3D is the software used in the research of analysis for computational modeling. The work done by A.P. Ambily and S.R. Gandhi is taken into consideration for validating the finite element model. The load-carrying test was numerically analyzed. The taken model is 210 to 420 mm diameter and height of soft clay bed is 450 mm. The stone column diameter as 100 mm and height 450 mm is placed at the center of the soft clay bed with prescribed displacement of 30 mm. The validation result plotted between experimental model and numerical model. The load vs displacement graph obtained from PLAXIS 3D program (Fig. 1) shows well matched with experimental model.

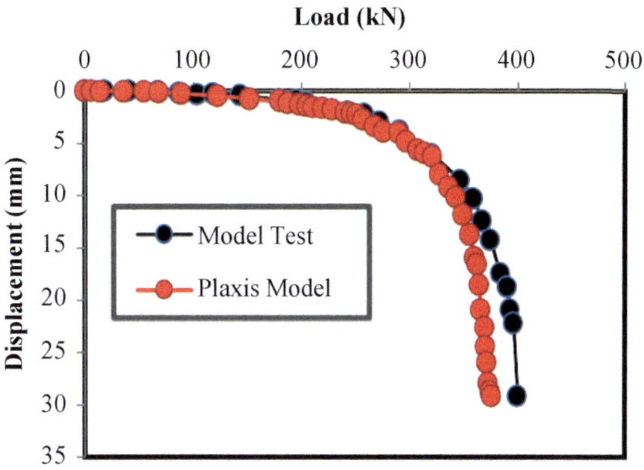

Fig. 1 Validation of PLAXIS 3D program

2.1 Numerical Model of Annular Stone Column 3D

A 3D program is used to analyze the singular annular stone with thickness variation of two soil strata. The properties of the clayey soils, annular stone column, and parameter properties were considered as per [7] and [5]. With the results obtained, the column's capacity for supporting loads can be assessed. The inner and outer diameters of the column were measured at 0.4 and 0.8 m, respectively, for the model construction. The diameter (D) of a stone column is taken into account in the modeling process along with variations in the overlying soil's thickness. The upper soil thickness varies in a dual soil system as 1D, 2D, 3D, and 4D.

The model is a unit cell model with dual soil strata and an annular stone column placed at the center of the soil body. In the procedure of numerical analysis for load identification, prescribed settlement is taken as 50 mm on the surface of a stone column. The column is covered with geo-reinforcement along the length of 10 m inside and outside of the annular portion. The geo-reinforcement stiffness (J) is 3000 kN/m. The dual soils considered for model testing are soft clay and stiff clay. A Mohr–Coulomb failure criterion model with drained and un-drained conditions in 3D analyses was done. The soil model material properties are taken for the analysis given below.

Properties:

The dual soil system is considered with two types of clay. The soil conditions with some specifications are taken with stone properties.

Table 1 represents the specifications of the soil model which are used in PLAXIS program for the load-carrying behavior of the annular stone column. Table specifications or properties are considered from literatures (5 and 7). The changes in the thickness of dual soils will affect the stone column performance that will be presented in the outcome of the analysis. The parametric study was also done by considering the internal friction angle of the stone column. The established outcomes will be presented in graphical representation with a comparison of conventional and

Table 1 Specifications for model

Considerations	Soft clay	Stiff clay	Annular stone column
γdry	17	19	20
γwet	17	19	20
Ec	2500	15,000	45,000
μ	0.4	0.35	0.15
su	25	–	–
C	–	75	0
ϕ	0	2	40
ψ	0	0	10

annular columns. As the internal friction angle is considered, the specifications are in the ranges given as and considered values are 35, 40, and 45 (ϕ). The established outcomes will be presented in graphical representation with a comparison of conventional and annular columns. The outcomes will generate a conclusion at last.

3 Outcomes and Colloquy

The specifications study outcomes present the load-carrying strength of the column in changing of dual soil width at top of each case. From the all, model verification plotted the result as graphical presentation. The displacement and load are considered in *x* and *y* coordinates, respectively. The generated graphs will expose the stone column condition, according to given terms of soil characters.

3.1 Alteration of Thickness Effect on Annular Column in S1 Series

Figure 2 indicates S1 series load versus displacement graph. S1 series exhibit the soft clay as top soil and stiff clay is bottom one. For each case of thickness alteration of top soil, express the software result in the form of graphical curves. Outcome indicates the loading strength of the column gradually slowdowns, by enhancement of soft clay thickness from 1 to 4D. That is the alteration of loading strength due to low stiffness soft clay. The prescribed displacement given for the column and due to low stiffness of soft clay, loading will be slowdowns as the increment in thickness of top clay soil.

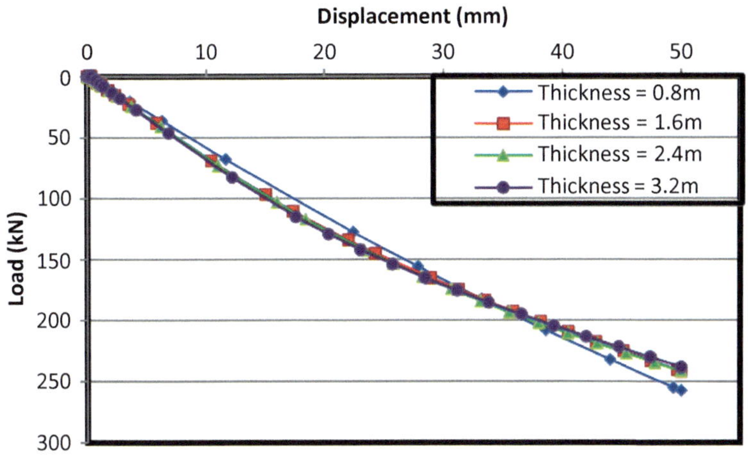

Fig. 2 Single annular column—load vs displacement for change in upper soft clay thickness covered with geo-reinforcement

3.2 *Alteration of Thickness Effect on Annular Column in S2 Series*

Figure 3 conveys the load-carrying strength of the annular column graph with prescribed displacement of 50 mm. The alteration of top soil is stiff clay and soft clay is bottom. Due to soil more stiffness of stiff clay exhibits the enhancement in loading strength of the column. In the same way, the thickness alteration from 1 to 4D the load gained up from 431 to 519 kN, respectively. The provided geo cover with stiff clay enhances the strength of the column loading ability.

The annular column condition varied for dual soil alteration with soils different specifications. Soft clay at upside of the stiff clay as series 1 presented the solution for loading strength of column has shrink for enhanced thickness of soft clay. Stiff clay at upside of the soft clay as series 2 represents the outcome for column loading strength has gained up for improved thickness of stiff clay.

3.3 *Alteration of Internal Friction Angle Effect on Annular Column in S1 Series*

Figure 4 presents the alteration of internal resistance angle of stone column. As the soft clay overlies on stiff clay, the loading strength of the annular column was increased with an enhancement of internal friction angle of stone. The friction increment in stones of column will catch the geo-reinforcement with dual soils. For each enhancement of friction angle from 35 to 45 degrees, the load-carrying strength also

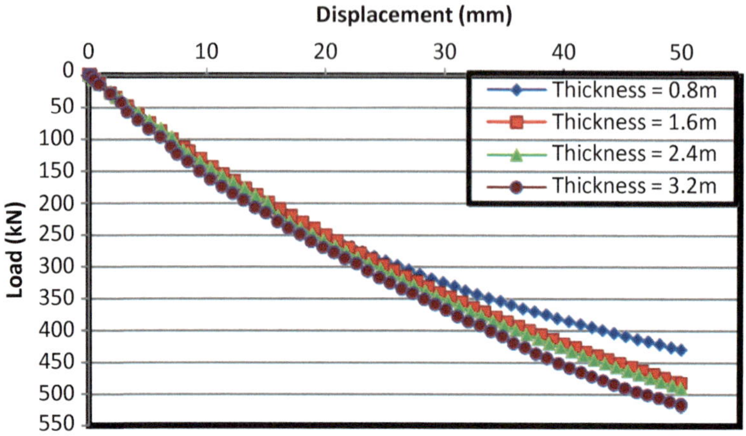

Fig. 3 Single annular column—load vs displacement for change in upper stiff clay thickness covered with geo-reinforcement

gained up 22% for 45 degrees friction angle than 35 degrees at 1D, 2D, 3D, and 4D thicknesses of soft clay.

3.4 *Alteration of Internal Friction Angle Effect on Annular Column in S2 Series*

Figure 5 shows the alteration of internal resistance angle of stone column. As the stiff clay overlies on soft clay, the loading strength of the annular column was increased with an enhancement of internal friction angle of stone same as in the series 1. The friction increment in stones of column will catch the geo-reinforcement with dual soils. For each enhancement of friction angle from 35 to 45 degrees, the load-carrying strength also gained up 30% for 45 degrees friction angle than 35 degrees at 1D, 2D, 3D, and 4D thicknesses of stiff clay. This reveals that stiff clay has more stiffness that loading values are higher than soft clay at up in dual soils.

For the confirmation of annular structure of column in the field application, the comparison is made analyzed for both conventional and annular stone columns with elastic clay soil model specifications.

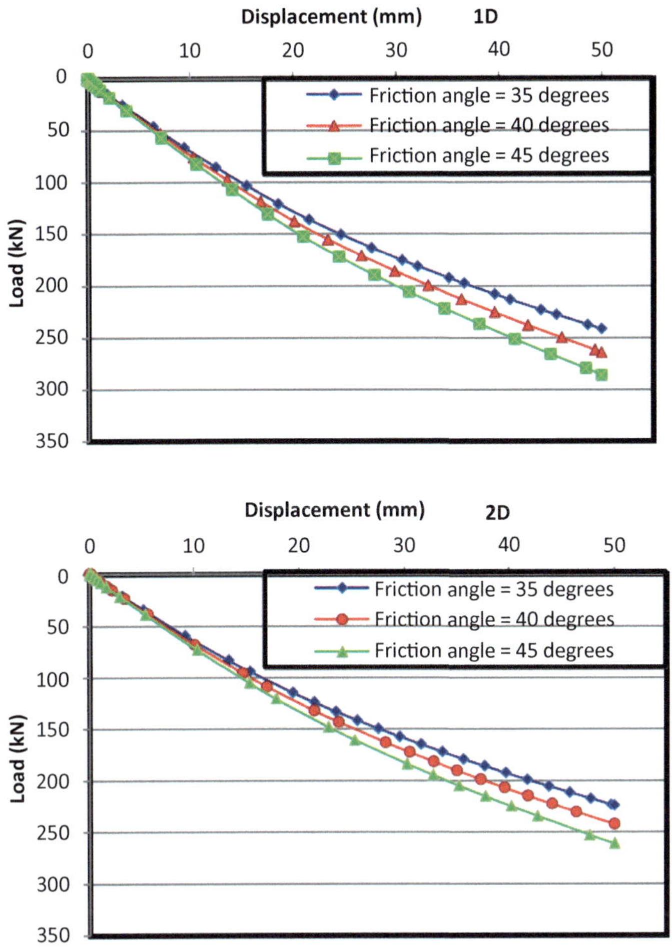

Fig. 4 Single annular column—load versus Displacement for change in internal friction angle of column covered with geo-reinforcement at 1D, 2D, 3D and 4D in series 1

3.5 Annular Column and Conventional Column Comparison

Figure 6 exhibits the comparison graph for both annular and conventional stone columns. The geometry for the 3D modeling taken the dimensions for both columns as mentioned below.

Conventional column:

Length of the column: 10 m Diameter of column: 0.7 m.

Soft clay: along full length of the column.

Annular column:

Length of the column: 10 m Diameter of column: 0.8 m.

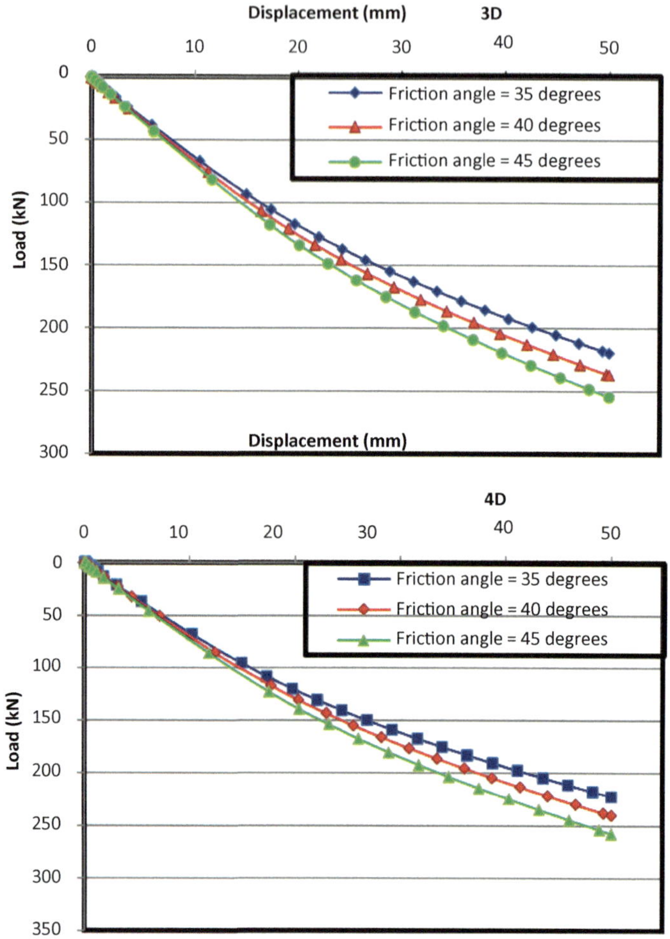

Fig. 4 (continued)

Soft clay: along full length of the column.

The stone and soft clay specifications are considered as same in the Table 1 for two models with covered geo-reinforcement of stiffness 3000 kN/m.

The above graph (Fig. 4) indicates loading strength of the columns for the prescribed displacement of 50 mm. The compared loading strength of two columns represents that the annular-structured column having more ability than the conventional column in soft clay soil.

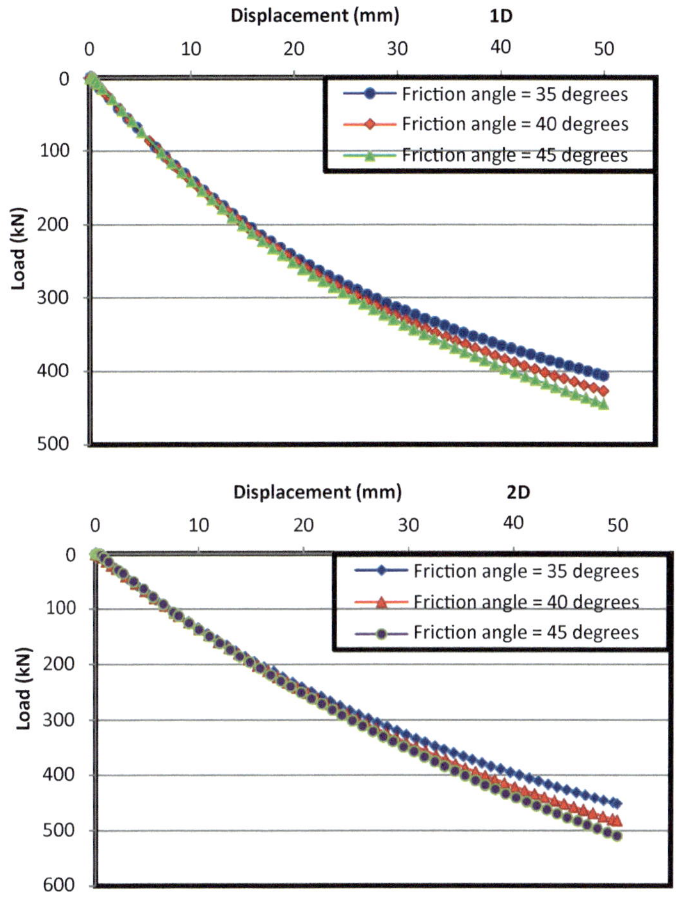

Fig. 5 Single annular column—load versus displacement for change in internal friction angle of column covered with geo-reinforcement at 1D, 2D, 3D, and 4D in series 2

4 Conclusions

The behavior of annular stone column by alteration of crest thickness in dual soil system by implementing cover around the column concludes that:

1. The column analysis in 3D exhibits accurate outcomes for the loading strength. Geometry for the model analysis considered as unit cell model for proper execution of stone column as in field.
2. As the soft clay soil at upside of stiff clay in dual soil system the annular stone column loading ability was slow downed due to low stiffness of soft clay in every case of alteration of thickness.
3. The loads of series 1 are in descending order of 7.8% decreased with enlargement of thickness of soft clay between 1 and 4D.

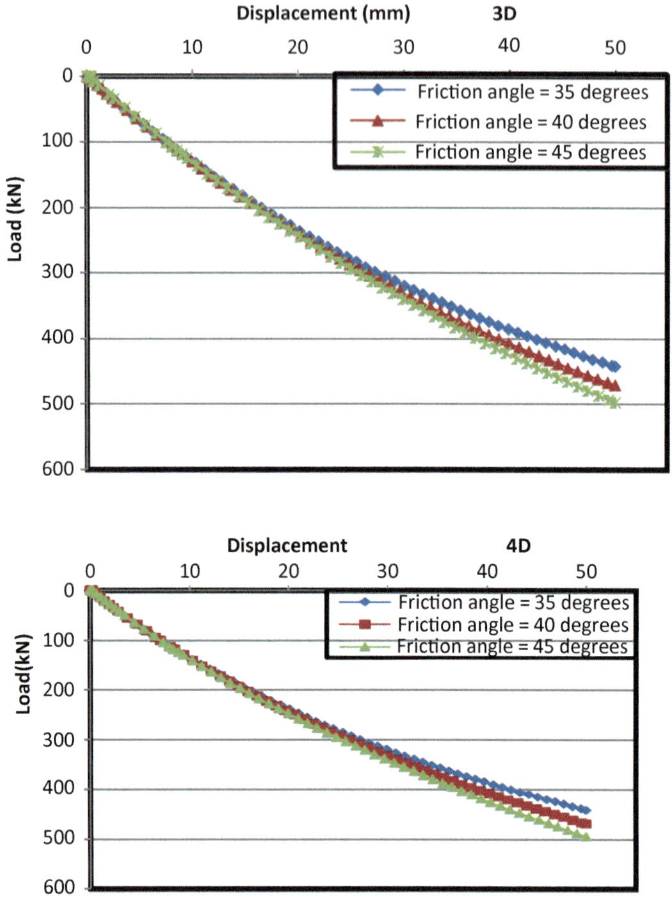

Fig. 5 (continued)

4. As the stiff clay soil at upside of soft clay the column loading ability was gained up due to thickness enlargement of top soil and stiff clay is stiffer than soft clay.
5. The loads of series 2 are in ascending order of 17% raised up between 1 and 4D thickness of stiff clay.
6. The comparison results shows that annular column can be executes better loading strength than conventional column due to the structure variation, that may leads to an isolator for moving loads on surface of annular column.
7. At last outcomes conclude that annular column can be applicable for load carrying in the field.

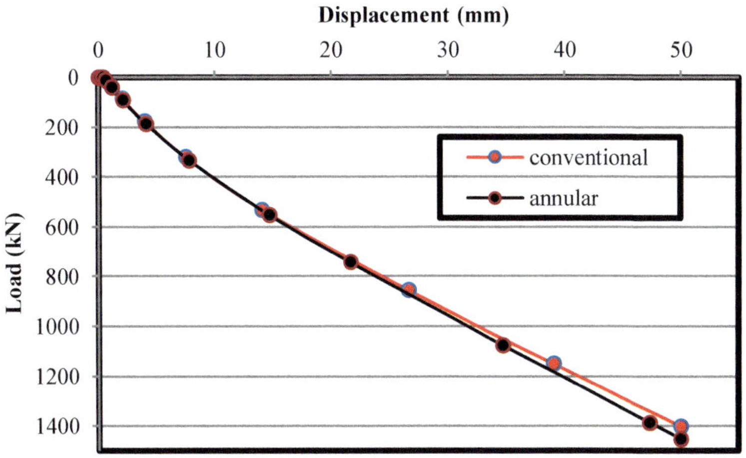

Fig. 6 Load versus displacement comparison for annular and conventional columns

References

1. Ambily AP, Gandhi SR (2004) Experimental and theoretical evaluation of stone column in soft clay. ICGGE-(2004)
2. Killeen MM, Mc Cabe BA (2014) Settlement performance of pad footings on soft clay supported by stone columns; a numerical study. The Japanese Geotech Soc (Soils and Foundations) 54(4):760–776
3. Frikha W, Bouassida M, Canou J (2015) Parametric study of a clayey specimen reinforced by a granular column. Int J Geo-mech.https://doi.org/10.1061/(ASCE)GM.1943-5622.000049
4. Al- Ani W et al. (2015) Numerical Modelling of stone columns in soft soil: Geotech Eng Infrastruct Developm. https://doi.org/10.1680/ecsmge.60678
5. Mohanty P, Samanta M (2015) Experimental and numerical studies on response of stone column in layered soil. Int J Geosynth Ground Enghttps://doi.org/10.1007/s40892-015-0029-z
6. Basack S, Indraratna B, Rujikiatkamjorn C (2016) Analysis of behavior of stone column stabilized soft ground supporting transport infrastructure: centre for geomechanics and railway. Engineering. https://doi.org/10.1016/j.proeng.2016.06.044
7. Tandel YK, Solanki CH, Desai AK (2013) 3D FE analysis of an embankment construction on GRSC and proposal of a design method: Hindawi publishing corporation. ISRN Civil Eng 348973:11. https://doi.org/10.1155/2013/348973
8. Das AK, Deb K (2019) Response of stone column—improved ground under c–φ soil embankment. Soils and Foundat. https://doi.org/10.1016/j.sandf.2019.01.003
9. Fernandez-Ruiz J, Miranda M, Castro J, Rodriguez LM (2021) Improvement of the critical speed in high-speed ballasted railway tracks with stone columns: a numerical study on critical length. Transport Geotech. https://doi.org/10.1016/j.trgeo.2021.100628
10. Miranda M, Fernandez-Ruiz J, Castro J (2021) Critical length of encased stone columns: geotextiles and Geomembrane: https://doi.org/10.1016/j.geotexmem.2021.05.003

Performance Assessment of Treated Jute Fiber Encased Stone Columns in Improving Soft Clay Beds

Venkateswarlu Dumpa, G. V. R. Prasada Raju, R. Dayakar Babu, and A. M. N. Kashyap

1 Introduction

The various potential problems that are associated with certain problematic soil deposits, such as peat, soft clay, marine clay, etc., are safe bearing capacity, higher compressibility, lateral flow tendency, etc. Such soil beds require proper treatment methods for the increment in the engineering behavior on par with the design requirements of the structure. One such method is utilization of stone columns for these problematic soils. Stone columns or granular piles are vertical elements created underground by compacting fragments of uncemented stone or gravel. When a vertical load is applied to a stone column, it experiences compression due to bulging of the aggregate. This downward movement of the column causes the granular material to compress the soft soil, transmitting stress through shear. Consequently, the bulge leads to an increment in lateral stress within the clay, providing additional confinement for the sand [1]. Geosynthesized encased stone columns are very efficient in enhancing the soft clay under moving vehicles cyclic loading [2]. The behavior and performance of the stone columns can be enhanced by adding the plastic fibers with sand and smaller amounts of gravel [3]. The impact of geogrid encased stone columns are studied and reported that extra lateral confinement is generated because of the column encasement which leads to enhanced load-carrying capacity [4]. To overcome these challenges, it is possible to mitigate the drawbacks by employing a jute fiber matrix to envelop each stone column. This approach offers numerous benefits,

V. Dumpa (✉) · A. M. N. Kashyap
Godavari Institute of Engineering and Technology (A), Rajahmundry, India
e-mail: dumpa.venkateswarlu@gmail.com

G. V. R. Prasada Raju
Jawaharlal Nehru Technological University, Kakinada, India

R. Dayakar Babu
Freelance Academician, Consultant, Kakinada, India

B. T. Jose et al. (eds.), *Proceedings of the Indian Geotechnical Conference 2022 Volume 4*, Lecture Notes in Civil Engineering 479,
https://doi.org/10.1007/978-981-97-1753-8_26

such as facilitating the formation of stone columns and enhancing their rigidity and durability. Wrapping the stone columns with jute fiber helps prevent the stones from being lost in the adjacent soft clay soil [5–12].

Nonetheless, it is important to recognize that this technique heavily relies on the neighboring clayey soil to mobilize the load-bearing capacity. The improvement in load capacity cannot exceed 25 times the strength of the soft clay [13]. Eventually, an equilibrium state is achieved, resulting in a reduction in vertical movement comparable to the unimproved soil [14].

In areas with soft clay, the installation of stone columns may present challenges due to inadequate confinement from the surrounding soft clay, which can lead to stone losses. Furthermore, the inclusion of soft clay soil may contaminate the stone aggregate, diminishing its frictional strength and impeding drainage within the column [15].

Numerous researchers have documented the positive effects of stone column techniques on enhancing bearing capacity and reducing settlements in soft clays [1]. This study focuses on investigating the behavior of load settlement in soft clay beds, comparing the performance of untreated stone columns with those encased in jute fiber as a reinforcing material. The aim is to assess the impact of jute fiber encasement on the load-settlement characteristics of the soft clay beds, thereby contributing to the existing body of knowledge in this field. Conventional stone columns and jute fiber encased stone columns at center of the model box were studied to assess the load-settlement behavior of soft clay beds. The effect of jute fiber encasement and gradation of the metal (6.3–2.0 mm) was also investigated.

2 Materials

The following section provides details regarding the different materials utilized in the experimental study.

2.1 Soil

The soil used in this study was marine clay collected from 'Kakinada Seaport Limited (KSPL), Kakinada', East Godavari District, Andhra Pradesh State, India. Inorganic clay of high compressibility (CH) is the categorized soil as per IS division. The properties of the marine clay assessed based on relevant I.S. code provisions are given in Table 1.

Table 1 Marine clay characteristics

S. No.	Property	Value
1	G	2.60
2	*Grain size distribution*	
	Sand (%)	9
	Silt (%)	20
	Clay (%)	71
3	*Compaction properties*	
	MDD (kN/m^3)	13.8
	OMC (%)	27.2
4	*Atterberg limits*	
	LL (%)	78
	PL (%)	33
	P I (%)	45
	S L (%)	14.8
	IS classification	CH
5	D F S (%)	42
6	S B C(kPa)	32.86
Shear parameters		
7	C (kPa)	30
8	Ø (degrees)	0

2.2 *Metal*

HBG (hard broken granite) metal with sizes between 2 and 6.3 mm was used as reinforcing material for preparing stone columns.

2.3 *Jute Fiber*

Jute fiber was procured from local market and is treated and used as encasing material for columns.

2.4 *Preparation of Stone Column*

Stone columns of size 5 cm in diameter and 50 cm height was prepared with granular material (from 2 to 6.3 mm at different combinations).These columns are installed in soft clay beds which were already prepared in the model tank [3, 4].

3 Experimental Study

A model box of dimensions 50 cm (L) × 50 cm (B) × 60 cm (H) is fabricated by using 8 mm thick transparent plates. Clay bed was prepared in the model box in layers each of 5 cm to its OMC and MDD up to full height of the model box.

3.1 Model Tests

Tests were carried out by installing the unencased and treated jute fiber encased stone columns at center in model box by adopting the standard procedures in the laboratory (Fig. 1; Table 2).

All the prepared models (Model 1 to Model 7) are mounted on the testing machine, a bearing plate of size 10 × 10 × 0.5 cm is placed centrally at the top of the box. The application of load is carried out incrementally, with a gradual rate of 10 N/s. To measure the corresponding settlements, two dial gauges are affixed to the top of the bearing plate, allowing for accurate recording of the settlement values. Load tests were conducted on unreinforced and reinforced soil model in the laboratory (Fig. 2).

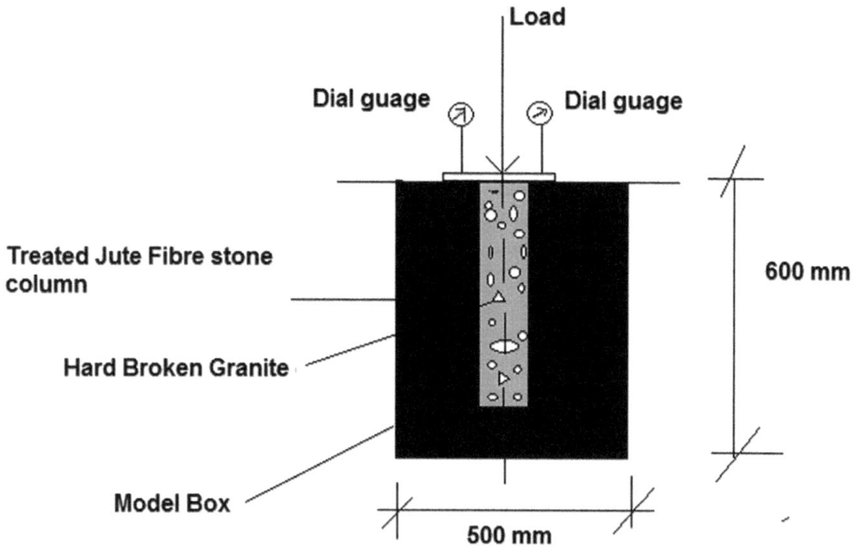

Fig. 1 Schematic diagram of model load test setup

Table 2 Description of models

Model no	Description of the model
Model I	Unreinforced soil model
Model II	Reinforced without encasement (4.75 mm passing and 2 mm retained metal)
Model III	Reinforced without encasement (6.3 mm passing and 4.75 mm retained metal)
Model IV	Reinforced without encasement (6.3 mm passing and 2 mm retained)
Model V	Reinforced with treated jute fiber encasement (4.75 mm passing and 2 mm retained metal)
Model VI	Reinforced with treated jute fiber encasement (6.3 mm passing and 4.75 mm retained metal)
Model VII	Reinforced with treated jute fiber encasement (6.3 mm passing and 2 mm retained metal)

Fig. 2 Model box with soft clay bed installed with treated jute fiber encased stone columns

4 Results and Discussions

The results obtained from the laboratory experiments have been organized into tables, presenting the collected data. These results are subsequently discussed in detail, providing a comprehensive analysis of the findings (Table 3).

Figure 3 shows the load–settlement curves for different soil models (Model: 1–7) constructed with and without reinforcement.

According to Fig. 3, there is a significant increase in the load-carrying capacity for Models 2–7. The installation of both unencased and encased stone columns can be attributed to the improvements in load-carrying capacity and the reduction in settlement.

Table 3 Ultimate load versus settlement

Model no	Ultimate load (N)	Settlement (mm)	Safe-bearing capacity (kPa)
Model 1	493.66	28	32.86
Model 2 (4.75mm ↓ 2.0mm ↑)	1840	21	122.66
Model 3 (6.3mm ↓ 4.75mm ↑)	2031.12	19	135.4
Model 4 (6.3mm ↓ 2.0mm ↑)	2274.3	17	151.6
Model 5 (4.75mm ↓ 2.0mm ↑)	2632.14	14	175.46
Model 6 (6.3mm ↓ 4.75mm ↑)	2860	11	190.66
Model 7 (6.3mm ↓ 2.0mm ↑)	3195.36	08	213

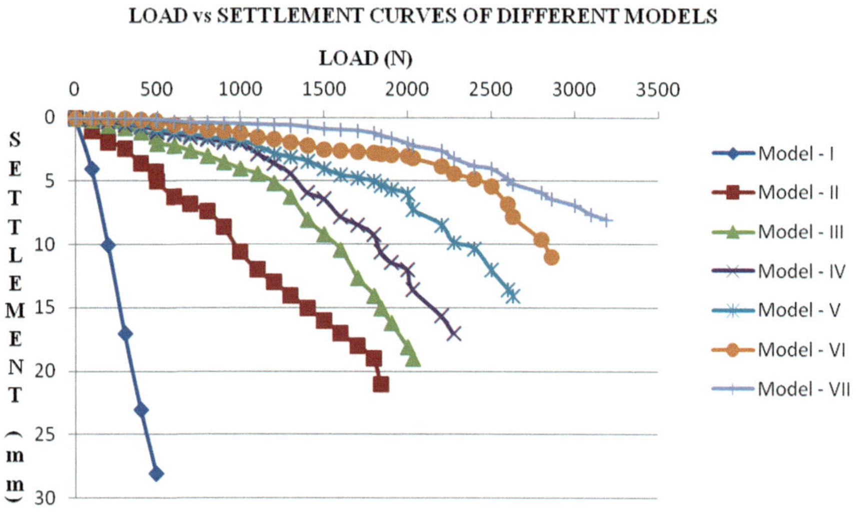

Fig. 3 Load versus settlement curves of reinforced and unreinforced models

The ultimate load-carrying capacity is enhanced by 2.73, 3.11, 3.61, 4.33, 4.80, and 5.47 times for the Models 2–7 with respect to unreinforced soil model (Model I). The settlements are decreased by 25%, 32%, 39%, 50%, 61%, and 72% for the Models 2–7 with respect to unreinforced soil model (Model I).

The improvement in load-bearing capacity and reduction in settlements can be attributed to the mobilization of lateral confinement in the models. The test results demonstrate that the load-carrying capacity of the jute fiber encased columns is further enhanced compared to the untreated ones, specifically for Models 5–7. The encasement of stone columns provides lateral confinement and transforms them into load-bearing elements.

The load-carrying capacity of unencased columns was observed less when compared to the treated jute fiber encased columns. However, the load-carrying capacity of both unencased and treated jute fiber encased columns is enhanced when compared to the Model-I (without reinforcement).

5 Conclusions

1. The load-carrying capacity of unencased reinforcement models, i.e., II–IV is enhanced when compared with unreinforced model, i.e., Model I, and the increment is in the range of 2.73–3.61 times, respectively.
2. The settlement of unencased reinforcement models, i.e., II–IV is decreased when compared with unreinforced model, i.e., Model I, and the decrement is in the range of 25–39%, respectively.
3. The load-carrying capacity of encased reinforcement models, i.e., V–VII is enhanced when compared with unreinforced model, i.e., Model I, and the increment is in the range of 4.33–5.47 times, respectively.
4. The settlement of encased reinforcement models, i.e., V–VII is decreased when compared with unreinforced model, i.e., Model–I, and the decrement is in the range of 50–72%, respectively.

Based on the laboratory model studies and the test results data, it can be concluded that the soft clay beds can be strengthened and the settlements can be decreased more efficiently by the technique of treated jute fiber encased stone columns.

References

1. Dumpa V, Kumar G, Rayi CS, Anjan Kumar M, Prasada Raju GVR (2021) Effect of soil slope on failure mechanism of soil-nailed structures by aluminum nails and bamboo nails. In: Patel S, Solanki CH, Reddy KR, Shukla SK (eds) Proceedings of the Indian geotechnical conference 2019 . Lecture notes in civil engineering, vol 136. Springer, Singapore. https://doi.org/10.1007/978-981-33-6444-8_71
2. Yoo C, Abbas Q (2019) Perfomance of geosynthetic-encased stone column-improved soft clay under vertical cyclic loading. ScienceDirect Soils and Found 59(2019):1875–1890
3. Mekkiyah HM, Al-Saadi SZ (2016) Experimental study for granular column and fiber granular column. J Civil Eng Res 2016(6950):119–127
4. Gniel J, Bouazza A (2009) Improvement of soft soils using geogrid encased stone columns. Geotext Geomembr 27(2009):167–175
5. Murugesan S, Rajagopal K (2008) Performance of encased stone columns and design guidelines for construction on soft clay soils. In: Proceedings of the 4th Asian regional conference on geosynthetics June 17–20, Shanghai, China, pp 729–734

6. Di Prisco C, Galli A, Cantarelli E, Bongiorno D (2006) Geo-reinforced sand columns: small scale experimental tests and theoretical modeling. In: Proceedings of the 8th international conference on geosynthetics. Yokohama, Japan, pp 1685–1688
7. Murugesan S, Rajagopal K (2006) Geosynthetic-encased stone columns: numerical evaluation. Geotext Geomembr 24(6):349–358
8. Dimiter A, Brokemper D, Lothspeich S (2005) Geo-textile encased columns (GEC): load capacity, geo-textile selection and pre-design graphs. Geotechnical Special Publication. No. 130–142, GeoFrontiers, pp 497–510
9. Raithel M, Kempfert HG, Kirchner A (2002) Geotextile-encased columns (GEC) for foundation of a dike on very soft soils. In: Seventh international conference on geosynthetics, Nice, pp 1025–1028
10. Kempfert HG, Gebreselassie B (2006) Excavations and foundations in soft soils. Springer, BerlinHeidelberg, Netherlands
11. Shivashankar R, Babu M, Nayak S, Rajathkumar V (2011) Experimental studies on behaviour of stone columns in layered soils. Geotech Geol Eng 29:749–757. https://doi.org/10.1007/s10706-011-9414-0
12. Li G, Chen Y, Tang X (2008) Geosynthetics in civil and environmental engineering: geosynthetics Asia 2008: proceedings of the 4th Asian regional conference on geosynthetics in Shanghai, China
13. Chummar AV (2000) Ground improvement using stone columns: problems encountered. In: An international conference on geotechnical and geological engineering. Geo Engineering, Melbourne, Australia
14. Greenwood DA (1970) Mechanical improvement of soils below ground surface. In: Ground engineering proceedings conference organised by the institution of civil engineers. London, June, pp 11–22
15. McKenna JM, Eyre WA, Wolstenholme DR (1975) Performance of an embankment supported by stone columns in soft ground. Geotechnique 25(1):51–59

Electrokinetic Remediation of Copper- and Zinc-Contaminated Soil Using Carbon Nanotube

Aiswarya Chandran, Sobha Cyrus, Anil Joseph, and G. Sanoop

1 Introduction

1.1 General

Leaded paint and petrol, mine tailings, the application of fertilizers to the ground, animal manures, sewage sludge, pesticides, wastewater irrigation, coal combustion residues, petrochemical spills, and atmospheric deposition all contribute to the contamination of soils by the accumulation of heavy metals and metalloids [2]. Across the world, soil contamination is a serious threat to the environment. Because of anthropogenic and natural sources as well as human activities, heavy metals get released into the soil. The amount of heavy metals currently mined worldwide is enormous. Nickel (Ni), lead (Pb), cadmium (Cd), chromium (Cr), copper (Cu), cobalt (Co), zinc (Zn), manganese (Mn), aluminum (Al), and mercury (Hg) are heavy metals that are frequently found in soils [3]. Among these heavy metals, As, Pb, Cd, and Hg are included in the top 20 Hazardous Substances of the Agency for Toxic Substances and Disease Registry (ATSDR) and the United States Environmental Protection Agency (US EPA).

An efficient new approach for removing up heavy metal-contaminated soil is electrokinetic remediation. For fine-grained soils, it can be applied as an in situ or ex situ remedy [4]. In this method, electrodes are installed in the soil, and a low

A. Chandran (✉)
EDC, Kochi 682035, India
e-mail: aishwarya96ce@gmail.com

S. Cyrus · G. Sanoop
Cochin University of Science and Technology, Kochi 682022, India

A. Joseph
Geostructurals Pvt Ltd, Kochi 682035, India

B. T. Jose et al. (eds.), *Proceedings of the Indian Geotechnical Conference 2022 Volume 4*, Lecture Notes in Civil Engineering 479,
https://doi.org/10.1007/978-981-97-1753-8_27

voltage DC is supplied across them. Due to the electrolysis process, this results in a generation of H^+ and OH^- ions at the anode and cathode. Contaminants migrate to the respective electrode chambers as a result of the voltage difference between the electrodes. Electromigration, electroosmosis, and electrophoresis are the three basic ways that contaminants are transported. Treatment of soil polluted with heavy metals and organic compounds has been accomplished using the electrokinetic process [5]. The electrokinetic technique has been used to successfully remediate soil polluted with heavy metals and organic chemicals [5]. Electrokinetic remediation is one of the most efficient methods for removing soil contamination, whether it is in situ or ex situ, as it has a high removal competence and time usefulness in soil with limited permeability. A novel and cost-effective physical technique for the removal of heavy metals from soil is soil electrokinetic remediation.

2 Experimental Procedure

2.1 Materials

Cochin Marine Clay. Marine clay collected from a site at Vallarpadam, Kochi, India. For uniformity, soil samples were collected and pooled together and thoroughly mixed into a uniform mass and preserved in polythene bags to maintain the water content. The properties of marine clay used in the study are given in Table 1.

Zinc Powder. Zinc is a transition metal with the following characteristics, atomic number 30, atomic mass 65.4, density 7.14 g cm^{-3}, melting point 419.5 °C, and boiling point 906 °C. The maximum allowable limit of Zn concentration in soil is 300 mg/kg. The zinc powder used in the study was procured from Alpha Chemicals, Kochi.

Table 1 Properties of marine clay used in the study

Property	Value
Water content (%)	112
Specific gravity	2.6
Liquid limit (%)	138
Plastic limit (%)	48
Plasticity index (%)	90
Shrinkage limit (%)	20
Grain size distribution	
Clay (<0.002 mm)	55
Silt (>0.002 mm< 0.075 mm)	35
Sand (>0.075 mm)	10

Copper Powder. Copper is a transition metal which belongs to period 4 and group IB of the periodic table with atomic number 29, atomic weight 63.5, density 8.96 g cm^{-3}, melting point 1083 °C, and boiling point 2595 °C. Copper is the third most used metal in the world. The maximum allowable limit of Cu concentration in soil is 100 mg/kg. The copper powder used in the study was procured from Alpha chemicals, Kochi.

Carbon Nanotubes (CNT). Carbon nanotube (CNT) can now be considered as the "king" of nanomaterials. Carbon nanotube is a tube-shaped material, made of carbon with a diameter measuring on the nanometer scale. As a group, Carbon Nanotubes typically have diameters ranging from < 1 nm up to 50 nm. Overall, carbon nanotubes show a unique combination of stiffness, strength, and tenacity compared to other fiber materials which usually lack one or more of these properties. Thermal and electrical conductivity are also very high, and comparable to other conductive materials. Carbon Nanotube (CNT) is selected for the study because of the exceptional sorptive ability which is driven by their characteristically large surface areas and highly hydrophobic nature. CNT exhibit many exceptional properties, including large surface area, can be easily modified chemical or physical, and they have ability of removing both organic and inorganic contaminants. The carbon nanotube used in the study was procured from Platonic Nanotech Pvt Ltd, Bangalore.

2.2 *Experimental Setup*

Figure 1 shows the schematic diagram of the experimental setup, and the Fig. 1 shows the experimental fabricated in the lab. The EK tests were performed in clear glass rectangular tank. Glass being chemically and electrically neutral material, the tank was fabricated with glass. The experimental tank has to be large enough to hold reasonable quantities of soil so that samples can be obtained to conduct the planned physicochemical tests at the specified regions between the electrodes during and after the EK treatment. Considering these requirements, it was decided to fabricate the glass tank with dimensions: 45 cm in length, 35 cm in width, and height of 25 cm. The glass tank was divided in to three compartments—two compartments for electrolyte, each having dimension 10 cm length and 35 cm width and one compartment for soil with dimensions 25 cm length and 35 cm width.

The tank was divided into three compartments using permeable layer made up of nylon mesh of 2 mm size and filter paper. The photograph of experimental setup made in the laboratory is as shown in Fig. 2. The selection of electrodes is important in EK treatments. Cost and chemical reactions are the main considerations in the selection of electrodes. The use of gold and platinum is impractical because they are very expensive metals. Non-metallic conductors such as carbon and graphite have been found to give very poor performance. Stainless steel was chosen as the electrode, due to the following considerations such as available at low cost, easy to fabricate, considerable strong, less likely to corrode, and also a common material that can be used in field situations. The electrodes used for the study were made in

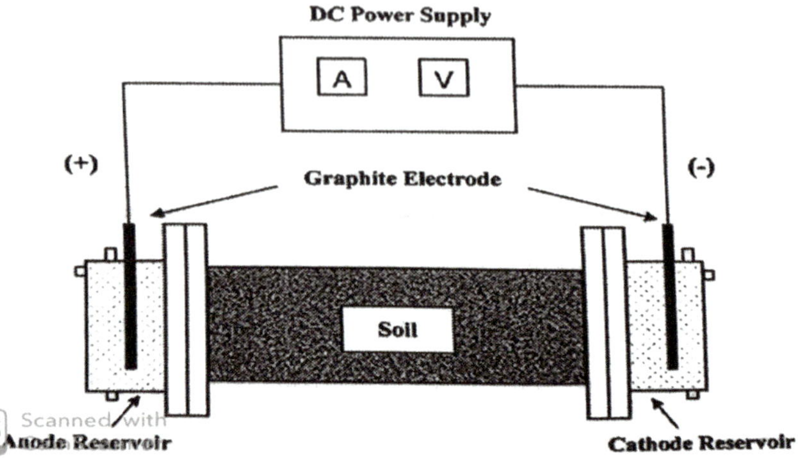

Fig. 1 Schematic view of the experimental model

Fig. 2 Photograph of experimental setup made in the laboratory

the shape of hollow tubes with 35 mm in diameter, 2 mm thickness, and 25 cm in height and were perforated. A 30 V DC supply is used for applying current to the soil. The DC supply is connected to the electrode via an electric probe.

2.3 *Preparation of Test Bed in the Tank*

Marine clay collected was mixed uniformly and stored in airtight bags. Before filling the tank, it is again uniformly mixed. The slurry sample was prepared by mixing the clay soil with deionized water to achieve a water content of 135%, which is equal to the LL of sample. Firstly a dissolved solution of Zn/Cu was prepared. For which 10%

of zinc/copper is taken and dissolved in deionized water using mechanical stirrer. This Zn/Cu solution is mixed with 5 kg soil sample in the mechanical mixer, achieving a uniform Zn/Cu spiked soil mix of water content of 135% for the test sample. The soil sample is poured into the soil compartment of tank and properly tamped to remove the trapped air inside the soil and the top is leveled. Filter paper and nylon mesh is attached at each end of soil column to avoid leakage of soil particles. The electrolyte compartment is filled with deionized.

water till level of soil. Then, two hollow perforated stainless steel tubes were placed apart in the electrolyte compartments. The electric probes are connected to the anode and cathode which in turn connected to the DC power supply, providing total voltage 30 V. During the main experimental program, the electric current and the voltage were maintained continuously throughout the test period. A voltage of 30 V was maintained throughout the test. Soil sample at different normalized distances from anode such as 0.15, 0.5, and 0.85 were collected on 5, 10, and 15 days and were tested. In order to study the effect of CNT on enhancement of removal efficiency of contaminant, the electrolyte used were deionized water mixed with 0.05% of CNT.

In the present study we determine the effect of electrokinetic remediation on different regions between the electrodes; therefore the soil between the electrodes in the soil compartment is divided into three regions with average normalized distance of 0.15, 0.5, and 0.85 cm, respectively, within the soil column. Normalized distance is the distance from the anode to the sample position divided by the distance between the electrodes. The soil at these sections was tested for variations in moisture content, index properties, and concentration of contaminant as a result of EK process for various periods 5,10, and 15 days. For testing, the change in concentration of contaminant soil where collected from the vicinity of anode and cathode. The soil samples collected were tested to find the final concentration of contaminant.

2.4 EK Treatment on Zinc- and Copper-Contaminated Soil

Electrokinetic remediation treatment was carried out on copper- and zinc-contaminated soil with and without CNT. This study was carried out to evaluate enhancement of the treatment efficiency of a heavy metal-contaminated soil by combined process of the injection of carbon nanotube (CNT) and electro kinetic remediation. The Cu/Zn-contaminated soil sample prepared is placed into the soil compartment of tank in three layers at and properly tamped to remove the trapped air inside the soil and the top is leveled. The electrolyte compartment is filled with deionized water. The electrodes are placed in electrolyte compartment and are connected to DC supply of capacity 30 V. During the experimental program, the electric current and the voltage were maintained continuously throughout the test period. A voltage of 30 V was maintained throughout the test. Soil sample at different normalized distances from anode such as 0.15, 0.5, and 0.85 were collected on 5, 10, and 15 days and were tested.

3 Results and Discussion

3.1 *Results of Physicochemical and Engineering Properties of Soil After EK Treatment*

Metal Concentration. Variation of metal concentration in the soil sample from anode to cathode is shown in Table 2. The samples collected were digested in HNO_3, and made into a 100 ml solution, then filtered and analyzed with ICP-MS system to obtain the concentration on Zn and Cu. For all test periods, experimental values show a similar trend from anode to cathode. A decrease in the concentration of Zn from anode to cathode was observed from the average values of the corresponding heavy metal concentrations in the soil at different treatment periods. The initial zinc concentration in the soil is 10.208% of soil. Total Zn removed from soil after 15 days treatment is 2.043% of soil. Hence, the removal efficiency is 20.0%. The initial copper concentration in the soil is 9.866% of soil. Total Cu removed from soil after 15 days treatment is 2.821% of soil. Hence, the removal efficiency is 28.6%.

Table 2 Variation in Zn concentration after EK treatment on Zn-contaminated soil

Normalized distance	Treatment duration (days)	
	Zn concentration (%)	
	0.15	0.85
0	9.981	10.436
5	9.484	8.301
10	9.303	7.518
15	8.432	7.898

3.2 *Results of Physicochemical and Engineering Properties of Soil After EK Treatment with CNT*

Metal Concentration. Variation of metal concentration in the soil sample from anode to cathode is shown in Table 3. The samples collected were digested in HNO_3 and made into a 100 ml solution, then filtered and analyzed with ICP-MS system to obtain the concentration on Zn and Cu. Experimental values show a similar trend from anode to cathode for all test periods. A decrease in the concentration of Zn from anode to cathode was observed from the average values of the corresponding heavy metal concentrations in the soil at different treatment periods. The initial zinc concentration in the soil is 9.9505% of soil. Total Zn removed from soil after 15 days treatment is 9.206% of soil. Hence, the removal efficiency is 92.517%. The

Table 3 Variation in Cu concentration after EK treatment on Cu-contaminated soil

Normalized distance	Treatment duration (days) Cu concentration (%)	
	0.15	0.85
0	10.121	9.610
5	8.98	8.509
10	8.015	7.958
15	7.981	6.109

Table 4 Variation in Zn concentration in Zn-contaminated soil after EK treatment with CNT

Normalized distance	Treatment duration (days) Zn concentration (%)	
	15	0.85
0	9.721	10.180
5	8.055	6.750
10	5.075	4.819
15	1.260	0.229

Table 5 Variation in Cu concentration in Cu-contaminated soil after EK treatment with CNT

Normalized distance	Treatment duration (days) Cu concentration (%)	
	0.15	0.85
0	9.809	9.519
5	6.591	4.230
10	3.984	3.019
15	0.907	0.127

initial copper concentration in the soil is 9.664% of soil. Total Cu removed from soil after 15 days treatment is 9.147% of soil. Hence, the removal efficiency is 94.650% (Tables 4 and 5).

3.3 *Comparison of Removal Efficiency of Copper and Zinc Through EK Treatment with and Without CNT*

The following equation was used to calculate the removal efficiency of copper and zinc from contaminated soil after the EK treatment.

$$\text{Removal efficiency } (\%) R_e = \frac{C_o - C_f}{C_o},$$

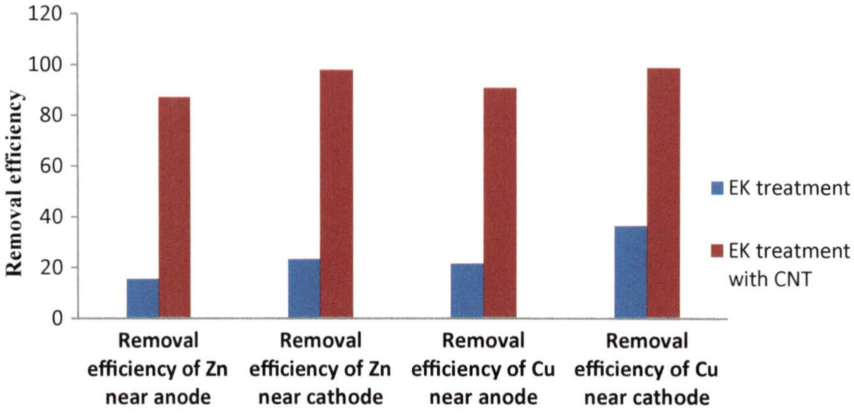

Fig. 3 Bar chart showing the comparison between the removal efficiency of EK treatment and EK treatment with CNT

where

C_o = the initial metal concentration

C_f = the final metal concentration

The removal efficiency of heavy metal from contaminated soil shown higher removal efficiency of Cu and Zn in EK treatment with CNT than EK treatment alone. The removal efficiency of zinc from the contaminated soil increased from 20.01% to 92.51%.The removal efficiency of copper from the contaminated soil increased from 28.56% to 94.65%. Figure 3 shows the removal efficiencies of EK treatment and EK treatment with CNT.

4 Summary and Conclusions

The present study enables a better understanding of physical and chemical changes of different properties of soil that occurs during the application of electrokinetics. The entire study carried out in the present work focused on the determination of enhancement rate in Zn and Cu removal from the contaminated soil using CNT. The important conclusions derived from the present work are as follows:

- From the study, it is concluded that electrokinetic remediation is a feasible treatment method for the remediation of fine-grained soil contaminated with zinc and copper.
- Use of CNT enhanced the contaminant removal efficiency of electrokinetic treatment.
- The electrokinetic treatment coupled with CNT, enhanced the removal efficiency of zinc from 20.0 to 92.5%.

- The electrokinetic treatment coupled with CNT, enhanced the removal efficiency of copper from 28.5 to 94.6%.

References

1. Gomes H, Ferreia DC, Ottosen ML, Riberio B (2015) Electroremediation of PCB contaminated soil combined with iron nanoparticles: Effect of the soil type. Centre Environ Sustain Res 6545–6535
2. Zhang MK, Liu ZY, Wang H (2010) Use of single extraction methods to predict bioavailability of heavy metals in polluted soils to rice. Commun Soil Sci Plant Anal 41(7):20–831
3. Cynthia EV, David DA (1997) Remediation of metals-contaminated soils and groundwater. Technology Report TE-97-01, GWRTAC, USA, GWRTAC-E Series
4. Reddy KR, Chinthamreddy S (2004) Enhanced electrokinetic remediation of heavy metals in glacial till soils using different electrolyte solutions. J Environ Eng (ASCE) 130(4):442–455
5. Malavika SJ, Deepthy LB (2017) Electrokinetic remediation of lead contaminated soil. Int J Eng Res Technol (IJERT). ISSN: 2278-0181, ETCEA

Parametric Studies on Cemented Stone Column Using Demolition Waste Materials

P. P. Dahale, S. S. Geete, T. N. Singh, and K. H. Singh

1 Introduction

In 1830, probably for the first time, stone column technique was used by the French Military Engineers as support system to the heavy foundation in soft marine sediments. Reviewed literature shows that, studies on stone columns use were limited till 1950s but later on the technique became widely popular especially in Europe and United States.

In the conventional constructions, processed stone or coarsely grained materials (sizes varying from 75 to 2 mm) are used with partial replacement of unsuitable soil ($\approx$ 20%–35%). Typically the stone column diameter is 60–90 cm and its length is 5–10 m depending upon the conditions and requirements at the site [1, 2]. The technique proved to be advantageous as it increases load carrying capacity of weak grounds. Also, reduction in total and differential settlements of structures is seen.

Depending upon the condition of soils, we can construct stone columns either as end bearing columns resting on firm stratum underlain the soft soils or floating column, wherein the tip of column embedded within the soft soil layer. In both the types, if the column length is more than five times the diameter, bulging is observed in the top portion (depth $\approx$ 2–4 times the shaft diameter) [2–6].

The magnitude of bulging depends upon applied loaded form, i.e., when load is applied through rigid foundation base over an area greater than column diameter, it increases the vertical and lateral stresses in the surrounding soils. The larger bearing

P. P. Dahale (✉) · S. S. Geete
Shri Ramdeobaba College of Engineering and Management, Nagpur, Maharashtra, India
e-mail: dahalepp@rknec.edu

T. N. Singh
Indian Institute of Technology, Patna, Bihar, India

K. H. Singh
Indian Institute of Technology, Bombay, Maharashtra, India

B. T. Jose et al. (eds.), *Proceedings of the Indian Geotechnical Conference 2022 Volume 4*, Lecture Notes in Civil Engineering 479,
https://doi.org/10.1007/978-981-97-1753-8_28

area together with the additional support of stone column results in less bulging and greater ultimate load capacity with less settlement. The local bearing type of failure (mostly punching type) is also reported in some cases especially when stone columns are placed in a group.

To improve effectiveness of stone column by restraining the lateral bulging under loading, some of the techniques proposed by the researchers are:

a. provision of skirting around the stone column or group of stone columns,
b. geotextile encasement of the column,
c. use of several reinforcing geofabric,
d. incorporation of randomly distributed reinforcing fibers in sandy fill material,
e. stone column with vertical nails (small-diameter steel bars) and
f. stone columns strengthened by additives.

Limited studies are available on the use of cement [7–9] and waste [10–12] materials for stone column construction.

In the present work, demolition wastes (DA) is used as replacement to the natural aggregate (NA) for stone column models with cement (4%). The test program was done in two phases, viz., material testing and model (small and large scale) tests. Here, results of large-scale model tests are discussed. The nominal mix proportion used in test model preparation is given in Table 1.

With the addition of cement (4% by weight) to the materials, the stone column behavior becomes semi-rigid. This has helped in minimizing the lateral bulging under axial loading and in turn increases the carrying capacity of unit cell of stone column.

The collected soil and aggregates (natural and demolition) was processed as per IS code provisions in the laboratory and their properties were examined. The properties of soil, natural aggregate and demolition aggregate are tabulated below in Tables 2 and 3, respectively. Total five combinations for model preparations were planned, namely—100% NA, 100% DA, 25%NA + 75% DA, 50% NA + 50% DA and 75% NA + 25% DA.

Table 1 Mix proportion adopted for stone column model [8, 9]

Sr. No.	Material	Specification	Mix proportion
1	Coarse aggregate	Passing through 20 mm IS sieve and retained on 10 mm IS sieve	50%
2	Fine aggregate	Passing through 4.75 mm IS sieve	50%
3	Coarse sand	Retained on 2 mm IS sieve	25% of (1 + 2)
4	Cement	Ordinary Portland cement	4% of (1 + 2 + 3)
5	Water	Potable water	5% by volume

Table 2 Physical and mechanical properties of soil

Sr. No.	Soil properties	Relevant IS codes	Results	Typical range[a]
1	Natural moisture content (%)	IS 2720 (Part 2): 1973	18.24	10–60
2	Specific gravity	IS 2720 (Part 3): 1980	2.40	2.4–2.9
3	Clay: % (< 0.002 mm)	IS 2720 (Part 4): 1985	80.12	–
4	Silt: % (0.002–0.075 mm)		19.88	–
5	Soil classification	IS 1498: 1970	CH	–
6	Liquid limit (%)	IS 2720 (Part 5): 1985	50.72	26–85
7	Plastic limit (%)		24.42	15–50
8	Shrinkage limit (%)		11.80	9–25
9	Plasticity index (%)		26.30	10–44
10	Free swell index (%)	IS 2720 (Part 40): 1977	35	15–40
11	Cohesion, C_u (kN/m^2)	IS 2720 (Part 13): 1986	32.35	20–65
12	Frictional angle, Φ_u (°)		13.30	7–22
13	Bulk density (kN/m^3)	IS 2720 (Part 8): 1983	18.88	14–20
14	Maximum dry density (kN/m^3)		15.38	14–20
15	Optimum moisture content (%)		21.15	12–40
16	Unconfined compressive strength, q_u (kN/m^2)	IS 2720 (Part 10): 1991	50.85	25–70
17	Activity	SP 36-1-1987	0.325	0.2–4
18	Modulus of elasticity (kN/m^2)	IS 9221:1979	3100	500–6000
19	Poisson ratio, μ		0.40	0.3–0.5
20	Test bed moisture content (%)	–	28	–

[a] Values are referred from past available literatures

2 Test Results

The axial compressive strength of stone column models prepared with different mix proportions on 7 days curing was tested. The maximum strength is observed for column with NA. However, result is promising for the mix 25% NA + 75% DA among all other combinations, where the observed strength is 97% of NA column strength.

All the tests on stone column models embedded in soil were done in a large tank of size 60 × 60 × 90 cm. Before test conduction, various steps followed are preparations of soils bed (desired density 90–95%) in layers, creation of borehole in soil bed and in place column casting, etc. The typical test setup is as shown in Fig. 1.

Table 3 Properties of natural and demolition aggregate [13, 14]

Sr. No.	Properties	Relevant IS Codes	NA	DA	Typical range[a]
1	Specific gravity	IS 2386 (Part 3): 1963	2.96	2.59	2.5–3.0
2	Water absorption (%)		1.28	3.69	1.65–6.0
3	D_{10} (mm)	IS 2386 (Part 1): 1963	10.64	9.72	2–11
4	D_{30} (mm)		14.35	14.36	7–15
5	D_{60} (mm)		19.90	19.31	14–20
6	Coefficient of curvature (C_U)		1.871	1.098	1.5–2.2
7	Coefficient of uniformity (C_C)		0.972	1.98	0.8–1.2
8	Gradation symbol		GP	GP	–
9	Size range (mm)	–	2–14	2–14	–
10	Maximum dry unit weight (kN/m^3)	IS 2386 (Part 3): 1963	18.46	15.69	17–22
11	Minimum dry unit weight (kN/m^3)		15.54	13.73	14–18
12	Minimum void ratio (%)		37.00	39.65	–
13	Maximum void ratio (%)		47.50	47.19	–
14	Los Angeles abrasion test (%)	IS 2386 (Part 4): 1963	12	32.52	11–15
15	Aggregate impact value test (%)		10.90	31.98	9–15
16	Aggregate crushing value test (%)		10.53	20.17	8–4

[a] Values are referred from past available literatures

The load-settlement observations of all test models for various L/D ratios are presented in Fig. 2. The method double tangent was used to determine the ultimate load intensity from the curve. It was observed that, on insertion of stone column, the load carrying capacity of the soil increases. The load-settlement curve for virgin soils during the test was somewhat parallel to the ordinate axis; on insertion of column the curve shifts horizontally and steadily becomes parallel to horizontal axis. This shows the effects of semi-rigid column insertion in the clay, which makes the system slightly stiffer than the earlier one.

The result for combinations when compared shows, the strength improvement is more for NA column. For the mix proportion, 25% NA + 75% DA, the improvement in load intensity is at par with NA column especially for L/D ratio 6 and 8.

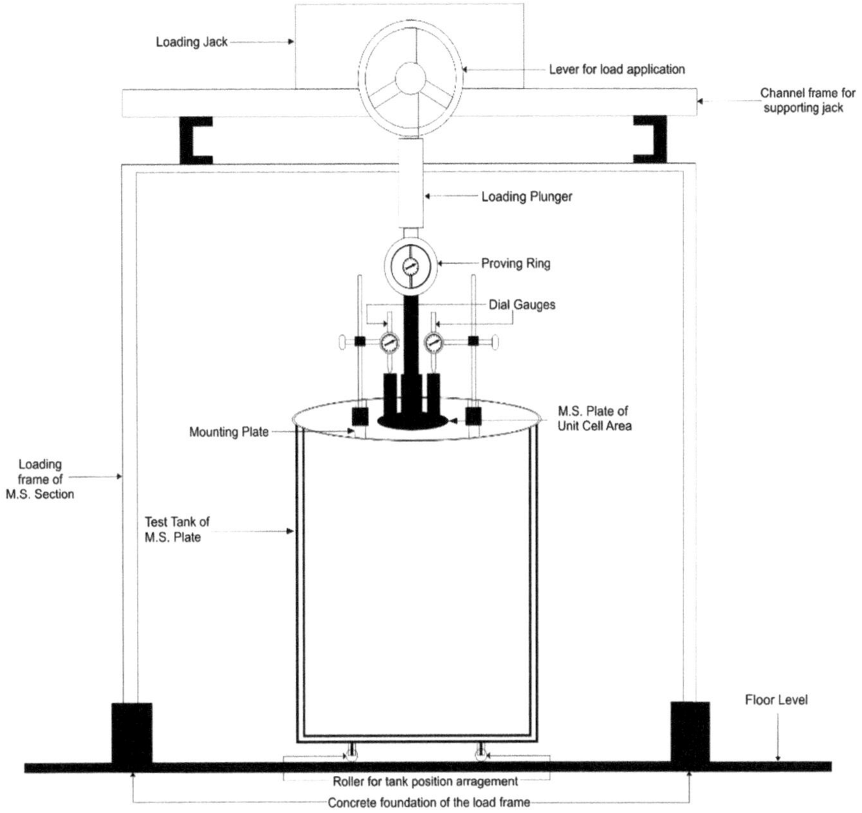

Fig. 1 Test setup

3 Conclusions

- An experimental program was carried out that shows noticeable increase in the load carrying capacity of soil with stone columns insertion.
- With cemented stone column, the lateral bulging reduces on axial loading as the mix column mix became semi-rigid which helps in increasing capacity of column.
- In case of column with demolition aggregates, the strength achieved is between 60 and 75% of that observed for column with natural aggregate. This observation confirms the possible use of demolition aggregates in stone column constructions.

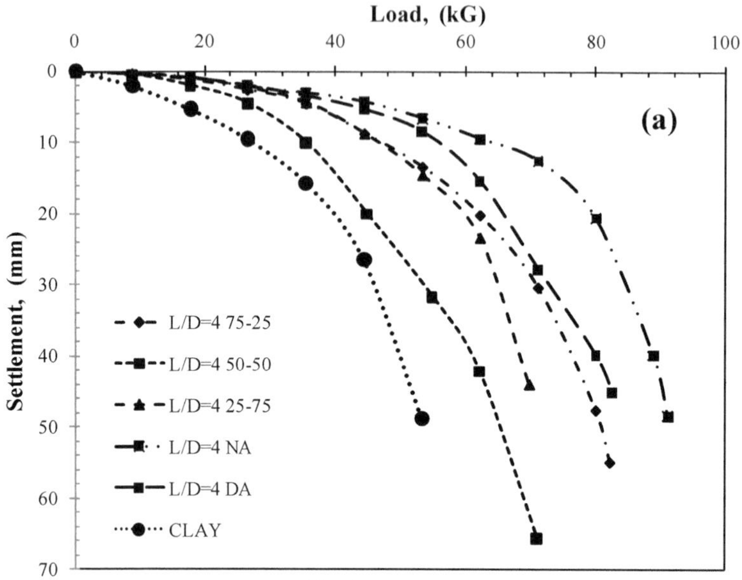

(a)L/D ratio 4

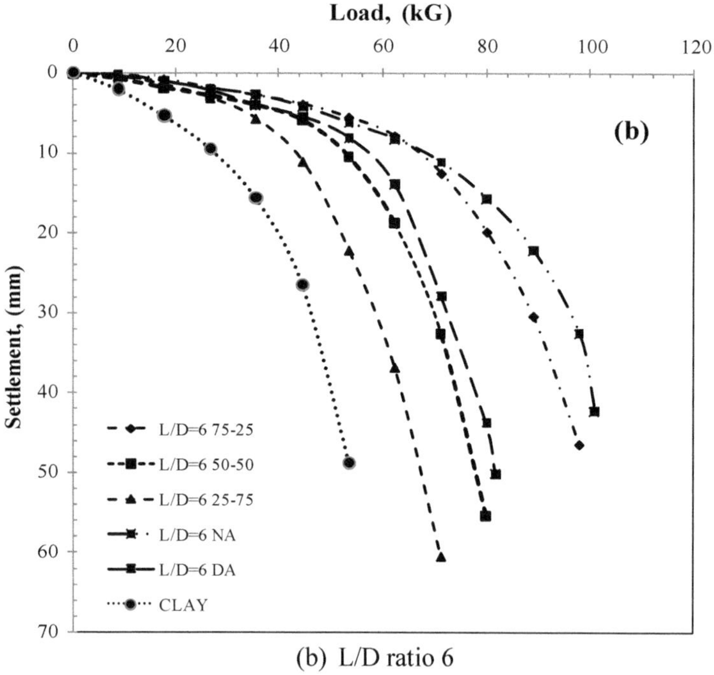

(b) L/D ratio 6

Fig. 2 Load-settlement curves for stone column tested in large tanks

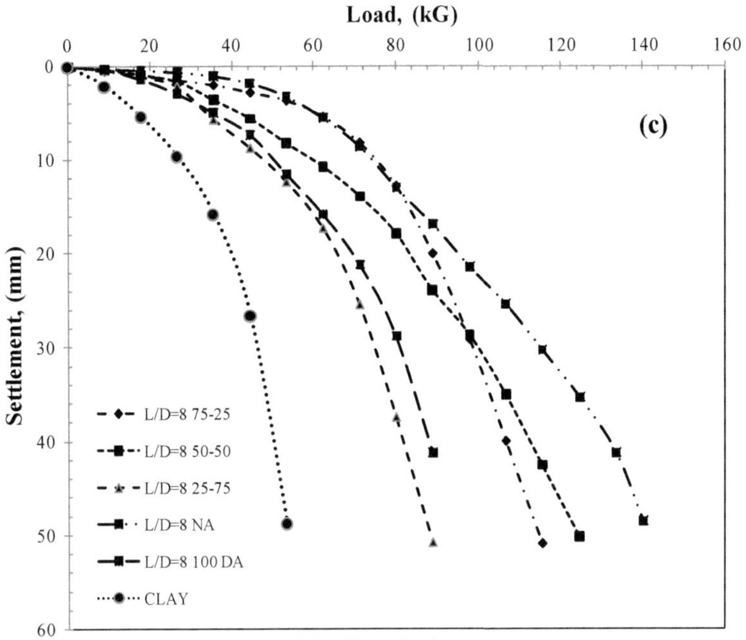

(c)L/D ratio 8

Fig. 2 (continued)

References

1. Balaam NP, Booker JR (1981) Analysis of rigid rafts supported by granular piles. Int J Numer Anal Methods Geomech 5(4):379–403
2. Barksdale RD, Bachus RC (1983) Design and construction of stone columns, vol 1. Report no. FHWA/RD 83/026, National technical Information Service, Springfield, Virginia
3. Lee JS, Pande GN (1998) Analysis of stone-column reinforced foundations. Int J Numer Anal Meth Geomech 22:1001—1020
4. Ambily AP, Gandhi SR (2007) Behavior of stone columns based on 1 experimental and FEM analysis. J Geotech Geoenviron Eng ASCE 133(4):405–4151
5. Murugesan S, Rajagopal K (2007) Model tests on geosynthetic-encased stone columns. J Geosynthetics Int 14(6):346–354
6. Shahidul M (2010) Stabilization of peat soil by soil column technique and settlement of the group column. Int J Phys Sci 5(9):1411–1418
7. Juran I, Riccobono O (1991) Reinforcing soft soils with artificially cemented compacted sand columns. J Geotech Eng https://doi.org/10.1061/(ASCE)073, 1042–1060
8. Golait YS, Satyanarayana V, Raju SSV (2009) Analysis of load carrying capacity of soft clay ground treated with cemented stone columns. In: Proceedings of Indian geotechnical conference, pp 356–360
9. Golait YS, Padade AH (2016) Analytical and experimental studies on cemented stone columns for soft clay ground improvement. ASCE Int J Geomech 17(4):04016100
10. Dipty SI, Girish MS (2009) Suitability of different materials for stone column construction. EJGE 14:1–12
11. Majeed AH, Al-Soud MS, Ali MS (2017) The behavior of soft clay treated with waste concrete column. Int J Eng Res Technol 6(02)

12. Shahverdi M, Haddad A (2019) Use of recycled material in floating stone column. Inst Civ Eng 173(2):99–108
13. IS 2386 (Part 1) (1963) Method of test for Aggregate for concrete: Particle size and shape. Indian Standards Institution, New Delhi, India
14. IS 2386 (Part 3) (1963) Method of test for Aggregate for concrete: Specific gravity, Density, Voids, Absorption and bulking. Indian Standards Institution, New Delhi, India
15. IS 15284 (2003) Indian standard code of practice for design and construction for ground improvement-guidelines
16. IS 2720 (Part 3) (1980) Methods for test for soil, determination of specific gravity of fine-grained soil. Indian Standards Institution, New Delhi, India
17. IS 2720 (Part 4) (1995) Method of test for soils: Grain size analysis. Indian Standards Institution, New Delhi, India
18. IS 2720 (Part 5) (1995) Method of test for soils: determination of liquid and plastic limit. New Delhi, India
19. IS 2720 (Part 7) (1995) Method of test for soils: determination of water content—dry density relation using light compaction. Indian Standards Institution, New Delhi, India
20. IS: 2720 (Part 10) (1995) Method of test for soils: unconfined compressive strength. Indian Standards Institution, New Delhi, India

A Comparative Study on the Uplift Capacity of Single Helix Helical Anchor and Granular Pile Anchor in Sand

Jerin Joseph, Shailendra Kumar, Jignesh B. Patel, Yogendra Tandel, and S. Ravi Teja

1 Introduction

1.1 Background

Helical piles and anchors have been in use for more than 170 years [20]. Helical piles were first put into use in the coastal regions of England lighthouse foundation in 1833 by an English inventor Alexander Mitchell. Later, they were adopted for the same purpose in North America. Helical anchors find application in various building foundations like dock foundation, solar panel foundation, communication, power transmission tower foundation, underpinning of settling structures, etc. They are foundation systems that can be used to support or resist any loading like compressive, tensile, or lateral. They are relatively easy to install and can be readily put into usage after the installation. The helical anchor consists of a hollow shaft made of steel and a helical plate fixed to this shaft. The number of helical plates can vary depending on the resistance needed. The helical shape helps in the installation of the helical anchor by acting as a cutting edge when it is being screwed into the ground by applying a torque at the top end. The torque is applied using a motor that may be fixed to the hydraulic arm extending from a vehicle. Granular pile anchor (GPA) was introduced as a modification to the currently existing stone column technique to carry

J. Joseph (✉) · S. Kumar · J. B. Patel · S. Ravi Teja
Department of Civil Engineering, Sardar Vallabhbhai National Institute of Technology, Surat, Gujarat 395007, India
e-mail: jerinphd@gmail.com

J. Joseph
Department of Civil Engineering, Albertian Institute of Science and Technology, Kalamassery, Kochi, Kerala 682022, India

Y. Tandel
Applied Mechanics Department, Government Engineering College, Valsad, Gujarat 396001, India

B. T. Jose et al. (eds.), *Proceedings of the Indian Geotechnical Conference 2022 Volume 4*, Lecture Notes in Civil Engineering 479,
https://doi.org/10.1007/978-981-97-1753-8_29

tensile loading [15]. The conventional stone column technique consists of making a cylindrically shaped hole in the ground and depositing granular material, which is then compacted to form a stiff column inside the soil mass. The compaction also increases the bearing capacity of the adjacent soil mass. In a granular anchor pile, a tensile element in the form of an anchor plate and an anchor rod is inserted inside the cylindrical hole before the granular material is deposited and compacted. The anchor plate is a mild steel plate of the same diameter as that of the hole or a concrete pedestal. The anchor rod should be of sufficient thickness and strength. The anchor rod is connected to the footing of the structure, which needs anchoring. A review of the literature revealed that the granular anchor pile is currently underutilized as a tension resistance foundation technique despite its relative simplicity and low-cost installation. This paper aims to present a comparative study of the single helix anchor and GPA installed in cohesionless soil based on the small-scale laboratory uplift tests. The uplift load versus displacement behavior of single model helical anchor and GPA are studied. A comparison of the ultimate uplift load and the effect of embedment depth on the uplift performance is given in this paper.

2 Literature Review

A limited review of the helical anchor and the GPA is provided in this section. Meyerhof and Adams [11] conducted uplift tests on model circular anchor plates buried in sand by varying the anchor diameter, embedment depth, and sand relative density. They concluded that a complex mechanism is involved in providing the resistance to the footings against the uplift where soil weight and the soil strength act together to resist the uplift force. Ghaly et al. [2] carried out pullout test on different types of single helical anchor, which were installed in sand beds of various densities and presented the experimental and theoretical investigations. The depth of the anchor and friction angle of the sand have a significant effect on the failure mechanism and the pullout capacity. The helical anchors were classified based on the failure mechanisms as shallow, transit, and deep and a limit-equilibrium-based mathematical model was developed. Hanna et al. [3] presented analytical models that predicted the uplift capacity and the load-displacement behavior for shallow single helical and plate anchors. Empirical expressions were presented to determine the critical depth of anchors, which separated shallow and deep failure mechanisms. The critical depth was found to depend on the anchor diameter and the friction angle of the sand. Sakr et al. [16] carried out uplift tests on large-scale helical piles installed in dense to very dense sand. It was noted that the 5% failure criterion, which estimates the ultimate capacity as the load at displacement equal to 5% of the helix diameter, is practically reasonable. George et al. [1] conducted an experimental and numerical investigation on the uplift capacity of helical anchors installed in the sand by displacement and non-displacement methods. A finite element method analysis using PLAXIS 3D software was also carried out to understand and simulate the effect of installation on the uplift capacity. It was observed that the uplift capacity depended on parameters like the relative density of sand, embedment depth, and installation method. A semi-empirical equation was also proposed by the authors to predict the uplift capacity. Most of available literature on GPA deals with its performance in expansive soil [12, 13, 15, 19]. It was found that the GPA resists the uplift forces due to swelling of the expansive soil and, at the same time, improves its bearing capacity. Some of the studies like [14, 18] have compared the uplift performance of GPA with concrete pile in expansive soil and c-ϕ soil. A comparative study was also carried out to compare the swelling uplift performance of GPA and helical pile anchor installed in expansive soil by Muthukumar and Shukla [14]. The limited available literature gives an indication that GPA can also be adopted as an anchoring method in cohesionless soil, and an embedment ratio of 10 is the optimum, after which the gain in uplift capacity is minimal [7–9, 17]. It is noted in the study conducted by Liu et al. [10], tested the GPA system under compression and tension in field conditions, and found it fit to be used as an anchor under sewage tanks. A study conducted on laboratory models and finite element modeling using PLAXIS 3D has revealed that the GPA can perform better than bored concrete piles in loose sand [5]. The GPA has been found to perform better than multi-helix plate helical anchor in medium dense sand [6]. Even though GPA has the potential to be used as an economical anchoring method, the available literature on its field application is minimal.

3 Experimental Methodology and Materials

3.1 Experimental Program

The experimental study consisted of a total of six uplift tests carried out on a single helix model helical anchor and GPA installed in loose sand beds. The diameter (d) of the helical plate and anchor plate was kept as a constant of 70 mm. The embedment length (L) of the model anchors were varied as 350, 525, and 700 mm. It is to be noted that the embedment length of a helical anchor was considered as the depth of the helical plate from the sand bed surface. The helical anchor installed are displacement-type anchors, which means they are installed by penetrating the sand bed by applying a downward vertical force and torque.

3.2 Materials

Sand: Sand used in the experiment to prepare the sand bed was sourced from Surat city. The sand passing through a 4.75 mm sieve with a specific gravity of 2.65 was used, and before conducting the experiment, it was air-dried. The sand was classified as poorly graded as per IS 1498 (Bureau of Indian Standards 1970) with maximum and minimum dry unit weights were 14.65 and 17.29 kN/m^3, respectively, with an

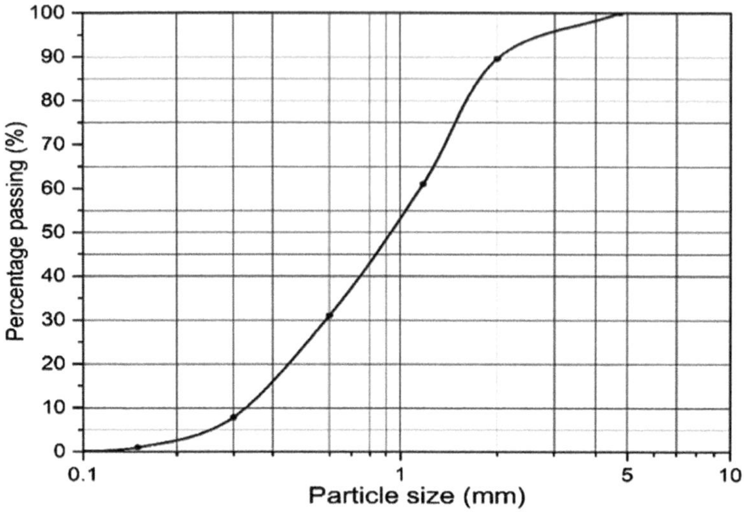

Fig. 1 Particle size distribution curve of sand

effective size of 0.33 mm, uniformity coefficient of 3.52, and coefficient of curvature of 0.94. The loose sand bed of relative density, $Dr = 25\%$, had a dry density of 15.23 kN/m^3 and an angle of friction equal to 31.250. The particle size distribution of the curve is shown in Fig. 1.

Helical anchor and GPA: The helical anchor consisted of a single helical plate of diameter 70 mm and a hollow shaft of diameter 25 mm. The pitch of the helical plate was 20 mm. The bottom end of the helical anchor was closed and pointed to facilitate penetration during installation. The helical plate and hollow shaft pipe were fabricated using 3 mm thick mild steel. The GPA consisted of an anchor plate of thickness 10 mm and an anchor rod of 5 mm thickness. The helical anchor and the GPA plate rod assembly used in the current study are shown in Fig. 2. The GPA material, i.e., the granular material, consisted of sand and gravel mixed at a ratio of 1:1.85. The sand used in the mixture passed through a 4.75 mm sieve, and the gravel size ranged from 4.75 mm to 8 mm. The GPA material mixture was decided based on the procedure adopted by Kumar [9] and Kranthikumar et al. [8].

3.3 *Sand Bed Preparation and Installation of Model Anchors*

The sand bed was prepared using the sand rainfall method. A hopper with a perforated mild steel plate at the bottom was used to deposit the required quantity of sand from a predetermined height to form sand layers of 50 mm each. The helical anchor was installed by manually applying torque to the shaft and pushing it inside the sand bed

Fig. 2 Helical anchor and anchor plate with rod

using a screw jack with the load frame acting as a reaction frame. Special care was taken to penetrate the helical anchor at the rate of 1 pitch (20 mm) per rotation of the helical anchor. The GPA installation was carried out simultaneously with sand bed preparation. The GPA plate and rod assembly were initially placed on a 50 mm sand layer. A hollow plastic pipe was placed on the GPA plate with the anchor rod passing centrally. The sand layers were placed around the hollow pipe, and simultaneously, the required quantity of GPA material was placed in the hollow pipe as layers of 50 mm each and compacted. The GPA material layers were compacted using a specially fabricated cylinder-shaped annular mild steel hammer, and at the same time, the hollow pipe was lifted. The process of sand layer deposition, filling of GPA material, compaction, and lifting of the hollow pipe was repeated until the required sand bed and GPA length were formed. The installation method adopted in the installation of the GPA is similar to Kranthikumar et al. [7, 8], and Kumar [9].

3.4 *Uplift Test Apparatus and Testing Procedure*

The uplift test apparatus consisted of a square tank of side length of 1000 mm and a depth of 750 mm, the load frame on which a multispeed load testing machine was mounted, and load/displacement measuring devices (Fig. 3). The square tank boundaries were well outside the influence of the model anchors. The multispeed load testing machine was used to apply the pullout load at a constant pullout displacement of 0.5 mm/min. The strain rate was fixed after a review of the work done by Ilamparuthi and Muthukrishnaiah [4] on the pullout of plate anchors in sand. The measure devices consisted of load cell and two linear variable displacement transducers, which were connected to a data logger. The data logger was connected to a laptop in which the data logging software was installed and gave real-time monitoring of the uplift load-displacement behavior.

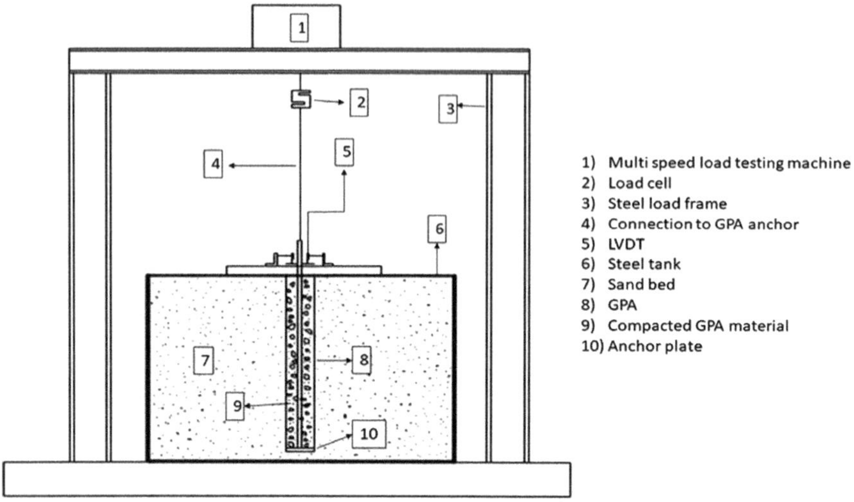

Fig. 3 Schematic diagram of uplift test apparatus

3.5 *Results and Discussions*

The uplift test results in the form of uplift load-displacement curves and the variation of ultimate uplift capacity with embedment length are given and discussed in the following sections.

Uplift load-displacement behavior: The uplift load-displacement behavior of the helical anchor and the GPA are given in Figs. 4, 5 and 6. The load-displacement curve can be divided into approximately three parts. In the first part, there is a linear increase in the uplift load resistance with respect to the displacement, which is rapid in the case of the GPA with respect to the helical anchor. In the second part, the increase in uplift is accompanied by a rapid increase in the displacement for both methods. In the third part, the uplift load becomes a constant with comparatively larger fluctuations in the helical anchor's case. The rapid increase in the uplift resistance in case of the GPA in the initial part of curve can be attributed to the confining pressure of the surrounding sand at the bottom portion of the GPA. The bottom portion of the GPA just above the anchor plate undergoes deformation during uplift which is resisted by the surrounding soil mass. As the uplift load increases, this confining support offered by the surrounding sand is overcome and the GPA material starts to deform which is reflected in the form of a flatter uplift load-displacement curve in the second part. The large fluctuations in the third part indicate plastic rearrangement sand/GPA material occurring above the helical/anchor plate. Similar uplift load-displacement behavior was also noted by Ilamparuthi and Muthukrishnaiah [4] for plate anchors.

Ultimate uplift load capacity: The ultimate uplift capacity of the helical anchor/ GPA was estimated as the uplift load at 10% of d uplift displacement of the helical

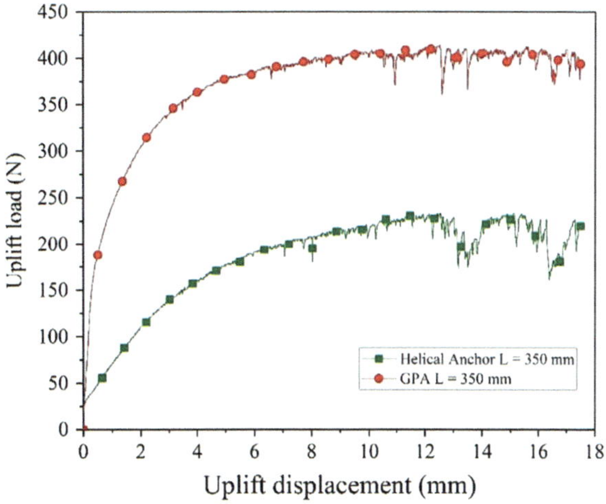

Fig. 4 Uplift load–displacement behavior of helical anchor and GPA at $L = 350$ mm

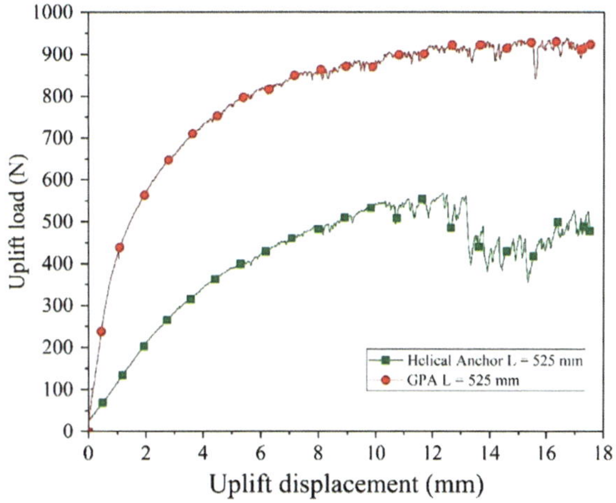

Fig. 5 Uplift load–displacement behavior of helical anchor and GPA at $L = 525$ mm

plate/anchor plate (ISSMFE criterion). The change in ultimate uplift capacity with an increase in *L*/*d* ratio is shown in Fig. 7 and tabulated in Table 1. It was observed that the ultimate uplift capacity increases with an increase in the embedment length for both the methods.

In the case of helical anchor, the percentage increase in uplift capacity when the embedment ratio changes from 5 to 7.5 is 130%, and 7.5–10 is 51%. Similarly, in GPA, the percentage increase in uplift capacity is 116% and 43% for the previously mentioned embedment ratio change. The decrease in the percentage increase in the uplift capacity is due to the change in the uplift resistance mechanism. In the lower embedment ratio, the uplift failure is shallow type in which the failure plane reaches the sand bed surface. In a larger embedment ratio, the failure is a deep anchor type in which the failure plane is confined inside the sand bed. A comparison of the uplift

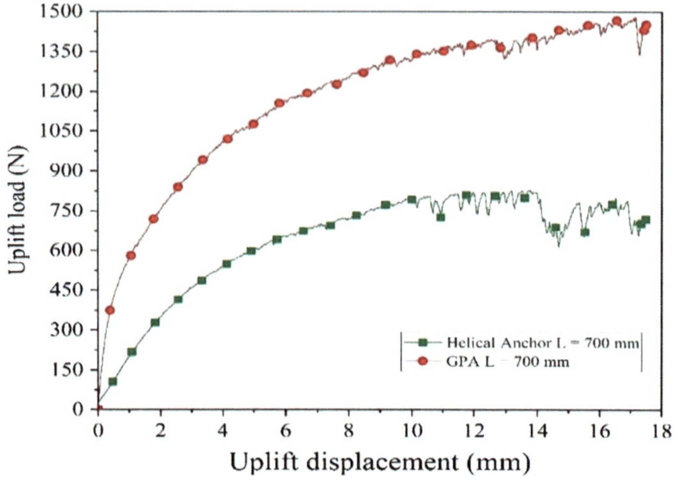

Fig. 6 Uplift load–displacement behavior of helical anchor and GPA at $L = 700$ mm

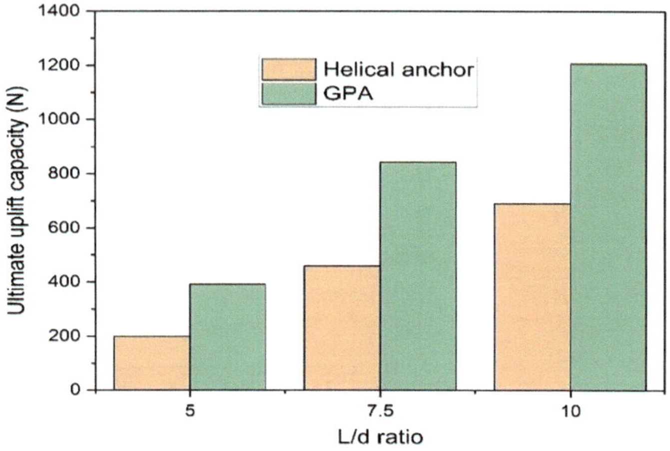

Fig. 7 Comparison of ultimate uplift capacity

Table 1 Geotechnical properties of the soil collected from the slope

L/d ratio	Ultimate uplift capacity (N)	
	Helical anchor	GPA
5	199	391
7.5	458	844
10	691	1207

capacities is illustrated in Fig. 7. The GPA gives a significantly larger uplift capacity compared with the helical anchor. The GPA's ultimate uplift capacity is found to be almost 2, 1.8, and 1.75 times that of the helical anchor for the embedment ratio 5, 7.5, and 10, respectively. The larges values of uplift resistance may be due to two reasons, firstly a part of the uplift being resisted by the compacted GPA material column above the anchor plate, and secondly, the confining pressure that is provided by the surrounding soil mass to the GPA material just above the anchor plate. Whereas the uplift resistance in case of the helical anchor is provided only by the sand mass above the helical plate.

4 Conclusions

A comparison of the load-displacement behavior and the ultimate uplift capacity of two methods that can be used to resist uplift loading is made in this study using small experimental modeling that is closely monitored in laboratory conditions. Laboratory-based small-scale testing cannot be relied upon to reflect field conditions fully. Nevertheless, the current study offers some limited insights into the comparative behavior of the two methods. The GPA resists uplift more uplift load compared with the helical anchor even at lower uplift displacement. The uplift capacity of both the methods increases with the increase in embedment ratio, but the percentage increase in the uplift capacity was observed to decrease. This decrease owing to the failure mechanism points to an optimum embedment ratio for both the methods. The GPA had 2–1.75 times the ultimate uplift capacity of the helical pile, which can be attributed to the compacted granular column above the GPA plate. Thus, the GPA is a much efficient method to resist uplift load than single helix helical anchors.

References

1. George BE, Banerjee S, Gandhi SR (2019) Helical piles installed in Cohesionless soil by displacement method. Int J Geomech 19(7):1–18
2. Ghaly A, Adel H, Hanna M (1991) Uplift behaviour of screw of screw anchors in sand. I: Dry sand. J Geotech Eng 117(5):773–793
3. Hanna A, Ayadat T, Sabry M (2007) Pullout resistance of single vertical shallow helical and plate anchors in sand. J Geotech Geol Eng 25(5):559
4. Ilamparuthi K, Muthukrishnaiah K (1999) Anchors in sand bed: delineation of rupture surface. Ocean Eng 26(12):1249–1273
5. Joseph J, Kumar S, Patel JB, Sawant V, Tandel Y (2022) Model tests on granular pile anchor and helical anchor: a comparative study. Int J Geosynthetics Ground Eng 8(3):1–12
6. Joseph J, Kumar S, Sawant VA, Patel JB (2022) An experimental and numerical comparative study on the uplift capacity of single granular pile anchor and rough pile in sand. Int J Geotech Eng 16(4):499–513
7. Kranthikumar A, Sawant VA, Shukla SK (2016) Numerical Modeling of Gran- ular Anchor Pile System in Loose Sandy Soil Subjected to Uplift Loading. International Journal of Geosynthetics and Ground Engineering, Springer International Publishing 2(2):15

8. Kranthikumar A, Sawant VA, Kumar P, Shukla SK (2017) Numerical and experimental investigations of granular anchor piles in loose sandy soil subjected to uplift loading. Int J Geomech 17(2):4016059
9. Kumar P (2002) Granular anchor pile system under Axial Pullout Loads. Ph.D thesis I.I.T., Roorkee
10. Liu KF, Xie XF, Zhang JF (2006) Compression/tension load capacity of stone column anchors. Proc Inst Civ Eng-Geotech Eng 159(3):161–165
11. Meyerhof GG, Adams JI (1968) The ultimate uplift capacity of foundations. Can Geotech J 5(4):225–244
12. Muthukumar M, Phanikumar BR (2015) Shrinkage behaviour of GPA-reinforced expansive clay beds subjected to swell–shrink cycles. Geotech Geol Eng 33(3):475–485
13. Muthukumar M, Shukla SK (2020) Comparative study on the behaviour of granular pile anchors and helical pile anchors in expansive soils subjected to swelling. Int J Geotech Eng 14(1):49–54
14. O'Kelly BC, Brinkgreve RBJ, Sivakumar V (2014) Pullout resistance of granular anchors in clay for undrained condition. Soils Found 54(6):1145–1158
15. Rao AS, Phanikumar BR, Babu RD, Suresh K (2007) Pullout behavior of granular pile-anchors in expansive clay beds in situ. J Geotech Geoenvironmental Eng 133(5):531–538
16. Sakr M, Eng P, Core G, Manager S, Canada WP (2011) Installation and performance characteristics of high capacity helical piles in Cohesionless soils. DFI J J Deep Found Inst 5(1):39–57
17. Singh J, Mital A, Arora VK (2019) Load displacement behaviour of Granular Anchor Pile (GAP) and group piles under oblique pullout loads. Indian Geotech J 49(1):100–107
18. Sivakumar V, O'Kelly BC, Madhav MR, Moorhead C, Rankin B (2013) Granular anchors under vertical loading—axial pull. Can Geotech J 50(2):123–132
19. Srirama Rao A, Phanikumar BR, Suresh K (2008) Response of granular pile-anchors under compression. Proc Inst Civ Eng-Ground Improv 161(3):121–129
20. Tsuha CHC (2016) Physical modelling of the behaviour of helical anchors. In: 3rd Eu ropean conference on physical modelling in geotechnics (EUROFUGE 2016). IFSTTAR Nantes Centre, France, 1st-3rd June

Utilization of Steel Slag for Stabilization of Subgrade Soil

Aditya D. Ahirwar and Hemant S. Chore

1 Introduction

In India, unavailability of suitable material for construction of highway severely hampers the schedule and cost of the project. This problem can be solved by utilizing the waste by-products such as steel slag, copper slag, zinc slag, bottom ash. As per world steel association, 1951 million ton of steel is produced in year 2021. During the production of steel, two types of slag namely blast furnace slag and steel making slag. In India, annual rate of slag generation is higher than the world average [1], thus tremendous quantity of waste which is usually dumped in the premises of the integrated steel plant. The use of processed steel slag in road construction allows for sustainable waste management and reduces reliance on perishable natural aggregates.

Several studies have been carried out on utilization of waste by-products in construction of flexible pavement. Akinwumi et al. [2] concluded that steel slag can be effectively used for soil stabilization in order to meet Nigerian standards for the utilization of various component layers of pavements. Tiwari et al. [3] provided an Indian perspective on the use of steel slag. Aldeeky et al. [4] investigated the efficacy of steel slag in the context of highly plastic soil and discovered an improvement in the geotechnical properties of such soil when used in the pavement subgrade. C3S, C2S, C4AF, and C2F present in steel slag impart cementitious properties [5]. Athulya et al. [6], Ahirwar et al. [7] discussed the suitability of soil–slag mixtures for use in various layers of pavement.

Development of strength of the stabilized mix depends on the % of admixture used and curing period. Poh [8] identified that development of strength for soil

A. D. Ahirwar (✉) · H. S. Chore
Dr. B.R. Ambedkar National Institute of Technology, Jalandhar, India
e-mail: ahirwarad.ce.19@nitj.ac.in

H. S. Chore
e-mail: chorehs@nitj.ac.in

B. T. Jose et al. (eds.), *Proceedings of the Indian Geotechnical Conference 2022 Volume 4*, Lecture Notes in Civil Engineering 479,
https://doi.org/10.1007/978-981-97-1753-8_30

Table 1 Properties of dredged soil

Property	Value
SG: Specific gravity	2.45
LL: Liquid limit	68.1
PL: Plastic limit	32.8
PI: Plasticity index	35.3
Soil classification	CH
MDD: Maximum dry density (kg/cm3)	1.46
OMC: Optimum moisture content (%)	27.5
UCS: Unconfined compressive strength (kPa)	118.6

stabilized with basic oxygen steel slag fines requires long curing period. Sabat et al. [9] examined the effect of curing on an expansive soil stabilized with marble waste and rice husk ash. Okagbue [10] investigated the effect of 7 day and accelerated 24 h curing on soil stabilized with marble dust. Athulya et al. [6] identified that strength of the soil stabilized with steel slag, cement, nanosilica improves with curing period.

2 Material

The experimental investigation makes use of three distinct materials: dredged soil, steel making slag, and cement. The soil used in this study was dredged from banks of the Uran River in Maharashtra. The basic properties of the dredged soil are shown in Table 1. In this study, ordinary Portland cement was used. Steel making slag is provided by Metarolls and Commodities Pvt. Ltd, Maharashtra. Slag is classified as a silty material with a low plasticity.

Sieve analysis and hydrometer testing were used to determine particle size. The particle size distribution of clay and steel making slag is depicted in Fig. 1. The results of chemical analysis carried out using an X-ray fluorescence test shows that silicon dioxide, aluminum oxide, iron oxide, and calcium oxide are the most abundant elements in the clay. Steel making slag was primarily composed of silicon dioxide, aluminum oxide, ferrous oxide, calcium oxide, and magnesium oxide.

3 Experimental Program

The current study looks into the possible utilization of steel making slag for stabilisation of clayey subgrade. The geotechnical investigation involved determining the strength characteristics of composite mix in terms of unconfined compressive strength (UCS) and California bearing ratio (CBR). The dredged soil was mixed with various proportions of steel slag. The laboratory investigations were expanded

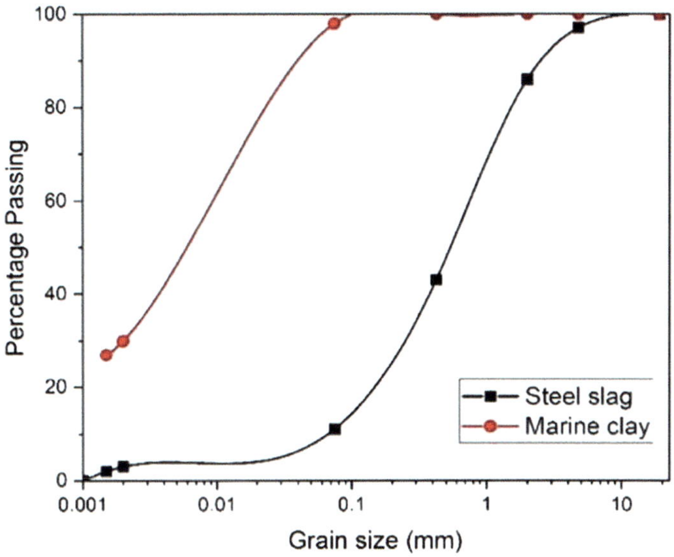

Fig. 1 Particle size analysis of dredged soil and steel making slag

to include the strength characterization of dredged soil stabilized with various proportions of steel slag and cement. Percentage of steel slag was varied in the range of 10–40%, whereas cement in the range of 4–8%.

Study is also stretched to evaluate the effect of curing period on unconfined compressive strength of the composite mix containing soil–slag–cement.

4 Proposed Model for Analysis of Flexible Pavement

The current analysis considers compacted subgrade with higher CBR and natural sub- grade with lower CBR as a two-layered system as shown in Fig. 2.

Elastic modulus of the both the layers are estimated using relationship given by Powell et al. [11]

Fig. 2 Two-layered system of compacted placed over natural subgrade

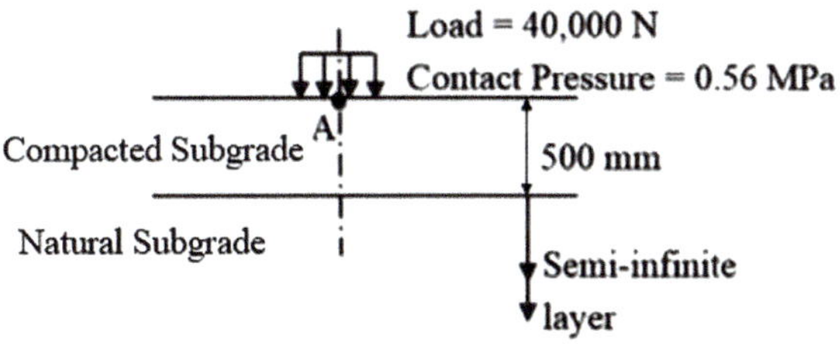

$$E = 10 * \text{CBR (Mpa) if CBR} <= 5 \tag{1}$$

$$E = 17.6\ \text{CBR}^{0.64}\text{(Mpa) if CBR} > 5 \tag{2}$$

The maximum surface deflection caused by a single wheel load of 40,000 N and a contact pressure of 0.56 Mpa for a two-layer system consisting of a 500-mm-thick subgrade laid over natural soil is calculated using the IITPAVE software. The equivalent single layer's resilient modulus is then calculated using equation

$$M_{RS} = \frac{2(1 - \mu^2)pa}{\delta} \tag{3}$$

where

p contact pressure
a radius of circular contact area
μ Poisson's ratio

According to IRC 37-2018, the design is based on the effective resilient modulus MRS and not the CBR. However, effective CBR can be calculated using Eqs. (1) and (2).

Composition of conventional pavement is made based on catalogue provided in IRC 37-2018 which is based on effective CBR. To run the analysis of flexible pavement on IITPAVE software, the modulus of different layers of pavement is calculated based on the IRC: 37-2018 guidelines. For the BC and DBM, VG 40 bitumen with air void content (V_a) of 3% and effective bitumen content by volume (V_{be}) = 13% is considered for the analysis. Based on the consideration, the modulus value of bituminous (MR) layer for all the cases is 3000 MPa [12]

Modulus of granular layer was calculated from the following formula given in IRC.

$$M_{\text{RGRAN}} = 0.2(h)^{0.45} \times M_{\text{RSUPPORT}}, \tag{4}$$

where

h thickness of granular layer in mm
M_{RGRAN} resilient modulus of the granular layer (MPa) MRSUPPORTresilient modulus of the supporting layer (MPa)

The rutting condition of the pavement is classified as critical or failure when the average rut depth is 20 mm or greater. The IRC recommends considering the equivalent number of standard axle load repetitions that the pavement can withstand before developing rut depths of 20 mm or greater.

The maximum strain at top of the subgrade is calculated on the basis of the rutting model for 90% reliability as per IRC:37-2018.

$$N_R = 1.4100 \times 10^{-8}[1/\varepsilon_v]^{4.5337}, \tag{5}$$

where

N_R subgrade rutting life (cumulative equivalent number of 80 kN standard axle loads that can be served by the pavement before the critical rut depth of 20 mm or more occurs)

ε_v vertical compressive strain at the top of the subgrade calculated using linear elastic layered theory by applying standard axle load at the surface of the selected pavement system

The maximum strain at the bottom of the bituminous layer is calculated on the basis of the Fatigue model for 90% reliability as per IRC 37-2018.

$$N_f = 0.5161 * C * 10^{-4}[1/\varepsilon_t]^{3.89} * [1/M_{RM}]^{0.854}, \tag{6}$$

where

$$M = 4.84(V_{be}/V_a + V_{be}) - 0.69 \tag{7}$$

V_a percent volume of air void in the mix used in the bottom bituminous layer considered 3

V_{be} percent volume of effective bitumen in the mix used in the bottom bituminous layer considered 13

Pursuant to the analysis of flexible pavement as per IRC:37–2018, the IITPAVE software analysis is carried out to determine the vertical and horizontal strain at the critical location of the traditional layer system for design traffic of 50 msa. The vertical compressive strain and horizontal tensile strain so obtained from IITPAVE software is then compared with the permissible strain to check the feasibility.

5 Results of Experimental Program

5.1 Unconfined Compressive Strength of the Composite Mix.

The unconfined compressive strength test was conducted on various composite mixes containing dredged soil and steel making slag. From the Fig. 3, it can be concluded that, unconfined compressive strength of the mix increases after addition of steel slag up to 30%. Akinwumi et al. [2], Aldeeky et al. [4] also observed that the unconfined compressive strength of subgrade soil improves with addition of steel slag. Similar pattern can be seen in case of samples tested on 7th and 28th day. Furthermore, the strength of the soil–slag mixes increases with the increase in curing period.

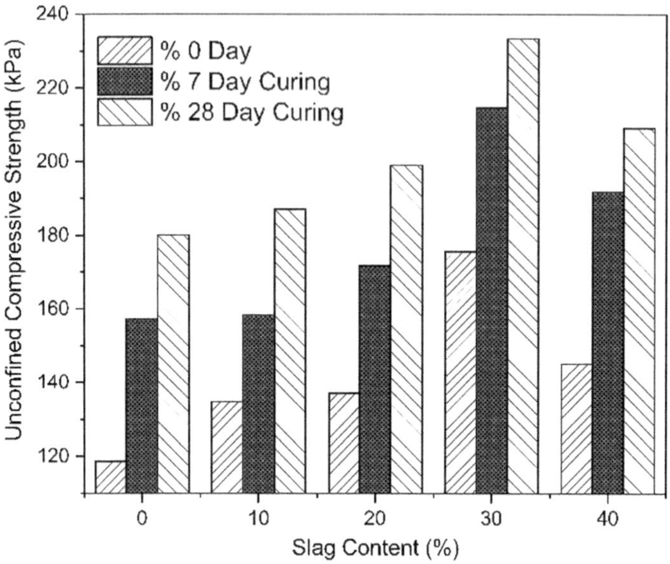

Fig. 3 Variation in UCS of the soil–slag mix for 0, 7, 28-day curing

UCS test were also conducted on mix containing soil–slag–cement. From the Fig. 4, it can be concluded that, UCS of the mix increases significantly after addition of cement. Furthermore, the strength of the mix improves with the curing period. Athulya et al. [6] observed significant improvement in strength of the soil–slag mix when admixed with the conventional stabilizer-like cement.

5.2 *California Bearing Ratio of Composite Mix*

The California bearing ratio test was conducted on various composite mixes containing, steel making slag and cement. Figure 5 suggests that there is no significant improvement in CBR when soil is admixed with the slag alone. Whereas after addition of cement, CBR of the mix increases significantly. Athulya et al. [6] also observed significant improvement in CBR of the soil–slag mix after addition of cement.

For a composite mix containing 4% cement, the CBR of a soil–slag–cement compo- site increases as slag content approaches 30%. Similar trend can be observed in case of mix containing 6% and 8% cement.

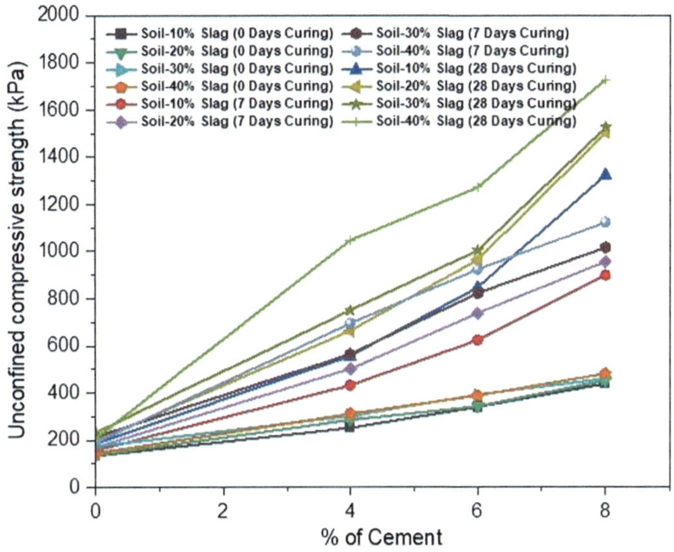

Fig. 4 Variation in UCS of the soil–slag–cement mix for 0-, 7-, 28-day curing

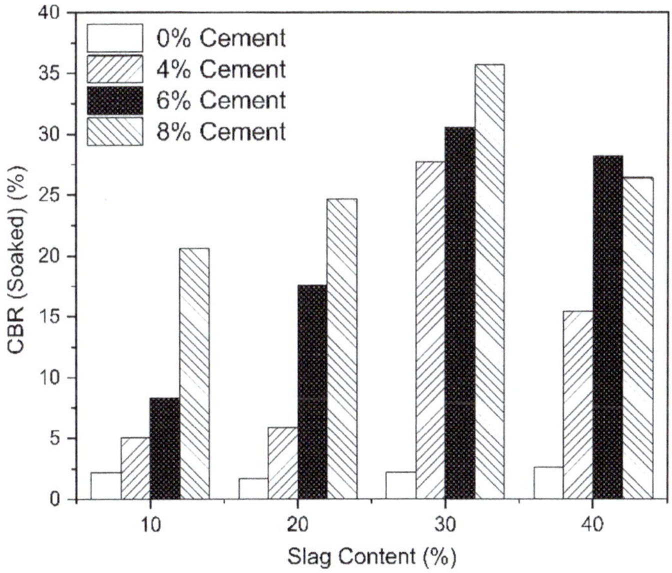

Fig. 5 California bearing ratio of composite mix

6 Analysis of Flexible Pavement on Stabilized Subgrade

Analysis of flexible pavement is based on traffic volume, wheel load, and characteristics of the materials used for the construction. Different methods of design of flexible pavement include the following: CBR method, Asphalt Institute method, Austroads, U.K. Road Note 29 and IRC: 37- 2001, 2012, and 2018. In the present study, IRC: 37-2018 guidelines based on the cumulative standard axial load is used.

Based on the laboratory investigations, the soaked CBR values of some of the combinations selected for the analysis is shown in Table 2.

From the laboratory investigation, it is observed that CBR of the natural subgrade is lower than the permissible limit prescribed by IRC standards. Thus, alternative 500-mm-thick capping subgrade should be laid having greater CBR. In such case, effective CBR of the single layer subgrade which is equivalent to combination of compacted subgrade and natural subgrade should be laid. In current investigation, it is proposed to construct the 500-mm-thick layer of the stabilized clay mixes having higher CBR. IRC standards suggests that it is effective resilient modulus value and not the CBR is used for design of the flexible pavement. Effective CBR of the single layer subgrade is equivalent to combination of subgrade layer and compacted subgrade (stabilized soil) is shown in Table 3.

Table 2 CBR values of dredged soil stabilized with different stabilizers

S. No.	Type of stabilized soil	Soaked CBR (%)
1	Dredged soil	2.15
2	Dredged soil + 30% steel slag + 4% cement	27.67
3	Dredged soil + 30% steel slag + 6% cement	30.53
4	Dredged soil + 30% steel slag + 8% cement	35.67

Table 3 Thicknesses of different layer for the conventional flexible pavement

Sr. No.	Stabilized soil mix	Effective CBR (%)	Total thickness (mm)	GSB (mm)	Base (mm)	Binder (mm)	Surface (mm)	References
1	Dredged soil	2.15	911	448	250	173	40	IRC 37-2012 [13]
2	Dredged soil + 30% steel slag + 4% cement	10.96	593	200	250	103	40	
3	Dredged soil + 30% steel slag + 6% cement	11.59	591	200	250	101	40	IRC 37-2018 [12]
4	Dredged soil + 30% steel slag + 8% cement	12.65	587	200	250	97	40	

The thickness of the flexible pavement is estimated using IRC 37-2012,18 standards considering conventional layer system and is based on effective CBR values of the various mixes for design traffic of 50 msa and is summarized in Table 3.

Thickness of the flexible pavement is reduced significantly when 500-mm-thick stabilized compacted subgrade is placed over the natural subgrade soil. When soil-30% and slag-4% cement mix is used as a compacted capping subgrade, thickness of the flexible pavement reduces by 34.90%. Whereas for mix containing 6% cement and 8% cement, reduction in thickness of about 35.12% and 35.56% can be observed, respectively.

The resilient modulus of the various layers of flexible pavement considering different mixes of subgrade soils is calculated as per IRC 37-2018 and is shown in Table 4.

Flexible pavements are subjected to fatigue and rutting damages due to the repetitions of heavy axle wheel load. Vertical compressive strain on top of the subgrade, corresponding to standard axle load repetitions, is indicative of failure of natural subgrade. Vertical compressive strain on top of the subgrade and horizontal tensile strain at the bottom of the bituminous layer are considered critical parameters that must be controlled to ensure satisfactory performance of the flexible pavement in terms of subgrade rutting and bottom-up cracking of bituminous layers. Based on the modulus of different layers and their thickness, the IITPAVE software analysis is carried out to determine the vertical and compressive strain at the critical location of the traditional layer system for design traffic of 50 msa.

In all the above-mentioned cases, the value of actual strain calculated using IIT Pave software is less than the allowable strain. From Table 5, it can be observed that conventional design by using IRC 37-2018 methodology can be considered safe rendering good serviceability.

Table 4 Resilient modulus for various layer of flexible pavement considering different stabilized soil subgrade

Sr. No.		Modulus values for different layers (MPa)		
	Stabilized soil	Subgrade	Granular sub-base and base	Bituminous layer
1	Dredged soil + 30% steel slag + 4% cement	81.477	254.689	3000
2	Dredged soil + 30% steel slag + 6% cement	84.448	263.976	3000
3	Dredged soil + 30% steel slag + 8% cement	89.335	279.252	3000

Table 5 Comparison between permissible strain and actual strain obtained using IIT Pave software

Sr. No.	Stabilized soil mix	Strain	Permissible limit (As per IRC 37)	Actual strain (IIT Pave)	Remark
1	Dredged soil + 30% steel slag + 4% cement	Horizontal ten sile strain	203.1×10^{-6}	173.0×10^{-6}	Safe
		Vertical compressive strain	371.69×10^{-6}	268.8×10^{-6}	Safe
2	Dredged soil + 30% steel slag + 6% cement	Horizontal ten sile strain	203.1×10^{-6}	171.7×10^{-6}	Safe
		Vertical compressive strain	371.69×10^{-6}	263.3×10^{-6}	Safe
3	Dredged soil + 30% steel slag + 8% cement	Horizontal ten sile strain	203.1×10^{-6}	171.5×10^{-6}	Safe
		Vertical compressive strain	371.69×10^{-6}	257.0×10^{-6}	Safe

7 Summary of Conclusions

The aim of the experimental research conducted for this paper was to explore the potential of using steel slag to stabilize subgrade soil. It involved characterizing the structural strength after it had been stabilized with steel making slag and cement. Pursuant to the analysis of flexible pavement as per IRC:37-2018, the IITPAVE software analysis is carried out to determine the vertical and compressive strain at the critical location of flexible pavement for design traffic of 50 msa. From the experimental and analytical work carried out in present study, following broad conclusions to be drawn.

- Unconfined compressive strength of the slag stabilized dredged soil im proves after addition of steel slag up to 30%.
- UCS of the mix increases with curing period.
- There is no significant improvement in strength characteristics like unconfined compressive strength and California bearing ratio, when soil is admixed with the slag alone. Whereas after addition of cement, CBR of the soil–slag–cement composite mix increases significantly.
- When soil stabilized with steel slag and cement is used as capping subgrade, thickness of the flexible pavement reduces significantly.
- When stabilized soil is used as a capping subgrade, the actual strain calculated using IIT Pave software falls within permissible limits thus can be considered safe rendering better serviceability.

The current study demonstrates the efficacy of using industrial waste such as steel slag to improve the engineering properties of weak soil when combined with a conventional stabilizer. The abundant use of steel making slag, in highway construction can be made in a sustainable manner, reducing environmental concerns.

References

1. TIFAC (2003) Management of steel plant solid wastes. Report prepared by Technology Information, Forecasting and Assessment Council (TIFAC), New Delhi
2. Akinwumi II, Adeyeri JB, Ejohwomu OA (2012) Effects of steel slag addition on the plasticity, strength and permeability of lateritic soil. In: Proceedings of second international conference of sustainable design, engineering and construction, fort worth, Texas, pp 457–464.https://doi.org/10.1061/9780784412688.055
3. Tiwari MK, Bajpai S, Dewangan UK (2016) Steel slag utilization—overview in Indian perspective. Int J Adv Res 4(8):2232–2246. https://doi.org/10.21474/IJAR01/1442
4. Aldeeky H, Al Hattamleh O (2017) Experimental study on the utilization of fine steel slag on stabilizing high plastic subgrade soil. Adv Civ Eng 2017 Article ID (9230279) 1–17. https://doi.org/10.1155/2017/9230279
5. Oluwasola EA, Hainin MR, Aziz MMA (2014) Characteristics and utilization of steel slag in road construction. Jurnal Teknologi 70(7):117–123. https://doi.org/10.11113/jt.v70.3591
6. Athulya GK, Dutta S, Mondal JN (2017) Performance evaluation of stabilised soil–slag mixes as highway construction material. Int J Geotech Eng 11(1):51–61. https://doi.org/10.1080/19386362.2016.1184874
7. Ahirwar AD, Chore HS (2022) Strength characterization of steel slag–cement-stabilized marine clay mixes for highway construction. Innov Infrastruct Solut 7:254. https://doi.org/10.1007/s41062-022-00846-4
8. Poh HY, Ghataora GS, Ghazireh N (2006) Soil stabilization using basic oxygen steel slag fines. J Mater Civ Eng 18(2):229–240. https://doi.org/10.1061/(ASCE)0899-1561
9. Sabat AK, Nanda RP (2011) Effect of marble dust on strength and durability of Rice husk ash stabilised expansive soil. Int J Civ Struct Eng 1(4):939–948
10. Okagbue CO, Onyeobi TUS (1999) Potential of marble dust to stabilise red tropical soils for road construction. Eng Geol 53(3–4):371–380
11. Powel et al (1984) The structural design of bituminous roads. Laboratory Report 1132, Transport and Road Research Laboratory, Department of Transport
12. IRC:37 (2012) Guidelines for design of flexible pavements. (Third Revision) Indian road congress, New Delhi
13. IRC:37 (2018) Guidelines for design of flexible pavements. (Fourth Revision) Indian road congress, New Delhi

Strengthening of the Foundation Using Deep Mixing and Stone Column Techniques for Large Earth Cum Rock Fill (ECRF) Dam for an Irrigation Project

Vidyaranya Bandi, Karsten Beckhaus, Cemal Can Maslak, and Vijaya Bhushan Rao

1 Introduction

The Government of Andhra Pradesh Irrigation and Command Area Development (I&CAD) Department proposed to construct a multipurpose dam across the river Godavari near Polavaram village, Andhra Pradesh, India (Fig. 1). Bauer is the subcontractor of the project to build Earth Cum Rock Fill (ECRF) dam for a length of 1750 m across the Godavari River. Ground improvement using deep soil mixing (DSM) is proposed by Bauer along the existing plastic concrete diaphragm wall (upstream side and downstream side) to ensure the overall stability of the ECRF dam by improving the stiffness of the stiffness of soils at GAP I and GAP II locations. Deep soil mixing (DSM) is introduced for the first time by Bauer in India at this project and it's a very sustainable solution replacing the traditional ground improvement techniques [1].

V. Bandi (✉)
Bauer Engineering India Pvt. Ltd., Chennai, India
e-mail: vidyaranya.bandi@bauer-india.com

K. Beckhaus · C. C. Maslak
Bauer Spezialtiefbau GmbH, Nordhausen, Germany
e-mail: Karsten.Beckhaus@bauer.de

C. C. Maslak
e-mail: Cemal-Can.Maslak@bauer.de

V. B. Rao
Bauer Spezialtiefbau GmbH (Dubai Office), Dubai, UAE
e-mail: Vijay.Bhushan@baueruae.ae

B. T. Jose et al. (eds.), *Proceedings of the Indian Geotechnical Conference 2022 Volume 4*, Lecture Notes in Civil Engineering 479,
https://doi.org/10.1007/978-981-97-1753-8_31

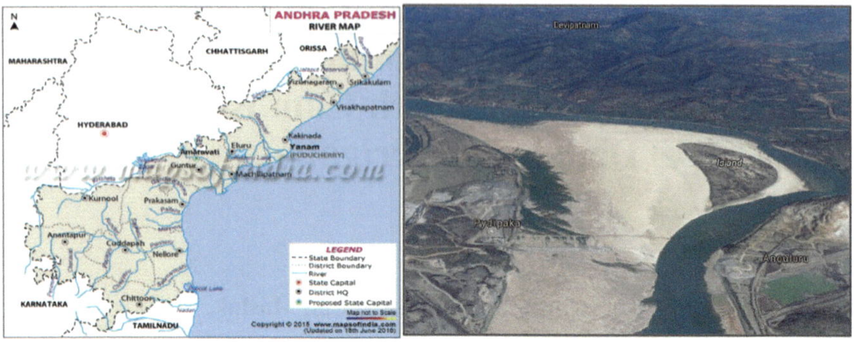

Fig. 1 Location of ECRF dam at Polavaram

2 Need for the Installation of the DSM Column Below the ECRF Dam

The DSM columns will form a transition treatment between the existing the cut-off wall and proposed ground improvement by stone columns below the footprint for ECRF Dam. The DSM columns act as transition of the deformation modulus of the ground from the center of the dam to the toe of the dam embankment. The soil improvement works consists of single column deep soil mixing (DSM) and shall be carried out along the installed plastic concrete diaphragm wall/cut-off wall in the clay sections as shown in the Fig. 2. As the stiffness of diaphragm wall is higher than the stone columns, so this DSM treatment will provide intermediate stiffness by transition. Installation of DSM columns shall confirm improvement of the initial soil stiffness by five times.

3 Designing the DSM Columns for Protection Works

For soil treatment beneath the rock fill embankment, stone columns (deep vibro floatation method) and rigid columns (DSM method) are used for ground improvement. Field trials are conducted near the GAP I area to establish the design parameters and mix design of the DSM columns to be installed at the site. The data obtained from the trials shall be used for execution of DSM column production at GAP I and GAP II.

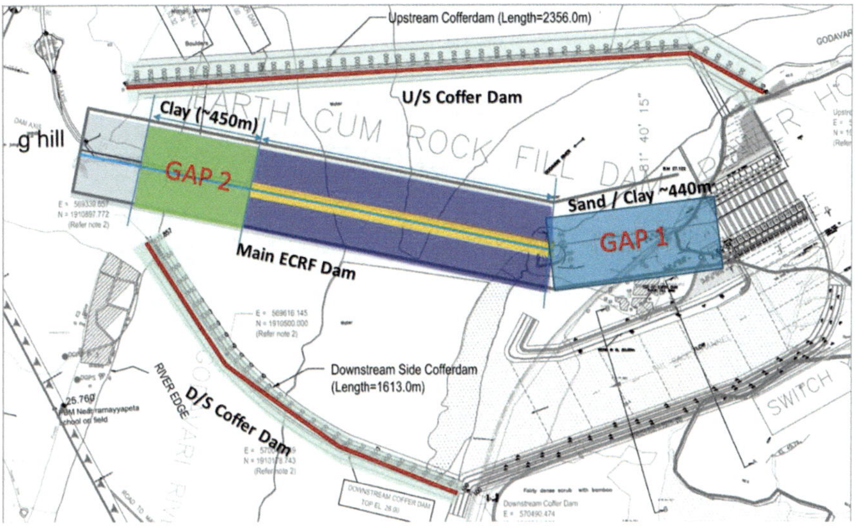

Fig. 2 Plan of deep soil mixing works at GAP I and GAP II

4 Summary of the Geotechnical Data at the Job Site

Based on the available geotechnical information for the site, the summary of the grain size distribution at the locations of improvement by deep soil mixing (DSM) are summarized below in Table 1.

4.1 Methodology for Trial Works for Deep Soil Mixing

Field trials are conducted prior to commencement of the main works to ensure performance of the proposed ground improvement scheme for both vibro stone columns and deep soil mixing columns. The tentative trial locations at the GAP I and GAP II are as shown in the Fig. 3. The results of field trials establish correctness of the assumed material properties (elastic modulus) which were considered in the numerical modeling for the case of post ground improvement scenario, in dam foundation

Table 1 Summary of the soil profile of the site

Location	Chainage (m)	Clay (%)	Silt (%)	Sand (%)	Gravel (%)
GAP 2	1365–1195	34.3	49.8	7.5	8.4
GAP 2	1195–1025	24.4	46.1	22.2	7.3
GAP 2	1025–910	< 50	< 50	> 50	> 50
GAP 1	250–425	16.7	48.6	30.3	4.5

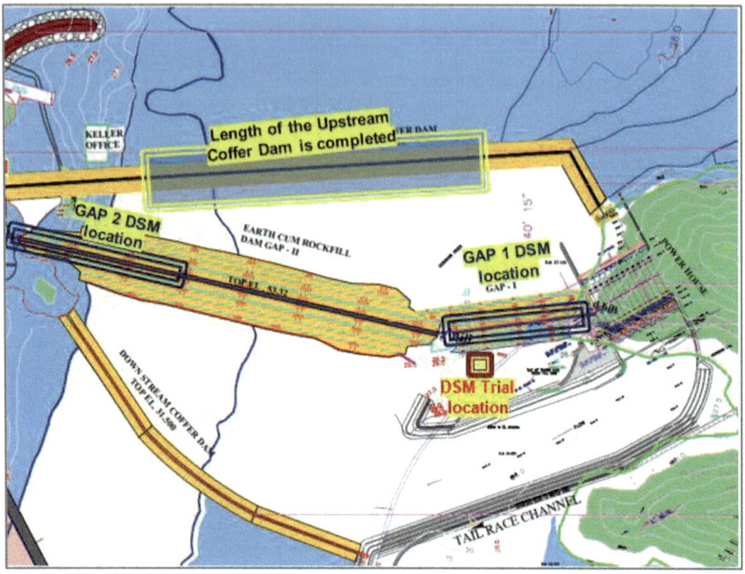

Fig. 3 Tentative location of the DSM works and proposed trial locations

design. The anticipated post treatment stiffness in case of stone columns is two times of the in situ stiffness whereas it is expected to be five times improvement for deep soil mixing.

4.2 Trial Columns

The field trial test section include installation of 18 nos of DSM columns. Layout of the DSM columns is shown below Fig. 4. For a treatment area ratio of 40%, with column diameter of $D_{\text{col}} = 1$ m and spacing between the columns in a squared grid is considered at $a = 1.$ m. However, the plan is to install trial columns with a 1.3 m spacing as per the tender specified for production. The quality control measures for the installation of the trial columns is specified in the Table 2.

4.3 Deep Soil Mixing Technology and Work Sequence at Field Trial Works and Main Works

The deep soil mixing technology implies mixing of existing soil with cement or compound binder (lime, flyash, slag, etc.) and aims to achieve significant improvement of mechanical properties of the soil. The improvement becomes possible by

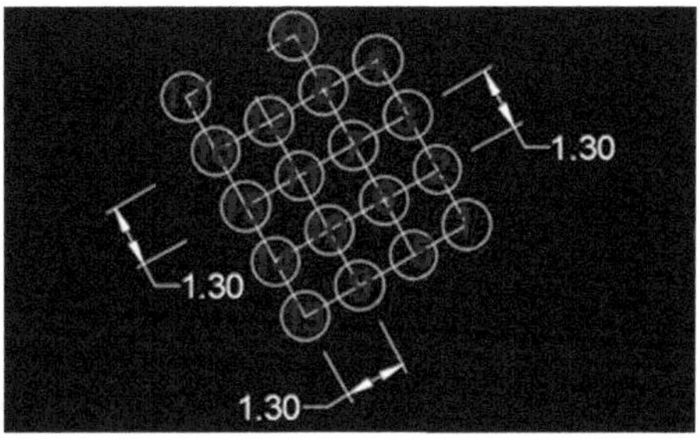

Fig. 4 DSM field trial columns setup

Table 2 DSM quality control requirements

Item	Description
Grout mix (W/C ratio)	1:1.2
Design stiffness	150 kPa
Target UCS of DSM columns	750 kPa
Amount of binding	150–300 kg/m^3 of soil
Methods of collection of samples (at various depths)	1 Wet grab sampling method 2 Three samples at various depths 3 Backflow sampling method
Tests conducted on DSM samples	1 UCS test 2 Stiffness measurement
Curing period of DSM samples	7, 14 and 28 days

strengthening with cement, bonding of soil particles and/or filling of voids by chemical reaction products. As a result of this mixing, soil–cement grout columns with higher strength than the natural soil are created. Along with improved strength, due to the known constitutive mechanical equation for cement-based materials, the DSM columns will have an improved stiffness [2].

Regarding the homogeneity, it shall be noted that soil particles of fist to cobble size will remain unmixed in the DSM material which is subject to the mechanical mixing with the DSM mixing tool but are not deemed to compromise the column stiffness significantly. One of the major advantages of deep soil mixing technology is an economical one in terms of no transportation and no purchasing of aggregates. The mass of excavated material and spoil that must be treated, transported, hauled, and dumped off site is also significantly reduced. This makes the method environmentally friendly as well. The complete process of the deep soil mixing technology is depicted in the Fig. 5. The deep soil mixing works are executed by single auger known single columns mixing (SCM). Installed DSM trial columns will not be removed from the site after testing.

The wet method of deep soil mixing is carried out by setting up drill rod with DSM mixing equipment (Fig. 6) and tool at column position. The mixing tool penetrates to the desired depth of treatment with simultaneous disaggregation of the soil by mixing tool and water or binder can be used for liquefaction. The process is continued till the

Fig. 5 Equipment for deep soil mixing technology

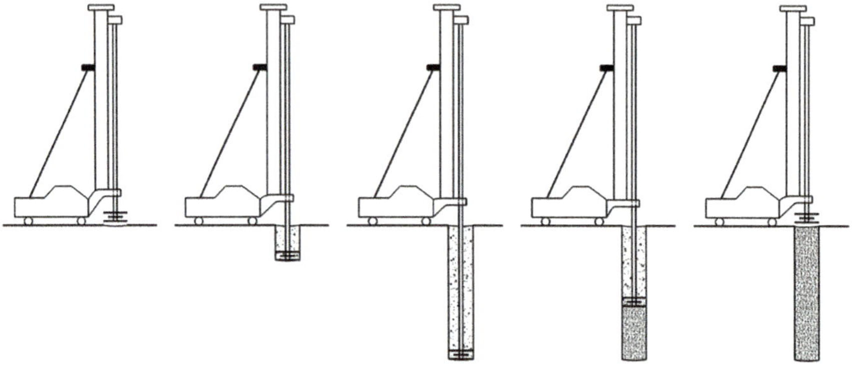

Fig. 6 Construction sequence of DSM columns

auger reaches the final depth of treatment. The binder slurry is injected into the soil during withdrawal, optionally by additional downwards and upwards strokes, where the mixing tool rotates in the horizontal plane and mixes the soil and the binder. The final column is formed in the ground as the auger is retrieved. The construction sequence of the DSM column installation is as shown in the Figs. 7 and 8.

4.4 Methodology of Collection of Samples at Various Depths

The samples of soil mixing will be collected using both wet grab sampling method and backflow sampling method for better quality. The process of wet grab samples is collected using a wet grab sampling box, typically secured to the excavator arm/

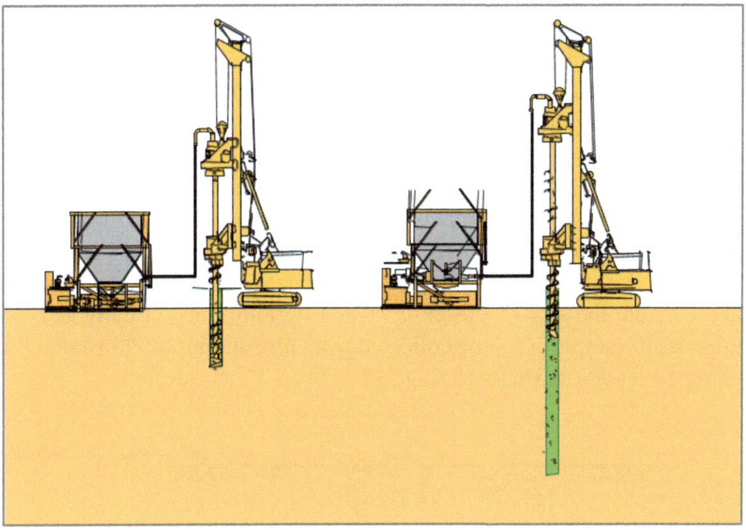

Fig. 7 **a** Down stroke—drilling down, pumping cement slurry (~70% of design) through the auger into the soil and mixing the soil with the slurry. **b** Up stroke—withdrawing auger, pumping slurry (~30% of design) through the auger and proving the more homogeneous mix

Fig. 8 DSM mixing tool

bucket, installation rig or crane and pushed into the freshly mixed column to collect an in-situ soil mix sample and casted in standard cylinder mounds (at least three samples at same depth). It is suggested to collect the samples from every column at three different depths, and it covers every 1 m depth among all columns (Fig. 9). The max depth of samples is able to collect by wet grab method is up to 2 m above the termination depth of DSM column Fig. 10 shows typical set up of wet grab sampler.

In any condition, wet grab samples are not able to collect from the desired depth, backflow samples shall be collected from surface for testing purpose. Back flow samples refer to the sludge generated during installation of wet mix column. Samples are collected from the sludge, casted in standard mold (cube or cylinder). At least, three numbers of samples shall be collected from all columns after installation DSM columns for necessary stiffness testing.

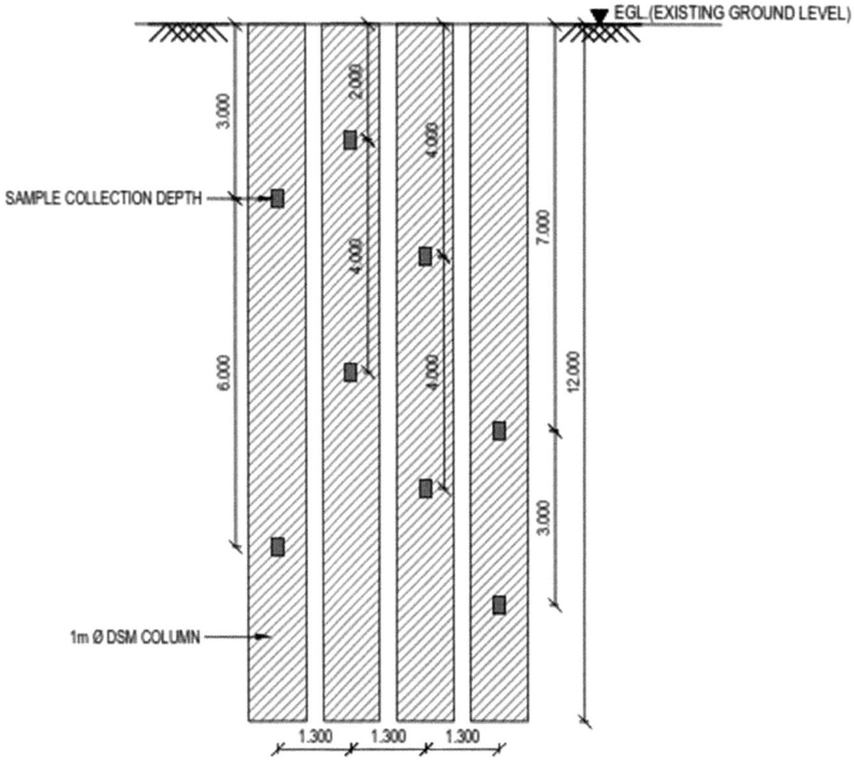

Fig. 9 Typical sample collection depths

Fig. 10 Typical pictures of wet grab sampler

4.5 Assessment of Improvement

The improved composite stiffness after treatment shall be calculated after achieving the actual DSM column stiffness values. An improved composite stiffness of treated ground were established using below set of formula [3].

$$E_{\text{comb}} = A_r E_{50,\text{col}} + (1 - A_r)\, E_{50,\text{soil}} \quad (1)$$

Estimation of degree of improvement in stiffness is as follows:

Step 1: Stiffness of original soils shall be established from either pre-treatment ECPTs or SPT N values from boreholes.
Step 2: Estimate the composite stiffness of treated ground as per formula shown in above.
Step 3: Compare the composite stiffness and stiffness of original soils. The degree of improvement in stiffness is the ratio of composite stiffness of treated ground to the stiffness of original soils.

4.6 Main Works

As a result of the performed field trials, a variation of mixing slurry and other construction parameters are examined, which cover the construction parameters necessary for the construction work, i.e., need to allow observing the dependence of the construction procedure and mechanical properties of installed columns. At main works, installing numerous DSM elements with previously established specific mixing parameters will be used to achieve the specified mechanical properties over the column depth and cross-section. For this, slurry mix designs and mixing parameters obtained from trial field shall be used.

In main works, same equipment proposed for trial field shall be used for the installation of the DSM columns. Based on the field trial results, the mix design and construction procedure shall be defined to achieve treated soil meeting the project specific acceptance criteria.

4.7 Post-performance Criteria for DSM Works

The whole DSM field trial program is aimed to establish all necessary DSM parameters to achieve the following post-performance criteria for the DSM material. Based on the results of trials field and considering aspects of required stiffness and compressive strength, following requirements on DSM material was suggested by project authority. The required unconfined compressive strength (UCS) shall be proved as a characteristic value, tested at cylindrical specimen with minimum dimensions of 100 mm, at the age of 28 days shall be within 0.5–1.0 MPa. The resulting average unconfined compressive strength shall correspond with an average design stiffness of the DSM columns, respectively, a deformation modulus of DSM material E within 40–135 MPa. This deformation modulus is supposed to be evaluated as a tangent modulus from the stress–strain graph from the UCS test. A correlation shall be assessed from a representative series of test samples taken from UCS samples.

5 Details on the Design and Execution Works

For the design study of deep soil mixing (DSM) columns in Polavaram project, layout of the field at GAP I and II locations, respectively (Fig. 11). The goal to achieve was calculating the required characteristics compressive strength f_{cm} of the DSM material in a defined grid. The general idea was to achieve a composite modulus for a lower bound of 50 MPa and upper bound of 150 MPa. A transition zone that allows for a smooth and gradual transition of settlement between the areas treated with DSM columns and treated with stone columns is also implemented along with lower and upper bound. To allow a mutual understanding of the effect of the soil improvement, a numerical model of the dam and an idealized underground situation has been set up in Plaxis 2D. Theoretical settlements have been calculated for the situations (1) Lower bound and (2) with upper bound of treated soils, considering the differently treated areas perpendicular to the dam (or cut-off wall) axis, i.e., across the DSM column zone, transition zone and stone-column zone. Although the total settlement values can be read only qualitatively, the comparison performed may help to decide for the necessary transition between DSM and vibro stone columns to smooth the incremental settlements across the dam.

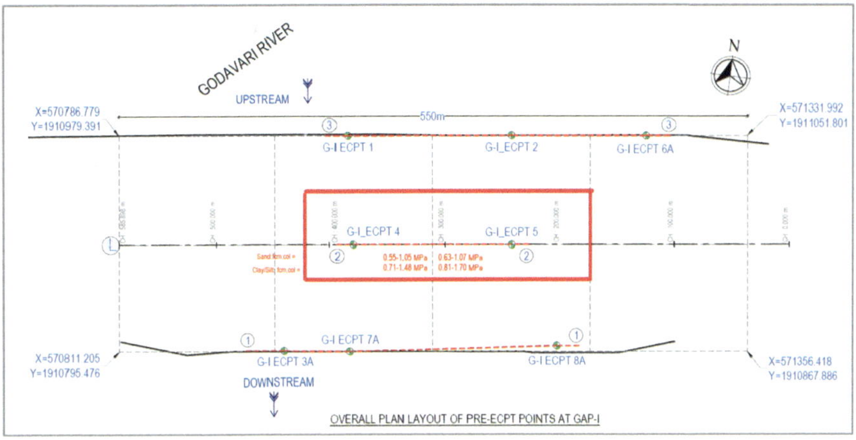

Fig. 11 Layout for GAP I with the range of f_{cm} values

5.1 Transition Zone

To provide a smooth transition of settlements between the DSM columns and stone columns, a consistent or gradual decrease in oedometric modulus ($E_{oed,composite}$) must be maintained. The average values of the lower and the upper bounds between the oedometric modulus of stone columns and DSM columns result in range of 32.5–82.5 MPa (Fig. 12).

The length of the transition zone is taken to be 10 m, which allows 5 rows of DSM columns between the stone columns. To utilize the transition zone to the optimal capacity and allow the stiffness modulus to shift along the profile, the first three layers of DSM columns in transition zone are taken at their original depth of 12 m, whereas the depths of last two layers are reduced to 6 m. The detailed summary of the DSM works in the main section and transition zone are provided in the Table 3 (Fig. 13).

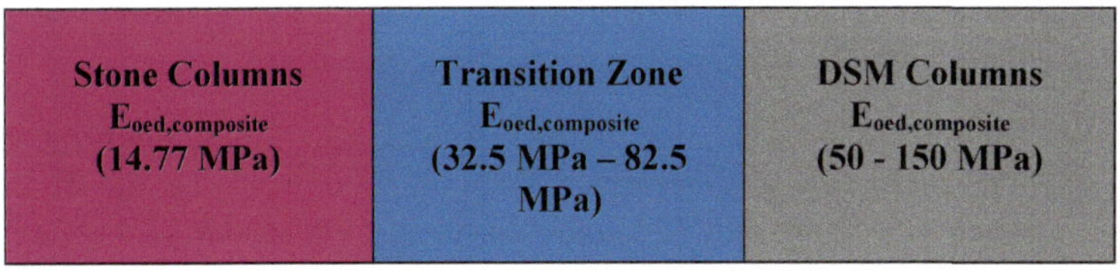

Fig. 12 Transition zone visualization

Table.3 Overview of soil improvement dimensions

Particulars	DSM up to 10 m wide	Transition zone				
		1st	2nd	3rd	4th	5th
Spacing, m	1.3	2	2	2	2	2
Diameter, m	1	1	1	1	1	1
Length of the DSM column, m	12	12	12	12	6	6
Length of the VSC column, m	17	17	17	17	17	17
Area/column, m	0.79	0.79	0.79	0.79	0.79	0.79
Ratio of replacement, %	46.47	19.63	19.63	19.63	19.63	19.63

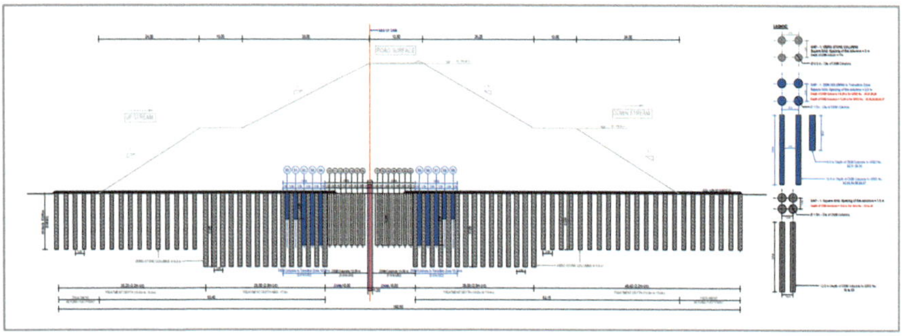

Fig. 13 Section detailing of DSM, transition zone (DSM + stone columns) and stone columns for GAP I

6 Conclusion

In this paper, use of the deep soil mixing (DSM) technique for the ground improvement beneath the dam foundation is presented. Detailed design analysis is carried out to estimate the settlements under different conditions like the dam being empty and flooded to the high flood levels. DSM acts as reinforcement element in providing protection to the plastic concrete cut-off wall along the axis of the dam thus reducing the settlements and improving the deformation modulus of the ground. The paper describes the installation of the deep soil mixing (DSM) methodology for ground improvement introduced for first time in India. The ground improvement by DSM is a sustainable solution and offers an alternative solution for the traditional ground improvement solutions by stone columns and other techniques.

References

1. Federal Highway Administration Design Manual. Deep mixing for embankment and foundation support. Publication No. FHWA-HRT-13-046
2. Bauer.de/export/shared/documents/pdf/bst/print/905_044_2_Deep_Soil_Mixing.pdf
3. Kitazume M, Terashi M (2013) The deep mixing method (1st edn). CRC Press. https://doi.org/10.1201/b13873

Drilled Displacement Columns as Ground Improvement in San Francisco

Akhil Anil and Daniel Wagstaffe

1 Introduction

When native soil cannot support the loads from a proposed development, ground improvement can be used to densify the soil, transfer loads to deeper layers, and/or help mitigate against the liquefaction pot. Ground improvement methods vary depending on the soil conditions and the site constraints. Some methods densify the soil using vibration or displacement, which have been shown to be an effective means of improving granular soils. Other methods reinforce the soil by constructing stiff elements within a soil mass that can transfer loads to more competent soil/bedrock. Still, other methods improve the soil by using additives, such as cementitious materials, and either (1) mixing the additives with the soil to create a composite material or (2) permeating them into granular soil to create a stiff matrix. This paper discusses the benefits of drilled displacement columns (DDCs) as a ground improvement method and provides a case history of their use at a site in San Francisco.

DDCs are a ground improvement method that provides densification and reinforcement. DDCs are grouted columns with a high modulus and controlled stiffness. During installation, DDC drilling equipment displaces soil as it advances, which increases the density of surrounding granular soil. The cavity created by the drilling equipment is filled with grout, and steel reinforcement can be added to the column while the grout is still fluid. DDCs are commonly installed through weak, compressible soils to transfer loads to more competent layers, reducing settlement, and increasing bearing capacity for overlying foundations. The displacement and

A. Anil (✉)
Texas A&M University, College Station, TX 77843, USA
e-mail: akhilanil07@tamu.edu

D. Wagstaffe
Langan, 135 Main St 1500, San Francisco, CA 94105, USA
e-mail: dwagstaffe@langan.com

B. T. Jose et al. (eds.), *Proceedings of the Indian Geotechnical Conference 2022 Volume 4*, Lecture Notes in Civil Engineering 479,
https://doi.org/10.1007/978-981-97-1753-8_32

subsequent densification of granular soil can also increase the soil's resistance to liquefaction. The soil reinforcement and densification provided by DDCs allow for increased bearing capacities, reduced settlement, and increased soil resistance to liquefaction.

2 Drilled Displacement Column Installation Procedure

DDCs are constructed using a displacement drill tool shaped to laterally displace soil as the tool advances, which densifies the granular surrounding soil. The drill tool consists of auger flights that drill into the soil and a bevelled section that displaces the soil. It is hollow, with the internal cavity opening to a port at the tip of the tooling. Once the drill tool reaches the DDC's design tip elevation, grout is pumped through the hollow centre and out of the port at the tip of the tooling. Grout fills the cavity and flows up around the augers until a pressure head is created. After a grout pressure head is created, the drill tooling continues to pump the grout while withdrawing from the cavity. The grout head helps prevent soil from caving into the DDC cavity, and grout typically surfaces from the hole before the tip of the drill tool reaches the ground surface. The auger and the pressurised grout result in a coarse-sided grout column. After the drill tool is withdrawn, steel reinforcement can be added to the wet grout. A gravel cushion is added to the top of the DDC to prevent hard points and to decouple the DDC from overlying structural elements. Figure 1 illustrates the DDC installation procedure.

Installation of DDCs produces low noise and minimal vibration in the soil, and the displacement of the soil reduces the number of spoils. The composite ground can support loads from footings, slabs, and mats with reduced settlement. DDCs can also

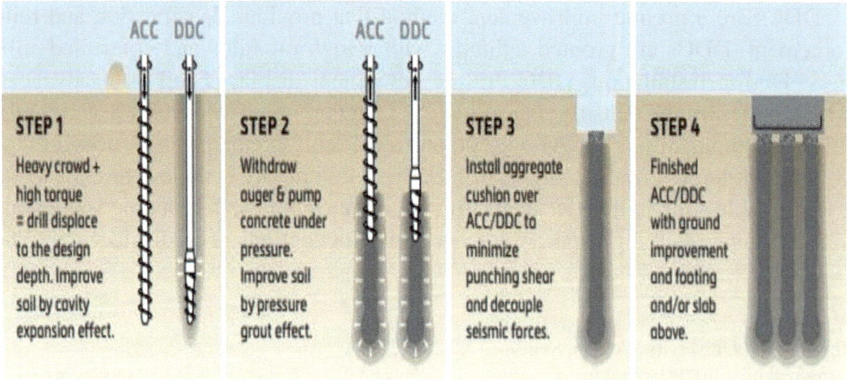

Fig. 1 Summary of DDC installation procedure (Farrell Design-Build [3]). Auger cast column (ACC) installation is also shown

cause settlement across a building footprint to be more uniform if they are installed systematically across the footprint.

The remainder of this paper will discuss a case study where DDCs were implemented at a site in San Francisco for foundation support. The site will be developed into a wholesale market and a parking garage for the market. The sections below describe the site, a summary of the subsurface conditions and geologic hazards at the site, considerations for foundation support, and an overview of the DDC installation at the site. The case study serves as an example of the successful implementation of DDC installation and ground improvement.

3 Site Description

The site for the proposed wholesale market takes up approximately 3½ acres and is located just west of the Mission Bay development area. The existing site features include a warehouse and an at-grade parking lot. As part of the proposed development, the warehouse will be retrofitted to house the market, and the at-grade parking lot will be developed into a parking structure. DDCs were installed to support the proposed parking structure. The regional geology and seismicity of the site are discussed below.

3.1 Regional Geology and Seismicity

The City of San Francisco is located at the northern tip of the San Francisco Peninsula, along the western edge of the North American plate. The terrain is characterised by hills consisting of highly deformed and fractured sedimentary rocks. The topography is largely from the east–west compression of coastal California during the late Pliocene and Pleistocene eras. As discussed in a report by the Association of Engineering Geologists [1], "the geology of San Francisco is comparatively simple: Several fault-bounded northwest trending bands of Mesozoic and Paleogene basement overlain by Quaternary surficial deposits (and artificial fill), and lacking any of the Late Cretaceous and Tertiary overlap sequences found elsewhere in the region, with the exception of a narrowband of Plio–Quaternary Merced Formation outcrops in the far southwest corner of the city. Mesozoic rocks are exposed in sea cliffs and as resistant knobs and hills throughout the city. Quaternary surficial deposits are poorly exposed, filling the low-lying areas between hills. Much of the land in the north and east side of the city is artificial fill overlying bay mud deposits." An example cross-section from a geologic map by Schlocker [5] can be seen in Fig. 2.

Because the region is seismically active, the area is prone to experience moderate to large earthquakes. The San Andreas, Hayward, and San Gregorio faults are the closest significant faults to San Francisco, as shown in Fig. 3. Figure 3 also displays the earthquake epicentres for previous events with a magnitude greater than 2.5.

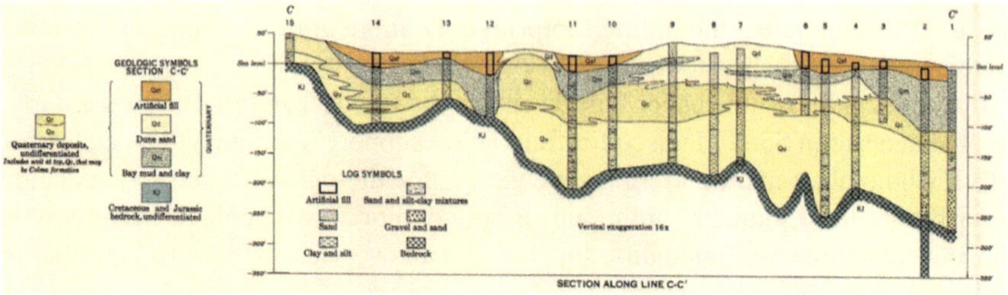

Fig. 2 Cross-section in the South of the Market area of San Francisco [5]

4 Subsurface Conditions

Multiple geotechnical and environmental investigations were performed at the site to characterise the subsurface conditions. As part of these investigations, fifteen borings and two cone penetration tests (CPTs) were performed within the limits of the proposed parking garage. The borings were drilled with both rotary-wash and hollow-stem-auger drilling equipment. The CPTs were performed by hydraulically pushing a cone-tipped probe into the ground, which measured tip resistance and frictional resistance. The CPT data were correlated to engineering properties, such as the soil types, apparent strength, and consolidation characteristics.

Based on the results of the investigations, the project geotechnical engineer concluded that the site was underlain by fill, soft clay (known locally as Bay Mud), layers of sand and clay, and bedrock. An example cross-section (based on a subsurface profile from the project geotechnical investigation report) is shown in Fig. 4, and the subsurface units underlying the proposed parking garage are discussed in more detail below.

Fill	The area of the proposed parking garage is blanketed by heterogeneous fill, extending to depths up to about 20 feet below the ground surface. The fill consists of gravel, sand, silt, and clay mixtures with varying amounts of construction debris and rubble, including concrete, rebar, brick, and timber fragments. Generally, it is not suitable for support of heavy building loads Based on the environmental assessment, the fill is contaminated
Bay	A weak and compressible marine clay deposit, referred to as Bay Mud, is
Mud	Present beneath the fill in many areas within the extents of the proposed parking garage. This layer is on the order of 5 to 10 feet thick. Similar to the fill, it is not suitable to support heavy building loads
Sand	Layers of sand and clay were encountered below the Bay Mud in most of
and	The borings and the CPTs. The sand has various clay and silt content, and
Clay	The clay has various sand content. These layers are more dense/stiff than the overlying fill and Bay Mud layers and can provide support for the proposed building loads

Bedrock	Within the extent of the proposed parking garage, the top of bedrock was encountered at depths ranging from roughly 20–70 feet below the ground surface. The depth to the top of the bedrock generally increased from south to north. Typical of bedrock in San Francisco, the strengths were generally weak, and the rock exhibited deep weathering

5 Geologic Hazards

The project site is expected to experience intense ground shaking during a major earthquake, which can cause various types of geologic hazards such as liquefaction and cyclic densification.

Liquefaction can occur in saturated sand and gravel and low-plasticity silt when an earthquake inhibits strong ground motion and causes the soil to behave like a liquid [6]. When liquefaction occurs, the soil can lose its ability to support structures and cause ground fissures and sand boils. This can be accompanied by a heterogeneous and erratic settlement of the ground surface and supported structure.

The site falls within a liquefaction hazard zone, as mapped by the California Geological Survey's seismic hazard zone map for the city of San Francisco [2], and the San Andreas Fault is roughly 12 kms from the site, which exhibited a characteristic magnitude of about 8.1 during the 1906 earthquake. Based on an evaluation by the project geotechnical engineer, it was concluded that layers of saturated sand are susceptible to liquefaction and associated liquefaction-induced settlements following a major earthquake on a nearby active fault.

Cyclic or seismic densification can also occur from intense ground shaking. Cyclic densification can occur in loose, granular deposits above the water table during a strong earthquake. The densification leads to ground surface settlement, and differential settlement can occur when the densification is not homogeneous. The degree of susceptibility to this phenomenon is correlated to the relative density of the granular soil. The project geotechnical engineer concluded that the dry sand underlying the site is susceptible to cyclic densification.

Both liquefaction-induced settlement and cyclic densification can occur in unison, and the total post-earthquake settlement caused by seismic activity is the sum of these settlements. The project geotechnical engineer estimated that several inches of post-earthquake settlement could occur at the site if the soil was not improved.

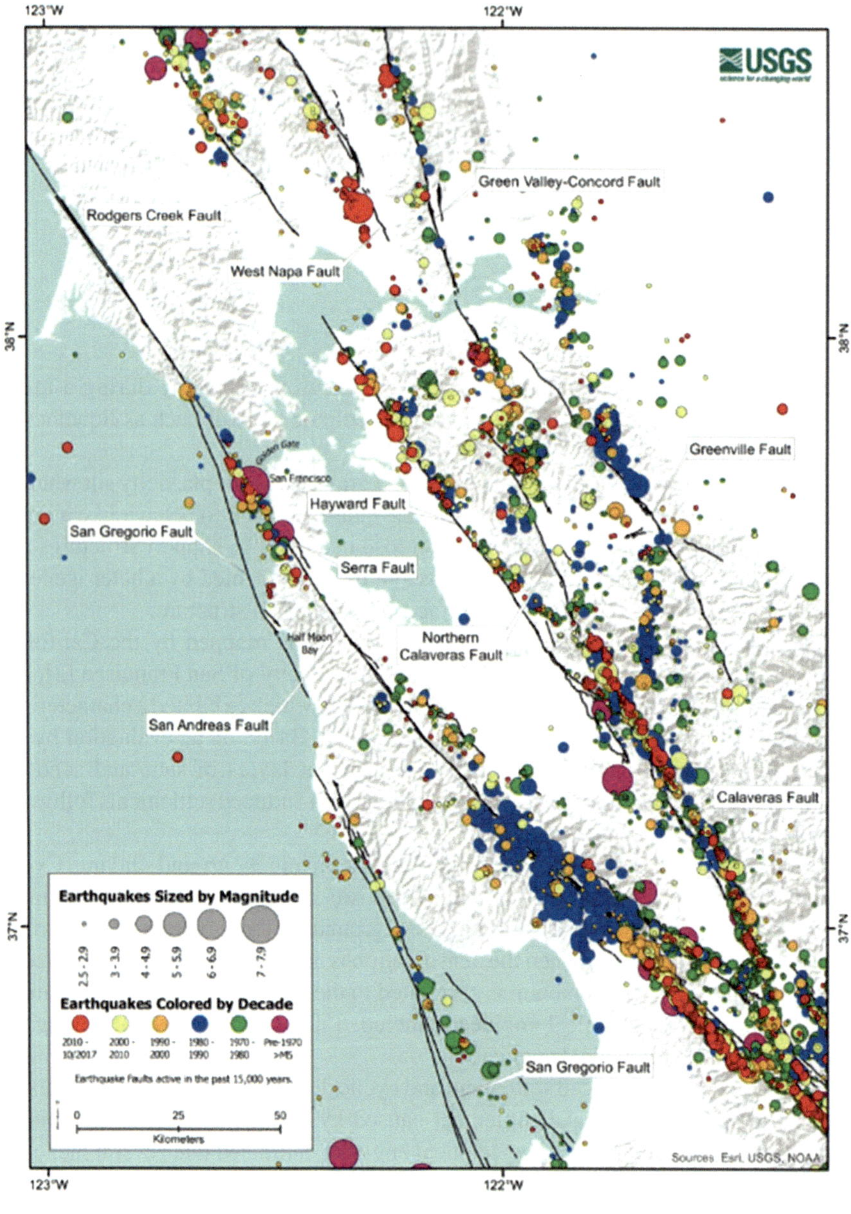

Fig. 3 Map of historical (M > 2.5) seismicity from the Northern California Earthquake Data Centre (NCEDC) catalogue with known active faults of the San Francisco Bay Area [1]

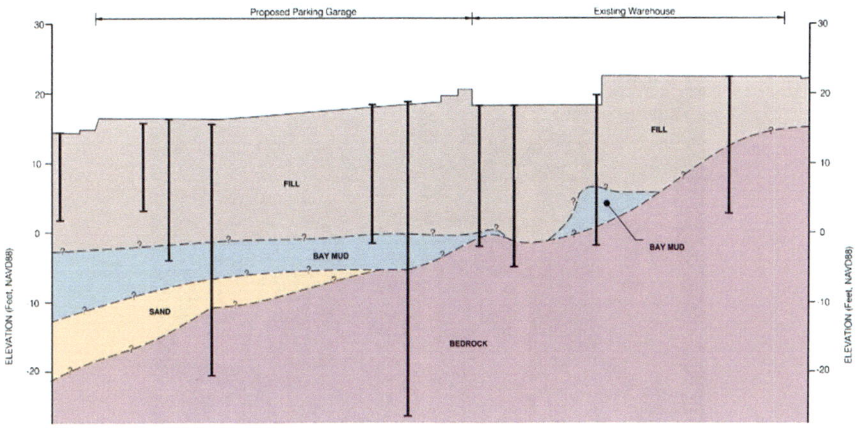

Fig. 4 Example cross-section based on an idealised subsurface profile from the project geotechnical investigation report. Brown colouring corresponds to fill, blue to Bay Mud, orange to sand, and purple to bedrock

6 Foundation Consideration

As discussed above, the fill and Bay Mud beneath the proposed parking garage footprint are unsuitable for foundation support. The fill below and above the groundwater level is subject to liquefaction and cyclic densification, respectively. The Bay Mud is weak and compressible. If a shallow foundation system was used to support the proposed parking garage, the project geotechnical engineer estimated that the Bay Mud could consolidate (settle) several inches, depending on the loads. Therefore, the design team considered using either deep foundations to bypass these poor soils or ground improvement that could improve the soil and extend building loads into competent soil and bedrock. The project owner determined that ground improvement would be the most economical solution for the project. Several ground improvement options were considered, including rigid inclusions, deep soil mixing, rammed aggregate piers, and DDCs. It was concluded that DDCs would be the most economical and appropriate ground improvement methodology for the site.

7 DDC Installation

In total, 509 DDCs were installed to transfer the loads from the proposed parking garage to competent soil and bedrock. The ground improvement installation contractor drilled each DDC using a Bauer BG-28 drill rig equipped with a 40-foot-long, 16-inch diameter displacement auger. Figure 5 shows DDCs installed at the site, and Fig. 6 shows the auger used for DDC installation. Figure 7 shows the planned DDC section.

Fig. 5 DDC installation at the wholesale market site

Fig. 6 View of the ground improvement installation contractor attaching the auger to their drill rig (left) and view of the auger on the right-hand side of the photo (right)

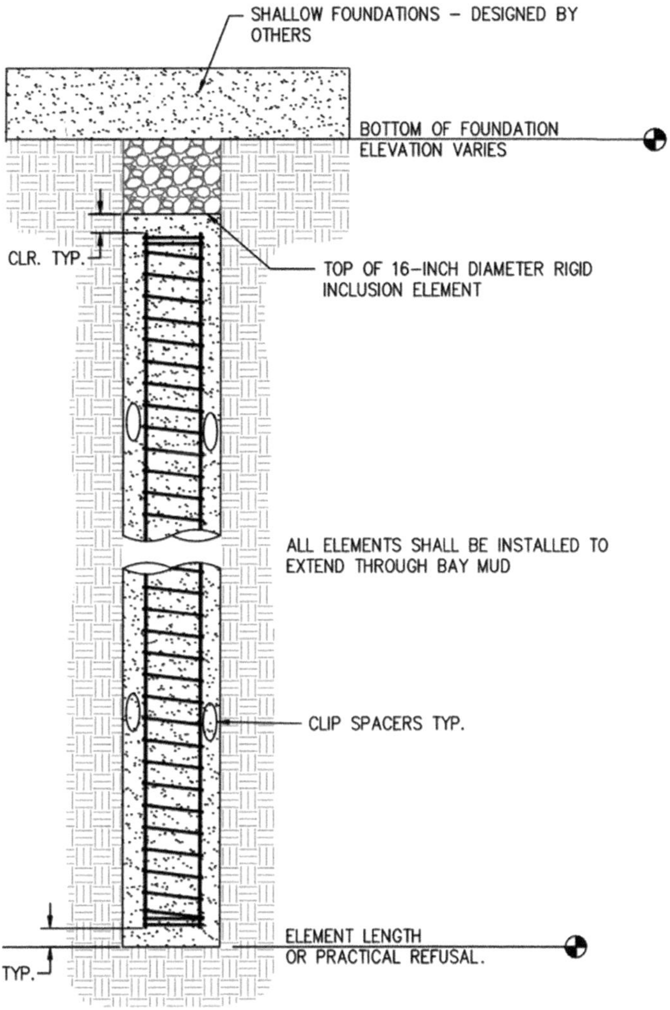

Fig. 7 Planned DDC section based on the section provided in the project DDC construction drawings (dimensions and references to details have been omitted)

To confirm the integrity of the DDCs, load tests were performed on four test DDCs to verify that they could support the anticipated loads with acceptable settlements. After the test program, the production DDCs were installed using the same criteria as the test DDCs. Production DDC drill depths were also estimated based on the results of the test program.

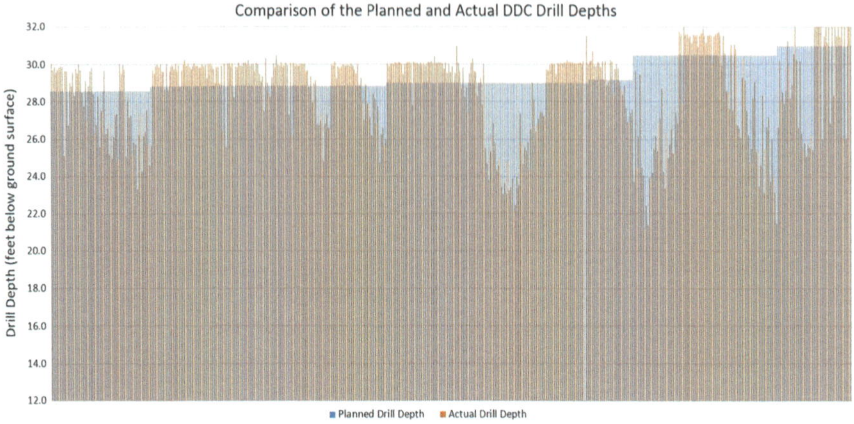

Fig. 8 Summary of the planned (blue) and actual (orange) DDC drill depths

During production DDC installation, the ground improvement installation contractor would cease drilling and begin grouting a DDC once they reached the target drill depth or once practical refusal was encountered during drilling. The practical refusal was defined as advancing the drilling equipment 12 inches or less for at least 1 min. A summary of the planned and actual DDC drill depths is shown in Fig. 8. As shown in Fig. 8, many DDCs encountered practical refusal.

During installation, the drill rig recorded installation parameters such as depth, auger torque, auger push pressure, grout pressure, and grout volume. The actual grout volume was compared to the theoretical grout volume to evaluate whether enough grout was placed into the DDC. To quantify this, a grout factor (the actual volume of grout divided by the theoretical volume of grout) was calculated for each DDC. The target grout factor was 1.28 for this project, which was based on the results of the DDC test program conducted prior to installing production DDCs. Figure 9 shows a summary of the grout factors for this project.

After the auger was withdrawn from the ground, a rebar cage was inserted into the DDC while the grout was still wet. Figure 10 shows examples of the ground improvement installation contractor inserting rebar cages into recently drilled DDCs. A one-foot thick aggregate cushion was installed at the top of each DDC, and shallow foundations for the parking garage were poured on top of these aggregate cushions.

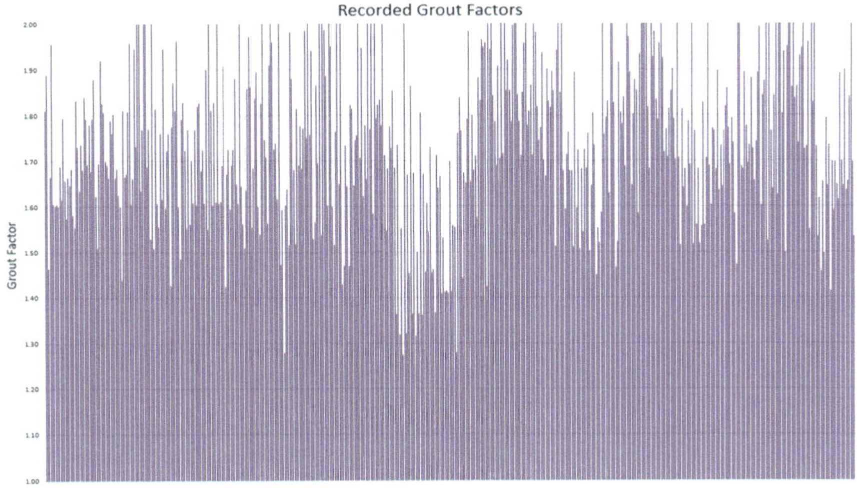

Fig. 9 Summary of DDC grout factors

Fig. 10 Rebar cages being lowered into DDCs

8 Conclusion

Improving soil with DDCs can be a cost-effective option to provide support to structures. They can be used to:

- reduce foundation settlement,
- mitigate liquefaction potential by replacing and densifying soil, and
- transfer building loads through weaker soil to a deeper bearing stratum.

Instead of requiring deep foundations for support, a structure can be supported on shallow foundations bearing on the firm composite ground created by DDC installation.

The case study we provided showcased DDC installation and presented an example of their use. DDCs were successfully installed in this project, and a parking garage was supported on the improved soil.

References

1. Baldwin J et al (2018) Johnson K, Bartow G (eds) Association engineering geologists, Geology of San Francisco. United States of America, California, pp 1–189
2. California Division of Mines and Geology (CDMG) (2000) Seismic hazard zone report for the City and County of San Francisco, California, CDMG Seismic Hazard Zone Report 043.
3. Farrell Design-Build Inc., 2023, Auger Cast Column™ & Drill Displacement Column™ - Design-Build Ground Improvement, Foundation Systems for California and West Coast, www.farrellinc.com/services/foundation-systems/auger-cast-columndrill-displacement-column.
4. Northern California Earthquake Data Center (NCEDC) (2014) UC Berkeley seismological laboratory. Dataset. https://doi.org/10.7932/NCEDC
5. Schlocker J (1974) Geology of the San Francisco North quadrangle, California, U.S. Geological Survey Professional Paper 782, 104 p., 3 plates, scale 1:24,000
6. Youd TL (1973) Liquefaction, flow and associated ground failure, U.S. Geological Survey Circular 688, 12pp

A Review on Biopolymers as an Eco-Friendly and Sustainable Solution for Soil Stabilization

Neha Raj and S. Selvakumar

1 Introduction

Since the dawn of human civilization, the technical discipline of geotechnical engineering has been practiced, particularly with regard to the handling and application of soil (or earth) in construction. Geotechnical engineering entailed stabilizing and enhancing the soil. The methods and supplies for doing this, like stabilizing the soil and combining it with conventional concrete binders, have indeed been applied to designed soil treatments. Approaches that are environmentally advantageous alternatives are essential. Possibilities, such as geotextiles, chemical monomers, geopolymers, microbial induction, and natural polymers, are being intensively investigated. Due to its widespread application as a realistic soil remediation agent, cement generates a substantial amount of global warming gases [14]. Subsequently, a variety of techniques and materials have been employed to enhance the characteristics of soils. Portland cement emerged as the foremost used substance for building and treating soil, especially following the Industrial Revolution [26, 35]. In the building sector, Portland cement is a substance that offers a number of well-known advantages, including affordability, toughness, workability, and durability. The manufacture and use of cement, however, are linked to some environmental concerns. The emissions associated with the usage of cement as one of main contributors of CO_2 emissions in the global level, accounting for roughly 8% of total CO_2 emissions worldwide [14, 35].

N. Raj (✉) · S. Selvakumar
Department of Civil Engineering, Vel Tech Rangarajan Dr. Sagunthala R&D Institute of Science and Technology, Avadi, Chennai, Tamil Nadu 600062, India
e-mail: neharaj1403@gmail.com

S. Selvakumar
e-mail: selvakumars@veltech.edu.in

B. T. Jose et al. (eds.), *Proceedings of the Indian Geotechnical Conference 2022 Volume 4*, Lecture Notes in Civil Engineering 479,
https://doi.org/10.1007/978-981-97-1753-8_33

The top priorities of soil application and betterment, also referred to as designed soil, are to enhance a particular soil's engineering characteristics, like its internal resistance, hydraulic gradient, and resilience under repetition drying and wetting circumstances, in addition to increasing environmental rejuvenation. Physical augmentation and chemical modification have historically been the two basic methods for producing customized soil. Physical enhancement is the method of improving the soil's material characteristics, such as deformation, permeability, exterior loading (such as overlay), consolidation, and perhaps other methods. When soil is treated chemically, interior chemical mechanisms such hydrolysis or cementitious reactions, which artificially link soil particles occurs together. One example of this is the usage of calcium silicate hydrate (C–S–H) [43].

For every ton of Portland cement produced, 0.55 tons of CO_2 are released during the calcination of calcium carbonate and released into the atmosphere. Additionally, as calcination requires heat at a temperature of up to 1450 °C, roughly 0.4 tons of CO_2 are released during the manufacturing of 1000 kg of cement as a consequence of the burning of carbon fuel. Every time a ton of cement is manufactured, 0.95 tons of CO_2 are released into the atmosphere [31]. Although the usage of cement contributes to numerous ecosystem problems, the building sector also uses about 40% of all the energy worldwide [4].

When it comes to soil engineering, biological approaches are currently being actively researched as a substitute to such traditional methods of soil treatment and enhancement. Particularly for soil treatment and enhancement, biopolymers have been presented as a novel kind of building binder [14]. Ordinary Portland cement and geopolymers have stronger fortifying effects than biopolymer techniques, but both utilize massive quantities of binders (10–50% of the soil's mass in binders, on average) nevertheless, biopolymers need 0.5–1.0% of the soil's weight in binders to reach strength levels comparable to those produced by intensive cement mixing. 0.5% biopolymer combinations are required to attain strengths that are comparable to or even greater compared to those for 10% ordinary Portland cement-soil combinations [11]. This study offers an overview of biopolymer in geotechnical engineering in response.

2 Biopolymers a Sustainable Approach

Polynucleotides (such as ribonucleic acid and deoxyribonucleic acid), carbohydrates and proteins (fatty acid molecules) that have already been historically among the most common types frequently used in diverse engineering procedures, three primary typical types of key biopolymers that are taken into consideration [8, 36, 47]. Organic polymers called biopolymers are created by living things. They are made up of monomeric pieces that are joined together to form larger forms. In fact, the utilization of biopolymers is not a completely new innovation in geotechnical engineering. In a broad sense, organic polymers like natural bitumen, straw, and sticky rice are also biopolymers and have been employed by ancient civilizations. As a binding agent,

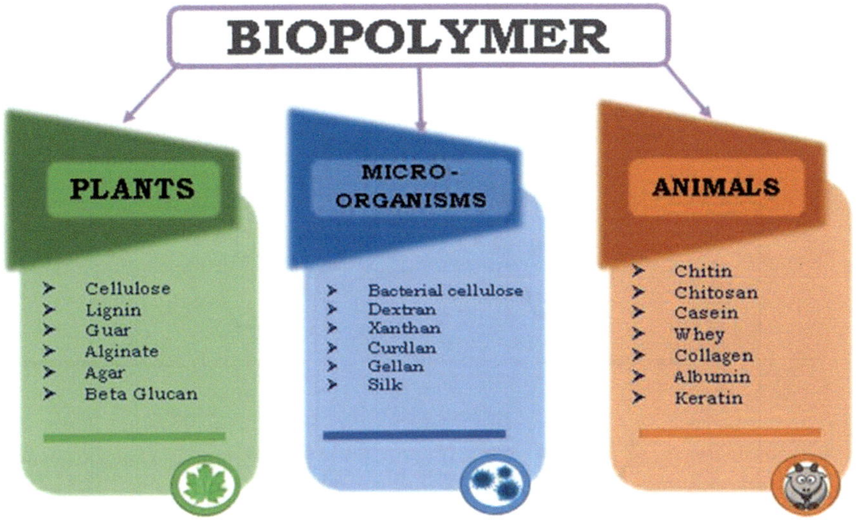

Fig. 1 Classification of biopolymer [25]

sticky rice mortar was utilized in ancient Chinese culture. Actinidia Chinensis cane liquid, lime, loess, and riverbed sediment were combined with glutinous rice broth to form a paste that has adequate resistance, high stiffness, and water-resistant properties [8, 50].

Biopolymers can be classified into groups as shown in Fig. 1, a list of biopolymers, organized by the production of source, that are mostly used for soil improvement. Plant-based biopolymers are produced using vegetation and agricultural waste. Biopolymers are generally made of plant-based materials, the majority of which are polysaccharides, and are used in geotechnics. Among the compounds in this category are cellulose, lignin, guar, alginate, agar, and beta-glucan. Microorganism- based biopolymers include xanthan gum, gellan gum, and dextran, which are made from bacteria through fermentation methods. Biopolymers originated from animals are those that are made from animal products. Waste crustacean shells, such as shrimp shells, are used to make chitin and chitosan, while milk and dairy products are used to make protein-based biopolymers [25].

3 Types of Biopolymers

Xanthan Gum: Due to its properties as a hydrocolloid rheology modifier, xanthan gum, a polysaccharide produced by aerobic degradation of sucrose by the bacterial species Xanthomonas campestris is frequently employed as a food ingredient [6, 20]. It is a D-uronic acid, D-mannose, pyruvylated mannose, 6-0-acetyl D-mannose, and 1, 4-linked glucan anionic polysaccharide [45]. Two tri-saccharide components

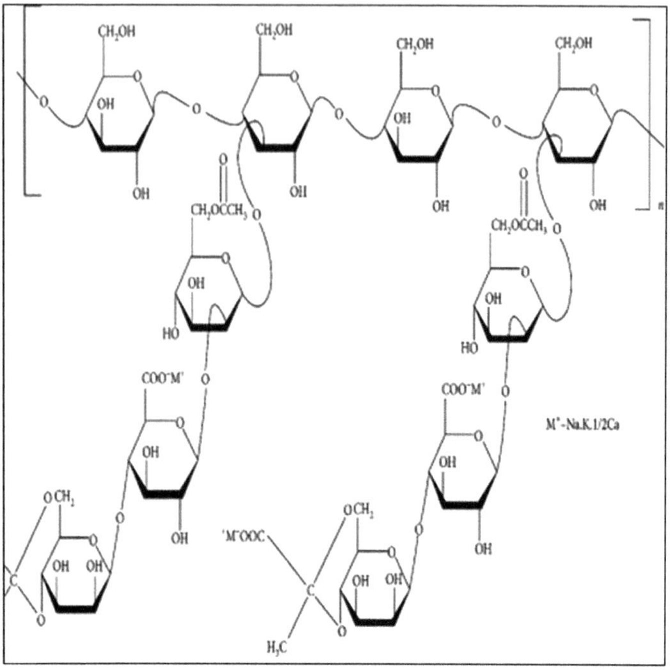

Fig. 2 Xanthan gum [20]

pair with each glucose element in the chemical structure to form a linear chain 1, 4-linked-D- glucose backbone. Between two D-mannose components, a D-glucuronic acid component makes up the other side chain [10, 45]. Figure 2 shows the chemical structure of xanthan gum where the aerobic fermentation of Xanthomonas campestris on glucose and has a cellulose backbone with two mannose and one glucuronic acid side chains [20].

Guar Gum: The leguminous plant seeds endosperms are processed to produce the non-ionic polysaccharide known as guar gum [21]. They work well as a soil supplement since they are totally soluble in water and non-toxic. The primary component of guar gum is a linear chain of high molecular weight galactomannans (1 → 4)-linked β-D- mannopyranosyl units with (1 → 6)-linked α-D-galactopyranosyl residues as side chains as shown in Fig. 3 [34].

Chitosan: The degraded form of chitin, known as chitosan (P-(1, 4)-2-amino-2-deoxy-D-glucose), is present in exoskeletons of crustaceans and cell walls of fungus. Typically, alkali sodium hydroxide treatment is used to remove it. Numerous applications have been made use for this hydrophilic biopolymer. Nourishment, material science, health sciences, microbiological, immunological, in addition to other sectors including farming, effluents, and degrading goods have all used and researched chitosan. Chitosan is a bio renewable, biocompatible, biodegradable, and bio functional biopolymer [37, 42, 49]. Chitin is a copolymer of α-(1 → 4) glucosamine

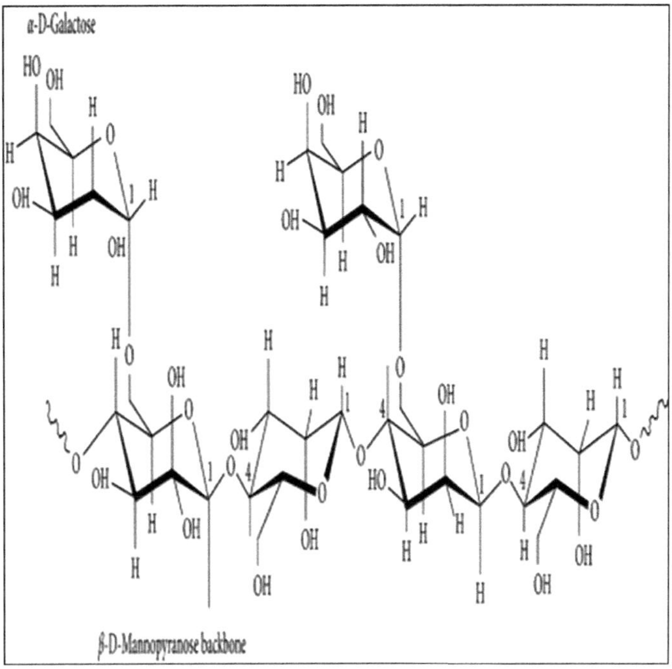

Fig. 3 Guar gum [34]

(C6H11O4N) n, with varying numbers of N-acetyl groups, while chitosan is its deacetylated derivative as presented in Fig. 4 [32].

Beta-Glucan: Biopolymers called beta-glucans are made of D-glucose monomers that are joined by glycosidic linkages [3]. Natural sources of glucan include plant material, cereal, and the cellular membranes of bacterial, yeast, and fungus, as well as other forms [15]. A polymer found in layers of the embryo of legumes crops called β- glucan as shown in Fig. 5, it is composed of straight glucose chains with β-(1 → 3) and β-(1 → 4) connections. The β-glucan found in yeast and fungus, in contrast, was made up of (1 → 3) connections and (1 → 6) connected branches [44].

Fig. 4 Chitosan [32]

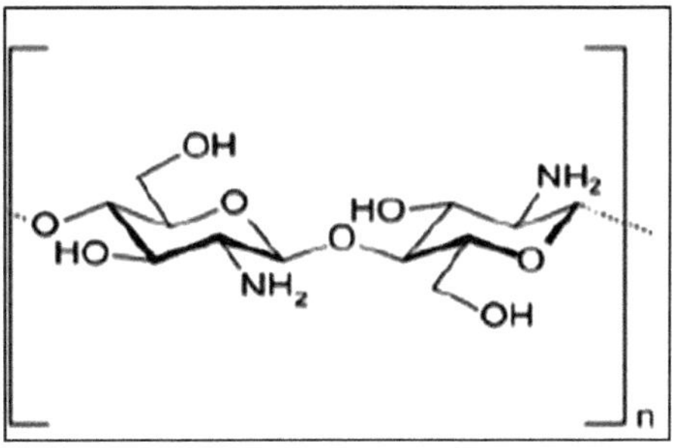

Fig. 5 Beta–Glucan [44]

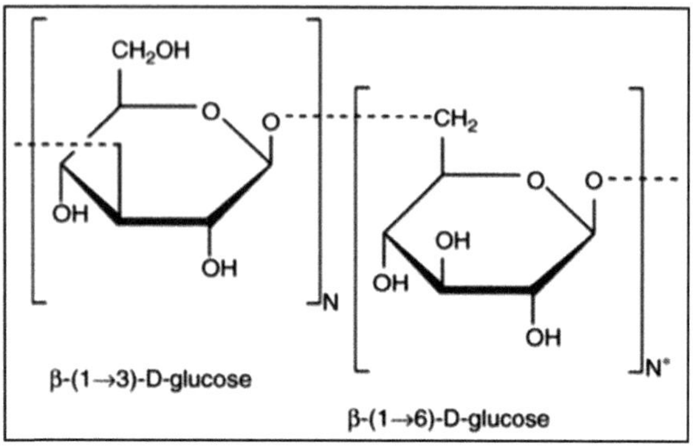

Agar Gum: Rhodophyceae are the source of the polysaccharide known as agar gum, which is made up of linked galactose molecules [26]. The rheological characteristics of agar gum make it effective as a thickening, stabilizer, and emulsifier. Agar gum also has a thermo-mechanical emulsification ability that allows it to form solid solutions once reduced down to ambient degree after becoming dispersed in hot water. [33]. Agar refers to gelling polysaccharides as presented in the Fig. 6 having an alternating 1,3- linked β -D-galactose and 1,4-linked 3,6-anhydro- α -L-galactose disaccharide called agarobiose or neo-agarobiose repeating unit as its primary backbone structure [30].

Gellan Gum: A high molecular weight polymer called gellan gum is produced by microorganisms called Sphingomonas elodea (formerly known as Pseudomonas elodea) [27]. Agar gum is typically replaced with gellan gum because of its many similarities to that substance, particularly its thermogelation capabilities [24, 46]. A tetra-saccharide repeating unit of two β -D-glucose, one β-D-glucuronic acid, and one α-L-rhamnose residue is connected together in this structure as given in Fig. 7 [22].

Fig. 6 Agar gum [30]

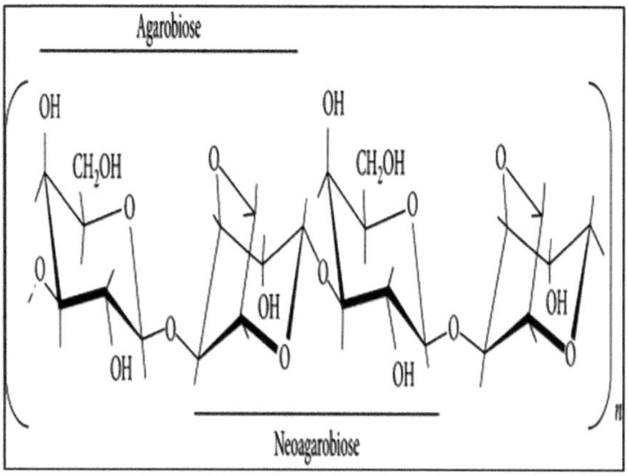

Fig. 7 Gellan gum [22]

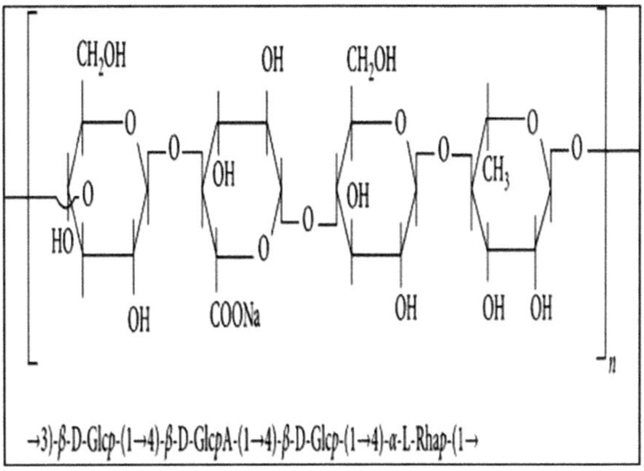

4 Biopolymers Used in Soil

Chitosan has been demonstrated to have the ability to decontaminate groundwater that contains pollutants like copper (II) (Cu^{2+}) and phosphorus in geotechnical and geoenvironmental situations (P-) [16, 41]. Practical applications for chitosan-coated layer sand particles in purifiers for groundwater clean-up [49] and can reduce the hydrostatic conductivity of sandy soils sediments to produce an appropriate plugging effect [28]. **β-Glucan** mix when compared to the strength of a 10% cement mix (2170 kPa), a modest amount of 0.25% of beta- glucan (i.e., ratio of mass to soil) in soil, produces a compressive strength of 2650 kPa, whereas 0.5% beta-glucan mixing results in soil strength of up to 4310 kPa [11]. According to geotechnical analysis, beta- glucan improves soil's shear stiffness (G) and Atterberg limit values while having little to no impact on the soil's constrained modulus (M) [12, 13].

Agar Gum without harming the environment, significantly increases the strength of sand (particularly cohesiveness) [29]. Researchers recently tried to use agar gum to increase soil strength. They found that using 3 percent (to soil mass) thermally treated agar gum increased soil's unconfined compressive strength up to 10 MPa when the soil was dried [10]. **Gellan Gum** due to its strong strengthening effect, gellan gum has a lot of potential for use in engineered soil [19]. According to a recent study, clayey soil containing 3% gellan gum has a maximum unconfined compressive strength of 12.6 MPa [10].

Xanthan Gum for the purpose of pore filling in geotechnical engineering has been introduced. This reduces the hydraulic conductivity of silty sand [6]. Additionally, raising the liquid limit will improve the soil's undrained shear strength [38]. Another current study identified the possibility for xanthan gum to be used as a soil strengthener and demonstrated that the substance preferentially creates hard xanthan gum-clayey soil matrices through hydrogen bonding [9]. **Guar Gum** helps building work in arid regions, attempts have been made to employ viscosity-controlled guar gum with additives such acrylamide, ammonium persulfate, and formaldehyde as injected grout for sand stabilization [23]. Guar gum can be used to prevent cracking at shallow depths by stabilizing expansive soils, such as desert sand, on slopes [1, 40]. Additionally, guar gum slurry has reportedly been used to make vertical barrier walls (such as cut-off walls) [17, 48].

4.1 Biopolymer Soil Interactions

To increase the overall effectiveness and dependability of soils treated with biopolymers, different types of biopolymers and/or workable application techniques might be combined. Because most biopolymers are highly sensitive to water, it is imperative to find ways to make biopolymer-treated soils more successful in it. This will increase the soils' dependability for ground stabilization and improvement. A number of approaches had been researched and also developed to address these problem, such as the usage of biopolymers with thermo-gelation capabilities. As was previously mentioned, biopolymers with such thermo-gelation capabilities include gellan gum and agar gum. When the gel is cooled to room temperature, the biopolymers begin to stiffen and becomes highly soluble in water at a temperature of around 85° to 90 °C. Even when exposed to water, these thermo-gelating biopolymers maintain a significant amount of strength [24, 46] (Table 1).

Cross-linking of biopolymers might provide an effective method of stabilizing soil in addition to current methods. Cross-linking is a technique, which is used to considerably enhance a material's properties by adding a component that promotes interactions between various polymer chains, increasing their overall strength. Cross-linking is widely employed when sulfur is introduced to rubber polymers, a process

Table 1 Soil biopolymer interaction

Soil	Biopolymer	Curing time	Observations	Remark
Sand	Xanthan gum, Guar gum	–	Copper metal absorption capacity to 85.6 and 60.7%, and reduce leachability to 10.3 and 22.8%	[18]
Sand with silt	Xanthan gum, guar gum	1–10 weeks	Cohesion of the sand was enhanced by 447 and 218 kN/m^2, and unconfined compressive strength value raised by 270 and 790 kPa	[2]
Sand	Xanthan gum	1, 3, 7,14 and 28 days	Hydraulic conductivity value was decreased	[7]
Sand bentonite mixture	Xanthan gum	–	Increase in permeability value	[5]
Kaolinite soil	Xanthan gum, guar gum	–	Liquid limit values of soil treated was 185.9%. With the addition of salts decreased the liquid limit of soil to 120.8% and 130%,	[38, 39]

known as vulcanization, which significantly increases the rubber's strength and stiffness.

5 Conclusions

A number of cement alternatives, various ideas, such as using biopolymers directly inside the soil profile, have been put forth, as well as the development of green cement. Recent research has demonstrated that biopolymers can enhance the soils. They have a number of benefits in these applications, including being environmentally friendly and efficient at shallow concentrations. According to numerous studies, the strengthening brought on by biopolymer treatment is greatest in the presence of fines, particularly clay particles.

Various investigations have demonstrated that in the presence of fines, particularly clay particles, maximizes the strengthening brought on by biopolymer treatment. Biopolymers, as an alternative to cement, have a significant capability to lower the emissions of carbon dioxide. Additionally, a number of biopolymers demonstrate the ability to support vegetative development and stabilize, which can be used as defense against threats to environmental conservation as well as the preservation of farmland and anti-desertification.

The ability to manufacture it ex situ and apply in situ with a greater level of quality control makes the biopolymer implementation superior to that of other biological soil approaches. Additionally, biopolymers may be mass manufactured in a commercial setting and react within the soil particles pretty quickly immediately after mixing, making them useful for quick or temporary support.

Limitations: Despite having many advantages of using biopolymers, there are currently numerous problems that must be fixed, such as their susceptibility to water, high cost, and potential biological degradation. Overall, the utilization of biopolymers in soil engineering seems to have a bright forthcoming given the large range of accessible biopolymers, their ability to be modified easily, and the multiple favorable qualities that they exhibit.

References

1. Acharya R, Pedarla A, Bheemasetti TV, Puppala AJ (2017) Assessment of guar gum biopolymer treatment toward mitigation of desiccation cracking on slopes built with expansive soils. Transp Res Rec 2657(1):78–88
2. Ayeldeen MK, Negm AM, El Sawwaf MA (2016) Evaluating the physical characteristics of biopolymer/soil mixtures. Arab J Geosci 9(5):1–13
3. Bacic A, Fincher GB, Stone B (eds) (2009) Chemistry, biochemistry, and biology of 1–3 beta glucans and related polysaccharides. Academic Press
4. Basu D, Misra A, Puppala AJ (2015) Sustainability and geotechnical engineering: perspectives and review. Can Geotech J 52(1):96–113

5. Biju MS, Arnepalli DN (2020) Effect of biopolymers on permeability of sand-bentonite mixtures. J Rock Mech Geotech Eng 12(5):1093–1102
6. Bouazza A, Gates WP, Ranjith PG (2009) Hydraulic conductivity of biopolymer-treated silty sand. Géotechnique 59(1):71–72
7. Cabalar AF, Wiszniewski M, Skutnik Z (2017) Effects of xanthan gum biopolymer on the permeability, odometer, unconfined compressive and triaxial shear behavior of a sand. Soil Mech Found Eng 54(5):356–361
8. Carvalho AJF (2011) Starch as source of polymeric materials. Biopolymers: Biomed Environ Appl
9. Chang I, Im J, Prasidhi AK, Cho GC (2015) Effects of Xanthan gum biopolymer on soil strengthening. Constr Build Mater 74:65–72
10. Chang I, Prasidhi AK, Im J, Cho GC (2015) Soil strengthening using thermo-gelation biopolymers. Construct Build Mater 77,:430–438
11. Chang I, Cho GC (2012) Strengthening of Korean residual soil with β-1, 3/1, 6-glucan biopolymer. Constr Build Mater 30:30–35
12. Chang I, Cho GC (2014) Geotechnical behavior of a beta-1, 3/1, 6-glucan biopolymer- treated residual soil. Geomech Eng 7(6):633–647
13. Chang I, Prasidhi AK, Im J, Shin HD, Cho GC (2015) Soil treatment using microbial biopolymers for anti-desertification purposes. Geoderma 253:39–47
14. Chang I, Im J, Cho GC (2016) Introduction of microbial biopolymers in soil treatment for future environmentally-friendly and sustainable geotechnical engineering. Sustainability 8(3):251
15. Chang YJ, Lee S, Yoo MA, Lee HG (2006) Structural and biological characterization of sulphated-derivatized oat β-glucan. J Agric Food Chem 54(11):3815–3818
16. Cheung WH, Ng JCY, McKay G (2003) Kinetic analysis of the sorption of copper (II) ions on chitosan. J Chem Technol Biotech: Int Res Proc Environ Clean Technol 78(5):562–571
17. Day SR, O'Hannesin SF, Marsden L (1999) Geotechnical techniques for the construction of reactive barriers. J Hazard Mater 67(3):285–297
18. Etemadi O, Petrisor IG, Kim D, Wan MW, Yen TF (2003) Stabilization of metals in subsurface by biopolymers: Laboratory drainage flow studies. Soil Sediment Contam: An Int J 12(5):647–661
19. Ferruzzi GG, Pan N, Casey WH (2000) Mechanical properties of gellan and polyacrylamide gels with implications for soil stabilization. Soil Sci Interdisc Approach Soils Res 165(10):778–792
20. Garcıa-Ochoa F, Santos VE, Casas JA, Gómez E (2000) Xanthan gum: production, recovery, and properties. Biotech Adv 18(7):549–579
21. Grasdalen H, Painter T (1980) NMR studies of composition and sequence in legume-seed galactomannans. Carbohydr Res 81(1):59–66
22. Grasdalen H, Smidsrød O (1987) Gelation of gellan gum. Carbohyd Polym 7(5):371–393
23. Gupta SC, Hooda KS, Mathur NK, Gupta S (2009) Tailoring of guar gum for desert sand stabilization
24. Huang M, Kennedy JF, Li B, Xu X, Xie BJ (2007) Characters of rice starch gel modified by gellan, carrageenan, and glucomannan: a texture profile analysis study. Carbohyd Polym 69(3):411–418
25. Ibrahim S, Riahi O, Said SM, Sabri MF, Rozali S (2019) Biopolymers from crop plants. Reference module in materials science and materials engineering
26. Ivanov V, Chu J (2008) Applications of microorganisms to geotechnical engineering for bioclogging and biocementation of soil in situ. Rev Environ Sci and Bio/Technol 7(2):139–153
27. Imeson A (2009) Food stabilisers, Thickeners and gelling agents, vol 368. Wiley-Blackwell Publishing, Chichester, UK
28. Khachatoorian R, Petrisor IG, Kwan CC, Yen TF (2003) Biopolymer plugging effect: laboratory-pressurized pumping flow studies. J Petroleum Sci Eng 38(1–2):13–21
29. Khatami HR, O'Kelly BC (2013) Improving mechanical properties of sand using biopolymers. J Geotech Geoenviron Eng 139(8)
30. Lahaye M (2001) Developments on gelling algal galactans, their structure and physico-chemistry. J Appl Phycol 13(2):173–184

31. Larson A (2011) Sustainability, innovation, and entrepreneurship; University of Virginia: Charlottesville. VA, USA
32. Mahmoud MG, El Kady EM, Asker MS (2019) Chitin, chitosan and glucan, properties and applications. World J Agric Soil Sci 3(1):1–19
33. McHugh DJ (2003) A guide to the seaweed industry; food and agriculture organization of the United Nations, Rome, Italy
34. Mudgil D, Barak S, Khatkar BS (2014) Guar gum: processing, properties and food applications—a review. J Food Sci Technol 51(3):409–418
35. Murphy EM, Ginn TR (2000) Modeling microbial processes in porous media. Hydrogeol J 8(1):142–158
36. Murthy VNS (2002) Geotechnical engineering: principles and practices of soil mechanics and foundation engineering. CRC press
37. Nascimento T, Oliveira H, Rego C (2007) Potential use of chitosan in the control of grapevine trunk diseases. Phytopathol Mediterr 46(2):218–224
38. Nugent RA, Zhang G, Gambrell RP (2009) Effect of exopolymers on the liquid limit of clays and its engineering implications. J Transp Res Board 2101(1):34–43
39. Nugent RA, Zhang G, Gambrell RP (2011) The effect of exopolymers on the compressibility of clays. In: Geo-frontiers 2011: advances in geotechnical engineering, pp 3935–3944
40. Puppala AJ, Pedarla A (2017) Innovative ground improvement techniques for expansive soils. Innovative Infrastruct Solutions 2(1):1–15
41. Renault F, Sancey B, Badot PM, Crini G (2009) Chitosan for coagulation/flocculation processes–an eco-friendly approach. Eur Polymer J 45(5):1337–1348
42. Sahoo D, Nayak PL (2011) Chitosan: the most valuable derivative of chitin. Biopolymers: Biomed Environ Appl 129–166
43. Sherwood PT (1993) Soil stabilization with cement and lime. HMSO, London, UK
44. Sima P, Vannucci L, Vetvicka V (2018) β-glucans and cholesterol. Int J Mol Med 41(4):1799–1808
45. Stewart TL, Fogler HS (2001) Biomass plug development and propagation in porous media. Biotechnol Bioeng 72(3):353–363
46. Tang J, Tung MA, Zeng Y (1997) Gelling properties of gellan solutions containing monovalent and divalent cations. J Food Sci 62(4):688–712
47. Van de Velde K, Kiekens P (2002) Biopolymers: overview of several properties and consequences on their applications. Polym Testing 21(4):433–442
48. Velimirovic M, Tosco T, Uyttebroek M, Luna M, Gastone F, De Boer C, Klaas N, Sapion H, Eisenmann H, Larsson PO, Braun J (2014) Field assessment of guar gum stabilized microscale zerovalent iron particles for in-situ remediation of 1, 1, 1- trichloroethane. J Contam Hydrol 164:88–99
49. Wan MW, Petrisor IG, Lai HT, Kim D, Yen TF (2004) Copper adsorption through chitosan immobilized on sand to demonstrate the feasibility for in situ soil decontamination. Carbohyd Polym 55(3):249–254
50. Yang F, Zhang B, Pan C, Zeng Y (2009) Traditional mortar represented by sticky rice lime mortar—one of the great inventions in ancient China. Sci China Ser E: Technol Sci 52(6):1641–1647

Characterization of Chicken Feathers Ash as a Potential Soil-Stabilizing Material

R. Kokilan and B. Divya Priya

1 Introduction

According to a National Chicken Council report, chicken is one of the most popular meats consumed around the world. India is one of the global highest producers of chicken meat. The demand for chicken has grown significantly over time, and consumption has risen significantly due to increasing average income and urban population. In 2021, it is found that over four million metric tons of chicken meat were consumed in India. Apart from the meat, chickens' feathers make up to 10% of their total weight, which indicates that every year, 8–9 million tons of waste chicken feathers (CFs) are generated worldwide [1–3]. The keratin and numerous cysteine bonds in CFs make them difficult to break down [4]. Traditionally, poultry litter (including feathers, poultry excreta, spilled feed, and other materials used as bedding in poultry operations) is spread on soils as an organic fertilizer. However, excess application of this material results in an overabundance of nutrients that eutrophicates water bodies, spreads diseases, produces phytotoxic chemicals, pollutes the environment, and emits greenhouse gases [5]. In developing countries, these CFs produced are commonly disposed of by open dumping which causes deposition and clogging of roadside drains and other open pits, thus posing a serious challenge in solid waste management and a high threat to people's health. In most cases, only a small portion of (CFs) are converted into useful byproducts like fertilizer and feather meal, while the rest is disposed of in a landfill. The utilization of feathers as a meal will affect the digestive system due to the presence of pepsin, and leachate produced through the landfilling of feathers will result in groundwater contamination. Therefore, there is a need for adequate waste management facilities and treatment of the CFs for their disposal. Adversely, presence of high p and k in the chicken litter ash

R. Kokilan · B. Divya Priya (✉)
School of Civil Engineering, Vellore Institute of Technology, Vellore 632014, India
e-mail: divyapriya.b@vit.ac.in

B. T. Jose et al. (eds.), *Proceedings of the Indian Geotechnical Conference 2022 Volume 4*, Lecture Notes in Civil Engineering 479,
https://doi.org/10.1007/978-981-97-1753-8_34

which includes chicken manure may affect the pozzolanic activity with the soil and it can be controlled by segregating the chicken feathers and turns to ash which can be used as a soil-stabilizing agent. Around the world, CFs are disposed of by two major methods, i.e., incineration and landfill disposal. Disposal of waste in landfills is a concept that has been practiced for many decades but bio-accumulation of heavy metals in feathers may lead to soil and groundwater contamination. Moreover, one of the effective methods for destroying potentially infectious pathogens and controlling leachate production is incineration process. Thus, using waste chicken feather ash as binding material will minimize environmental pollution. Municipal solid waste (MSW) can be incinerated to dispose of a significant volume of waste and generate energy at the same time, with a reduction in mass and volume to 70% and 90%, respectively [6–8]. But incinerating at high temperatures will affect its physicochemical composition and becomes useless as trash. So, CFS must incinerate at optimum temperature for high retention of cementitious properties.

According to Luyckx et al. [9], poultry litter in ash form contains elements such as Ca, Si, Mg, Fe, and Al which can be effectively used as a building material. However, the conversion of poultry litter into ash may increase the fertility when added to soil than the increase in strength properties due to the presence of high p and k components. Studies suggest that replacing a percentage of chicken feather ash (CFA) in cement production with other organic ash compounds (stimulating pozzolanic activity) reduces the percentage of greenhouse gases released into the atmosphere [10–12]. Hence, an attempt has been made to characterize the CFA (in powder and flakes forms) including morphological, mineralogical, and chemical composition and examine the presence of cementitious properties. Based on the results and literature analysis the feasibility of using chicken feather ash as a potential stabilizing agent is discussed.

2 Experimental Methods

As the Namakkal district in Tamil Nadu is one of the major poultry hubs in the country, the chicken feathers are collected from various farms and the poultry meat processing industry. The collected feathers are rinsed several times with warm water and soap to remove impurities and air-dried for 24 h. The cleaned feathers are burned at a different temperatures through an incinerator which is designed to a capacity of 150 g. Through various trials, it is found that the cementitious properties are retained at a high percentage at an optimum temperature of 600 °C. At this stage, the chicken feathers are incinerated at 600 °C at different time intervals (i.e., 30 and 10 min) to convert into two forms, i.e., ash powder and ash flakes. As a preventive measure, the filter to reduce the emission of sulfur dioxide is fixed in the furnace to control the air pollution and loss of ignition is calculated. The collected ash powder and flakes are kept inside the desiccator and allowed to cool for 30 min. The ash powder is pulverized and sieved to fine material. Now, CFA in both powder and flake forms is taken to a laboratory for further analysis. The entire process is depicted in (Fig. 1).

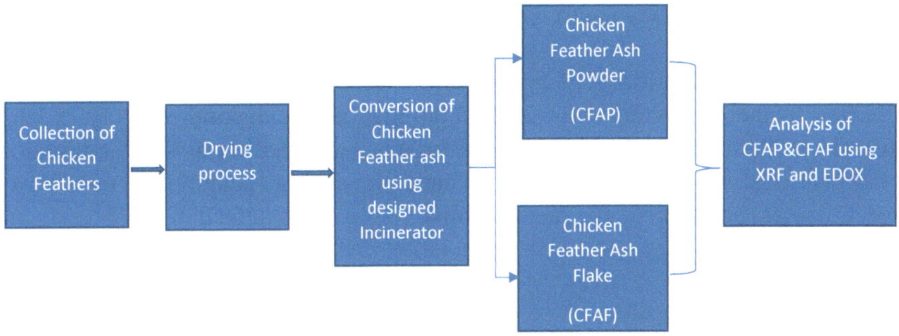

Fig. 1 Methodological process of conversion of chicken feather to ash

2.1 *Characterization of Samples*

More than 90% of a chicken feather's composition is beta keratin, an insoluble, fibrous structural protein that is extensively cross-linked by disulfide bonds. This study is conducted to assess the chicken feather compositions and the cementitious properties of feather ash (flake and powder). The chemical, mineralogical and morphological composition of CFs are analyzed as follows.

2.1.1 X-Ray Diffraction (XRD)

Samples are dried for 24 h in an oven at a temperature of 110 °C and then sieved through N60. The samples were stored in an air-tight bag. XRD analysis is conducted using XRD-diffractometer (Model: D8 advance). The samples were inspected by 2 at a rate of 1.5°/min in the 10–60° angstrom range. The qualitative analysis is found to have distinct peaks in the spectrum.

2.1.2 Scanning Electron Microscopy (SEM).

The morphological analysis is carried out by scanning electron microscopy (Model No: EVO/18 Research, Brand Name: Carl Zeiss). The materials for SEM analysis are prepared similarly to XRD analysis. The results of the SEM investigation are used to identify the surface features and size of the components. Energy dispersive spectroscopy (EDS) is carried out with SEM analysis to determine the characteristics and nature of the elements present in the sample.

2.1.3 X-Ray Fluorescence (XRF)

The chemical compositions of the ashes are determined by X-ray fluorescence. Chicken feather ash is dried for 24 h in an oven at a temperature of 110 °C and then sieved through a 60N sieve. XRF analysis is conducted at the NCES consisting of Bruker model S8 Tiger and S4 Pioneer sequential wavelength-dispersive x-ray spectrometers and sample preparation units.

3 Results and Discussion

Initially, SEM and EDS tests are performed on the chicken feather ash (powder and flake) to determine the cementitious properties and their morphological structure. The chemical composition of the CFA in powder and flake forms is given in Table 1.

3.1 SEM and EDS

3.1.1 SEM and EDS Analysis on Chicken Feather Ash Flake

SEM analysis on chicken flake is conducted to examine bonding structure, fabric arrangement, surface individuality, and particle orientation. EDOX is to determine the chemical composition of the chicken feather flake. The results show that the chemical composition of the flake structure contains more wollastonite, i.e., 44%. Wollastonite is a calcium inosilicate mineral with minor levels of iron, magnesium, and manganese that act as calcium replacements. It also contains silicon dioxide, i.e., nearly 42 and 21% of Feldspar as given in Table 1. The chemical formula SiO_2 (silicon dioxide), commonly referred to as silica, is an oxide of silicon that is most frequently found in nature as quartz mineral and organic waste materials. Silica makes up most of the sand in various places of the world. The morphological image of the CFAF indicates the presence of pores structure (see Fig. 2).

Table 1 Chemical composition of chicken feather ash: flake and powder

Element	CFAF (%)	CFAP (%)
$CaCO_3$	– 8.50	61.73
SiO_2	42.18	26.2
MAD-10 Feldspar	21.37	1.61
Wollastonite	44.95	4.18
Magnesium oxide	–	1.45
Albite	–	1.38
Iron disulfide	–	1.02

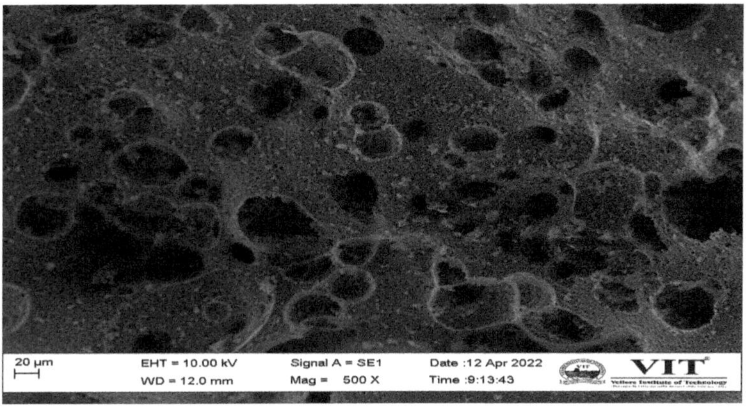

Fig. 2 Morphology of chicken feather flake

3.1.2 SEM and EDS Analysis on Chicken Feather Ash Powder

The internal structure and arrangement are inferred by examining the morphology of the chicken feather ash powder. It clearly shows that the chemical composition of the powder structure contains more calcium carbonate, i.e., of 61% a microfibrous material that has been extensively used in the reinforcement of cementitious material. The major constituent of feather ash is calcium carbonate which can be used as a stabilizing agent due to its cementitious properties. It also contains nearly 26% of silicon dioxide, 4% of wollastonite, 1% of magnesium oxide and albite, and of Feldspar as given in Table 1. The morphological image of the chicken feather ash powder clearly indicates that the presence of porous structure in the chicken feather ash powder is similar to CFAF but the structure is quite uniform when compared with chicken feather flake (see Fig. 3).

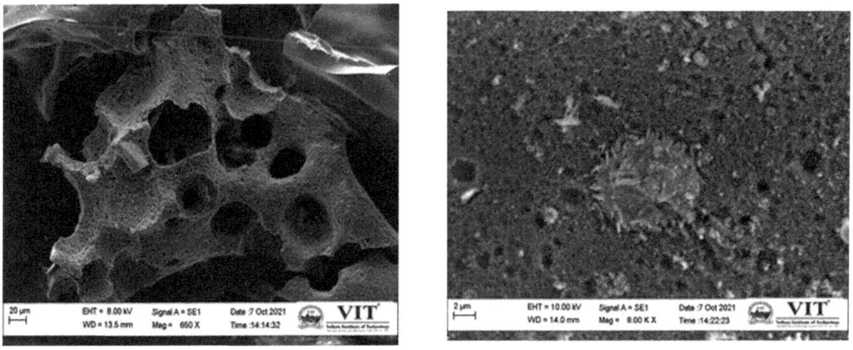

Fig. 3 Morphology of chicken feather ash

3.2 *Chemical and Mineralogical Composition of Chicken Feather Ash Powder:*

The results through SEM and EDS analysis for CFAP and CFAF indicates the presence of a porous structure and a high percentage of calcium carbonate is present in the chicken feather ash powder specifying the existence of cementitious properties which can be used to improve the engineering properties of soil and partial replacement in cement. The major constituent in CFAP is calcium oxide and in CFAF is calcium wollastonite. Chicken feathers converted into ash powder at different temperatures also indicate the presence of silicon and aluminum. Since calcium carbonate is the major constituent of CFAP, it can improve pozzolanic reaction in addition to soil with water, thus improving their engineering properties. It is found that the percentage of calcium carbonate varies with the process of incineration at different temperatures. Hence, the optimum temperature of 600 °C is established for incineration of CFs into CFAP with a higher proportion of calcium carbonate through various trial processes. The XRF analysis has been carried out and results are presented in Table 2.

It indicates that the major components of CFAP are CaO > 39%, P2O5 > 9%, SiO2 > 10%, K2O > 3%, MgO > 8%, SO3 > 12%, and Al2O3 > 3%. When added to water, it reacts with calcium oxide (CaO) and forms calcium hydroxide (Ca(OH)2), the calcium oxide compound is subsequently treated with carbon dioxide to create the calcium carbonate.

Thus improving the cementitious property and in turn, stimulating the pozzolanic activity increasing the engineering properties of soil when added to it, the results from EDS and XRF indicate that CFAP is rich in calcium component which is a cementitious material, whereas CFAF is rich in wollastonite.

Table 2 Chemical composition of chicken feather ash XRF

Sample (%)	CF Ash 600*
SiO2	10.20
TiO2	0.36
Al2O3	3.01
MnO	0.12
Fe2O3	4.55
CaO	39.20
MgO	8.63
Na2O	3.79
K2O	3.87
P2O5	9.78
SO3	12.30
Cl	2.69
ZnO	0.93
LOI	0.80

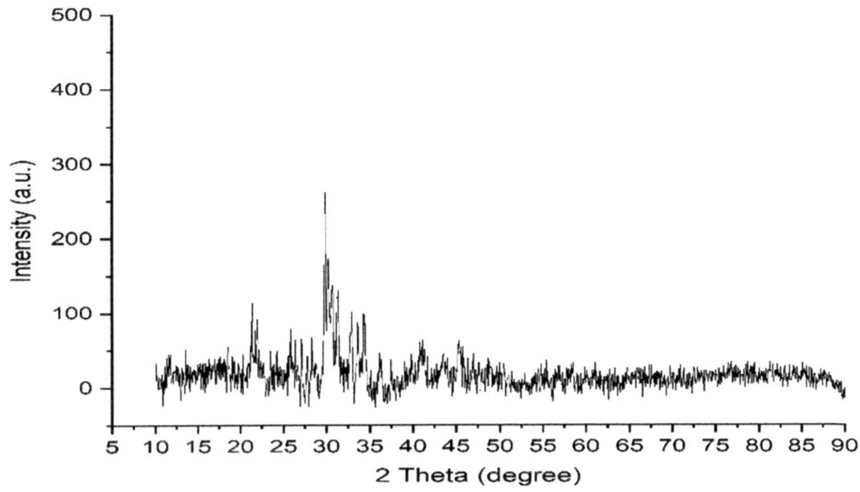

Fig. 4 X-ray diffract gram of the chicken feather ash

The chemical composition of EDS resembles the XRF results obtained for CFAP is rich in calcium carbonate a cementitious material whereas CFAF is rich in wollastonite. Cementitious minerals, like Quartz, Calcite, and Anhydrite present in the chicken feather ash powder are identified through the XRD which confirms that they can be used as a stabilizing agent [13]. Results are supported by micrographs of the ash material (see Fig. 4), depicting the trigonal shapes indicating the presence of calcite and anhydrite crystals as present like in cement similar to the reported literature [13].

4 Conclusion

In this study, a detailed characterization of the chemical properties of chicken feather ash has been performed with the aim to utilize the potential soil-stabilizing material. From the results, it is concluded that:

- An optimum temperature of 600 °C is derived for incineration to convert the chicken feathers into ash form with a high level of calcium carbonate (CFAP and CFAF) and silica.
- This research explores the efficacy of chicken feather ash as a suitable stabilizing material for soil when compared with traditional stabilizing agents where the calcium content in chicken feather ash can act as a binding material to enhance the engineering property of soil.
- Chicken feather ash powder contains lime content in the form of CaO, i.e., 39% which can be used as a potential stabilizing material in soil compared with the chicken feather ash flakes.

- CFAP is more likely to be used as a stabilizing material to enhance the engineering properties of the soil, due to the presence of cementitious properties which are established by examining its chemical, mineralogical, and morphology composition.

The use of chicken feather ash along with other silica-rich organic waste products as a stabilizing material to increase the strength properties of the soil is planned for future work.

Acknowledgements We acknowledge the X-ray Fluorescence Lab, National Centre for Earth Science Studies (Ministry of Earth Sciences, Government of India), Thiruvananthapuram, and Kerala for their support in XRF analysis.

We appreciate the support of the technicians in Scanning Electron Microscopy Laboratory, School of Biosciences and Technology, VIT Vellore for their assistance in morphological analysis.

References

1. Lasekan A, Abu Bakar F, Hashim D (2013) Potential of chicken by-products as sources of usefulbiological resources. Waste Manage 3(33):552–565
2. Peng Z, Mao X, Zhang J, Du G, Chen J (2019) Effective biodegradation of chicken featherwaste by co-cultivation of keratinase producing strains. Microb Cell Factories 1(18):11
3. ReddyN (2015) Non-food industrial applications of poultry feathers. Waste Manage 45:91–107
4. Callegaro K, Brandelli AD, Joner D (2019) Beyond plucking: feathers bioprocessing into valuable protein hydrolysates. Waste Manage 95:399–415
5. AcharyaB, Dutta A, Mahmud S, Tushar ML, Mathias (2014) Ash analysis of poultry litter, willow and oats for combustion in boilers. J Biomass Biofuel 1
6. MaB, Qiao X, Hou X, Yang Y (2016) Pure keratin membrane and fibers from chicken feather. Int J Biol Macromol 89:614–621
7. Qiang T, Heejong K, Kazuto E, Takeshi K, Toru I (2015) Size effect on lysimeter test evaluating theproperties of construction and demolition waste leachate. Soils Found 4(55):720–736
8. Rodier L, Bilba K, Onésippe C, Arsène MA (2017) Study of pozzolanic activity of bamboo stem ashes for use as partial replacement of cement. Mater Struct (Materiaux et Constructions)1(50)
9. Font-PalmaC (2012) Characterisation, kinetics and modelling of gasification of poultry manure and litter: an overview. Energy Convers Manage 1(53):92–98
10. Amah VE Ugwoha E, Macaulay J (2021) Chicken feather ash (CFA) as a partial replacementof cement in concrete. J Eng Res Rep 1(20):130–140
11. Cordeiro G,Tavares C, Toledo LM, Filho RD (2016) Improved pozzolanic activity of sugar cane bagasse ash by selective grinding and classification. Cem Concr Res 89:269–275
12. Irshad F, Yamamoto M,Honna S, Eneji T, Egrinya A (2009) Characterization of trace elements in chicken and duck litter ash. Waste Manage 1(29):265–271
13. Castillo D,Cruz JC, Trejo-Arroyo DL, Muzquiz EM, Zarhri Z, Gurrola MP, Vega-Azamar RE (2022) Characterization of poultry litter ashes as a supplementary cementitious material. Case Stud Constr Mater 17(23)

GPSR Compliance

The European Union's (EU) General Product Safety Regulation (GPSR) is a set of rules that requires consumer products to be safe and our obligations to ensure this.

If you have any concerns about our products, you can contact us on ProductSafety@springernature.com

In case Publisher is established outside the EU, the EU authorized representative is:

Springer Nature Customer Service Center GmbH
Europaplatz 3
69115 Heidelberg, Germany

Batch number: 10726133

Printed by Printforce, the Netherlands